AF725162

Pitman Research Notes in Mathematics Series

Submission of proposals for consideration

Suggestions for publication, in the form of outlines and representative samples, are invited by the Editorial Board for assessment. Intending authors should approach one of the main editors or another member of the Editorial Board, citing the relevant AMS subject classifications. Alternatively, outlines may be sent directly to the publisher's offices. Refereeing is by members of the board and other mathematical authorities in the topic concerned, throughout the world.

Preparation of accepted manuscripts

On acceptance of a proposal, the publisher will supply full instructions for the preparation of manuscripts in a form suitable for direct photo-lithographic reproduction. Specially printed grid sheets can be provided and a contribution is offered by the publisher towards the cost of typing. Word processor output, subject to the publisher's approval, is also acceptable.

Illustrations should be prepared by the authors, ready for direct reproduction without further improvement. The use of hand-drawn symbols should be avoided wherever possible, in order to maintain maximum clarity of the text.

The publisher will be pleased to give any guidance necessary during the preparation of a typescript, and will be happy to answer any queries.

Important note

In order to avoid later retyping, intending authors are strongly urged not to begin final preparation of a typescript before receiving the publisher's guidelines. In this way it is hoped to preserve the uniform appearance of the series.

Longman Group Ltd
Longman House
Burnt Mill
Harlow, Essex, CM20 2JE
UK
(Telephone (0) 1279 426721)

Titles in this series. A full list is available from the publisher on request.

201 Riemannian geometry and holonomy groups
S Salamon
202 Strong asymptotics for extremal errors and polynomials associated with Erdös type weights
D S Lubinsky
203 Optimal control of diffusion processes
V S Borkar
204 Rings, modules and radicals
B J Gardner
205 Two-parameter eigenvalue problems in ordinary differential equations
M Faierman
206 Distributions and analytic functions
R D Carmichael and D Mitrovic
207 Semicontinuity, relaxation and integral representation in the calculus of variations
G Buttazzo
208 Recent advances in nonlinear elliptic and parabolic problems
P Bénilan, M Chipot, L Evans and M Pierre
209 Model completions, ring representations and the topology of the Pierce sheaf
A Carson
210 Retarded dynamical systems
G Stepan
211 Function spaces, differential operators and nonlinear analysis
L Paivarinta
212 Analytic function theory of one complex variable
C C Yang, Y Komatu and K Niino
213 Elements of stability of visco-elastic fluids
J Dunwoody
214 Jordan decomposition of generalized vector measures
K D Schmidt
215 A mathematical analysis of bending of plates with transverse shear deformation
C Constanda
216 Ordinary and partial differential equations. Volume II
B D Sleeman and R J Jarvis
217 Hilbert modules over function algebras
R G Douglas and V I Paulsen
218 Graph colourings
R Wilson and R Nelson
219 Hardy-type inequalities
A Kufner and B Opic
220 Nonlinear partial differential equations and their applications: Collège de France Seminar. Volume X
H Brezis and J L Lions
221 Workshop on dynamical systems
E Shiels and Z Coelho
222 Geometry and analysis in nonlinear dynamics
H W Broer and F Takens
223 Fluid dynamical aspects of combustion theory
M Onofri and A Tesei
224 Approximation of Hilbert space operators. Volume I. 2nd edition
D Herrero
225 Operator theory: proceedings of the 1988 GPOTS–Wabash conference
J B Conway and B B Morrel
226 Local cohomology and localization
J L Bueso Montero, B Torrecillas Jover and A Verschoren
227 Nonlinear waves and dissipative effects
D Fusco and A Jeffrey
228 Numerical analysis 1989
D F Griffiths and G A Watson
229 Recent developments in structured continua. Volume II
D De Kee and P Kaloni
230 Boolean methods in interpolation and approximation
F J Delvos and W Schempp
231 Further advances in twistor theory. Volume I
L J Mason and L P Hughston
232 Further advances in twistor theory. Volume II
L J Mason, L P Hughston and P Z Kobak
233 Geometry in the neighborhood of invariant manifolds of maps and flows and linearization
U Kirchgraber and K Palmer
234 Quantales and their applications
K I Rosenthal
235 Integral equations and inverse problems
V Petkov and R Lazarov
236 Pseudo-differential operators
S R Simanca
237 A functional analytic approach to statistical experiments
I M Bomze
238 Quantum mechanics, algebras and distributions
D Dubin and M Hennings
239 Hamilton flows and evolution semigroups
J Gzyl
240 Topics in controlled Markov chains
V S Borkar
241 Invariant manifold theory for hydrodynamic transition
S Sritharan
242 Lectures on the spectrum of $L^2(\Gamma\backslash G)$
F L Williams
243 Progress in variational methods in Hamiltonian systems and elliptic equations
M Girardi, M Matzeu and F Pacella
244 Optimization and nonlinear analysis
A Ioffe, M Marcus and S Reich
245 Inverse problems and imaging
G F Roach
246 Semigroup theory with applications to systems and control
N U Ahmed
247 Periodic-parabolic boundary value problems and positivity
P Hess
248 Distributions and pseudo-differential operators
S Zaidman
249 Progress in partial differential equations: the Metz surveys
M Chipot and J Saint Jean Paulin
250 Differential equations and control theory
V Barbu

251 Stability of stochastic differential equations with respect to semimartingales
X Mao
252 Fixed point theory and applications
J Baillon and M Théra
253 Nonlinear hyperbolic equations and field theory
M K V Murthy and S Spagnolo
254 Ordinary and partial differential equations. Volume III
B D Sleeman and R J Jarvis
255 Harmonic maps into homogeneous spaces
M Black
256 Boundary value and initial value problems in complex analysis: studies in complex analysis and its applications to PDEs 1
R Kühnau and W Tutschke
257 Geometric function theory and applications of complex analysis in mechanics: studies in complex analysis and its applications to PDEs 2
R Kühnau and W Tutschke
258 The development of statistics: recent contributions from China
X R Chen, K T Fang and C C Yang
259 Multiplication of distributions and applications to partial differential equations
M Oberguggenberger
260 Numerical analysis 1991
D F Griffiths and G A Watson
261 Schur's algorithm and several applications
M Bakonyi and T Constantinescu
262 Partial differential equations with complex analysis
H Begehr and A Jeffrey
263 Partial differential equations with real analysis
H Begehr and A Jeffrey
264 Solvability and bifurcations of nonlinear equations
P Drábek
265 Orientational averaging in mechanics of solids
A Lagzdins, V Tamuzs, G Teters and A Kregers
266 Progress in partial differential equations: elliptic and parabolic problems
C Bandle, J Bemelmans, M Chipot, M Grüter and J Saint Jean Paulin
267 Progress in partial differential equations: calculus of variations, applications
C Bandle, J Bemelmans, M Chipot, M Grüter and J Saint Jean Paulin
268 Stochastic partial differential equations and applications
G Da Prato and L Tubaro
269 Partial differential equations and related subjects
M Miranda
270 Operator algebras and topology
W B Arveson, A S Mishchenko, M Putinar, M A Rieffel and S Stratila
271 Operator algebras and operator theory
W B Arveson, A S Mishchenko, M Putinar, M A Rieffel and S Stratila
272 Ordinary and delay differential equations
J Wiener and J K Hale
273 Partial differential equations
J Wiener and J K Hale
274 Mathematical topics in fluid mechanics
J F Rodrigues and A Sequeira
275 Green functions for second order parabolic integro-differential problems
M G Garroni and J F Menaldi
276 Riemann waves and their applications
M W Kalinowski
277 Banach C(K)-modules and operators preserving disjointness
Y A Abramovich, E L Arenson and A K Kitover
278 Limit algebras: an introduction to subalgebras of C*-algebras
S C Power
279 Abstract evolution equations, periodic problems and applications
D Daners and P Koch Medina
280 Emerging applications in free boundary problems
J Chadam and H Rasmussen
281 Free boundary problems involving solids
J Chadam and H Rasmussen
282 Free boundary problems in fluid flow with applications
J Chadam and H Rasmussen
283 Asymptotic problems in probability theory: stochastic models and diffusions on fractals
K D Elworthy and N Ikeda
284 Asymptotic problems in probability theory: Wiener functionals and asymptotics
K D Elworthy and N Ikeda
285 Dynamical systems
R Bamon, R Labarca, J Lewowicz and J Palis
286 Models of hysteresis
A Visintin
287 Moments in probability and approximation theory
G A Anastassiou
288 Mathematical aspects of penetrative convection
B Straughan
289 Ordinary and partial differential equations. Volume IV
B D Sleeman and R J Jarvis
290 *K*-theory for real *C**-algebras
H Schröder
291 Recent developments in theoretical fluid mechanics
G P Galdi and J Necas
292 Propagation of a curved shock and nonlinear ray theory
P Prasad
293 Non-classical elastic solids
M Ciarletta and D Ieșan
294 Multigrid methods
J Bramble
295 Entropy and partial differential equations
W A Day
296 Progress in partial differential equations: the Metz surveys 2
M Chipot
297 Nonstandard methods in the calculus of variations
C Tuckey
298 Barrelledness, Baire-like- and (LF)-spaces
M Kunzinger
299 Nonlinear partial differential equations and their applications. Collège de France Seminar. Volume XI
H Brezis and J L Lions
300 Introduction to operator theory
T Yoshino

301 Generalized fractional calculus and applications
V Kiryakova
302 Nonlinear partial differential equations and their applications. Collège de France Seminar Volume XII
H Brezis and J L Lions
303 Numerical analysis 1993
D F Griffiths and G A Watson
304 Topics in abstract differential equations
S Zaidman
305 Complex analysis and its applications
C C Yang, G C Wen, K Y Li and Y M Chiang
306 Computational methods for fluid-structure interaction
J M Crolet and R Ohayon
307 Random geometrically graph directed self-similar multifractals
L Olsen
308 Progress in theoretical and computational fluid mechanics
G P Galdi, J Málek and J Necas
309 Variational methods in Lorentzian geometry
A Masiello
310 Stochastic analysis on infinite dimensional spaces
H Kunita and H-H Kuo
311 Representations of Lie groups and quantum groups
V Baldoni and M Picardello
312 Common zeros of polynomials in several variables and higher dimensional quadrature
Y Xu
313 Extending modules
N V Dung, D van Huynh, P F Smith and R Wisbauer
314 Progress in partial differential equations: the Metz surveys 3
M Chipot, J Saint Jean Paulin and I Shafrir
315 Refined large deviation limit theorems
V Vinogradov
316 Topological vector spaces, algebras and related areas
A Lau and I Tweddle
317 Integral methods in science and engineering
C Constanda
318 A method for computing unsteady flows in porous media
R Raghavan and E Ozkan
319 Asymptotic theories for plates and shells
R P Gilbert and K Hackl
320 Nonlinear variational problems and partial differential equations
A Marino and M K V Murthy
321 Topics in abstract differential equations II
S Zaidman
322 Diffraction by wedges
B Budaev
323 Free boundary problems: theory and applications
J I Diaz, M A Herrero, A Liñan and J L Vazquez
324 Recent developments in evolution equations
A C McBride and G F Roach
325 Elliptic and parabolic problems: Pont-à-Mousson 1994
C Bandle, J Bemelmans, M Chipot, J Saint Jean Paulin and I Shafrir
326 Calculus of variations, applications and computations: Pont-à-Mousson 1994
C Bandle, J Bemelmans, M Chipot, J Saint Jean Paulin and I Shafrir
327 Conjugate gradient type methods for ill-posed problems
M Hanke
328 A survey of preconditioned iterative methods
A M Bruaset
329 A generalized Taylor's formula for functions of several variables and certain of its applications
J-A Riestra
330 Semigroups of operators and spectral theory
S Kantorovitz
331 Boundary-field equation methods for a class of nonlinear problems
G N Gatica and G C Hsiao
332 Metrizable barrelled spaces
J C Ferrando, M López Pellicer and L M Sánchez Ruiz
333 Real and complex singularities
W L Marar
334 Hyperbolic sets, shadowing and persistence for noninvertible mappings in Banach spaces
B Lani-Wayda
335 Nonlinear dynamics and pattern formation in the natural environment
A Doelman and A van Harten

A Doelman

Universiteit Utrecht, The Netherlands

and

A van Harten

Universiteit Twente, The Netherlands

(Editors)

Nonlinear dynamics and pattern formation in the natural environment

LONGMAN

Copublished in the United States with
John Wiley & Sons Inc., New York.

Longman Group Limited
Longman House, Burnt Mill, Harlow
Essex CM20 2JE, England
and Associated Companies throughout the world.

Copublished in the United States with
John Wiley & Sons Inc., 605 Third Avenue, New York, NY 10158

First published 1995

AMS Subject Classifications: 35Q30-55, 76E30, 35B32

ISSN 0269-3674

ISBN 0 582 27371 4

British Library Cataloguing in Publication Data

A catalogue record for this book is available from the British Library

Library of Congress Cataloging-in-Publication Data

Nonlinear dynamics and pattern formation in the natural environment / A. Doelman and A. van Harten (editors).
p. cm. -- (Pitman research notes in mathematics series, ISSN 0269-3674 ; ???)
1. Differential equations, Nonlinear--Congresses. 2. Fluid dynamics--Congresses. 3. Nonlinear theories--Congresses. 4. Pattern perception--Congresses. 5. Self-organizing systems--Congresses. I. Doelman, A. II. Harten, Aart van, 1949- III. Series.
QA377.N654 1995
500.2'01'175--dc20 95-9094
CIP

Printed and bound in Great Britain
by Biddles Ltd, Guildford and King's Lynn

Contents

Preface

The text of this book aims at providing insight in recent developments in the deep and beautiful theory of pattern formation. In the last decade there has been considerable progress in this field, both from a theoretical and a practical point of view. New applications of the theory arose in a multitude of scientific areas such as hydrodynamics, reaction-diffusion problems, oceanography, meteorology, combustion, geophysical and biological morphodynamics, semi-conductors, etc.. The book intends to show the interactions between the mathematical theory of nonlinear dynamics and the study of pattern generating phenomena in the natural environment. There is an intimate relationship between new insights in the mathematical aspects of nonlinear pattern formation and the comprehension of such phenomena. Therefore, there are two, partly overlapping, main themes: one in which the emphasis is put on generally applicable mathematical theories and techniques and one in which the phenomenology of pattern evolution in various areas is discussed. Although the emphasis changes considerably from paper to paper, one can say that in each contribution to this book the same two themes are present: all authors have aimed at achieving a suitable balance between the mathematical theory and the physical phenomena.

The growing interest of the scientific community in nonlinear dynamics and the phenomena of pattern formation in 'non-classical' application areas is reflected in the contributions by Komorova and Newell (river bed dynamics), Manneville (climate dynamics), Or-Guil et al. (semi-conductors), Segel (biological pattern formation), Seminara (river morphodynamics) and de Swart and Hulscher (coastal seas). On one hand, these papers open up rich new fields where the application of mathematical theories leads to better understanding of observations, on the other hand one encounters interesting phenomena which yield challenges for the further development of the mathematical theory. The papers by Doelman et al., Hărăguş and Kirchgässner, Grimshaw and Soto-Treviño and Kaper can be regarded as responses to such a challenge: model problems based on physical phenomena – ranging from meandering rivers or water waves to coherent structures in the ocean or nerve impulse transmission – are analysed in full mathematical detail.

Supplementary to the growing interest in 'new' application fields is the continuing evolution of the theory for the classical problems of pattern formation. The papers by Busse, Dijkstra and Manogg and Metzener are all addressing aspects of the very rich field of fluid convection. It is clear from these contributions that the recent developments in the mathematical theory of nonlinear dynamics provide important new insights. The same is certainly true for the other papers on extensions of classical problems: Clercx and van Heijst (coherent structures), Bayliss and Matkowski (cellular flames) and Renardy and Renardy (channel flows).

In many of the above mentioned papers the authors studied the nonlinear stability of a system at near-critical conditions by an appropriate (system of) modulation (or amplitude) equation(s). The papers by Bollerman et al., Knobloch and Mielke consider some of the foundations of these methods, respectively the validity and the limitations of the Ginzburg-Landau equation and the justification of the sideband-instability mechanism.

The contributions to this book are all selected from key presentations at the international conference on 'Nonlinear Dynamics and Pattern Formation in the Natural Environment', held in Noordwijkerhout, the Netherlands, July 4-7 1994. Altogether, the presentations at this conference, and in particular the contributions to this volume, show that the field of nonlinear dynamics and pattern formation is very much alive.

Therefore it is a pleasure to dedicate this book to one of the pioneers in this field: Wiktor Eckhaus, an early explorer of the basic principles in nonlinear stability theory, who has contributed much to the present understanding of the subject. On the occasion of his sixty-fifth birthday, many of his friends and colleagues want to recognise his merits through their contribution to this volume. It is very appropriate to quote from his book on the stability analysis of two-dimensional flows written in Paris, 1965: "I hope that a more unified theory thus has been achieved", from which it is clear that he recognised the importance and universality of the approach leading to what is nowadays called the Ginzburg-Landau equation at a very early stage. We mention here that he made substantial contributions to several other areas of applied mathematics as well, such as singular perturbation theory, perturbed Korteweg-de Vries equations and flame instabilities. Thus, it is particularly suitable that this volume, dedicated to him, addresses some of these topics as well.

Finally, we want to acknowledge the support of N.W.O. – the Dutch science foundation – , the European Union and the Mathematical Institute of the University of Utrecht for the above mentioned conference which lies at the basis of this text, and the invaluable help of Karen Schoenmaker with the technical editing of the manuscript.

Arjen Doelman
Aart van Harten

Utrecht, April 13, 1995.

Acknowledgements

We are grateful to the following for permission to reproduce copyright material:

Academic Press Ltd for Figure 2, p.260, Figure 3, p.261 and Figure 4, p.264; American Society of Civil Engineers for Figure 2, p.272; Cambridge University Press for Figure 2, p.56; Elsevier Science Publishers BV for Figure 1, p.271 and Figure 3, p.273; Faculty of Geographical Sciences, Utrecht University for Figure 1, p.316; Nederlandse Natuurkundige Vereniging for Figure 1, p.54; Springer-Verlag GmbH & Co KG for Figure 1, p.259.

Nonlinear evolution of cellular flame instabilities

A. BAYLISS and B.J. MATKOWSKY
Department of Engineering Sciences and Applied Mathematics, Northwestern University, Evanston, IL 60208, U.S.A.

Abstract

We describe cellular flames in a cylindrical geometry and focus on rotating waves traveling along the flame front. We consider two distinctly different parameter ranges, adiabatic flames with $Le < 1$ and nonadiabatic flames with $Le > 1$, where the Lewis number Le is the ratio of thermal diffusivity to mass diffusivity of a deficient component of the mixture. Both regimes exhibit transitions from uniformly rotating traveling waves (TWs) to modulated traveling waves (MTWs) and further transitions to more complex behavior, ultimately leading to chaos. We describe several different scenarios

1 Introduction.

Premixed flames often exhibit cellular structures, characterized by periodic arrays of crests pointing in the direction of the combustion products. The pointed crests are connected by smooth troughs that are convex toward the fresh fuel mixture. The temperature is higher at the troughs, which therefore appear brighter, and lower at the crests, which therefore appear darker [19].

In many instances the transition from laminar to turbulent combustion occurs as a smooth flame breaks up into cells which then undergo successive transitions leading to progressively more complex spatial and temporal behavior. Some of these transitions have been observed experimentally. For example the cells can exhibit traveling wave (TW) behavior (see e.g., [14, 19]). Alternatively the cells can exhibit non-periodic dynamics (possibly quasi-periodic or chaotic) [12, 13, 14]. The symmetry in which all cells are identical can also be broken so that different cells can have a different spatial structure [12, 14]. Finally, the cellular flame can exhibit an erratic motion, in which cells are randomly created and annihilated [13, 19].

One scenario for the development of cellular flames is that of thermo-diffusive cellular instabilities which arise as certain parameters are varied [18, 20, 24]. In fact, many of the complex spatio-temporal patterns observed in flames can be qualitatively described as thermo-diffusive instabilities. The thermo-diffusive model of flames was first introduced in [26, 27, 28, 29]. rather amazing that this simplified model is able to capture so many interesting effects. In this model the thermal expansion of the gas is assumed to be weak. As a result the density is taken to be constant and the velocity field is assumed to be unaffected by the chemical reaction and so can be specified beforehand from a calculation of the nonreacting gas flow problem. The model is described by a system of two nonlinear partial differential equations for the reduced temperature and the concentration of a deficient component of the fuel mixture. Analytical methods have been successful in describing primary transitions from uniform to cellular combustion (e.g. [1, 11, 18, 20, 22]). In some cases higher order transitions can be found by analytical methods as well [21, 22]. In most instances, however, secondary and higher order transitions are difficult to find analytically and are more readily found by numerical solution of the systems of partial differential equations modeling combustion.

In this paper we consider (i) adiabatic cylindrical flames stabilized by a line source of fuel and (ii) flames established in the region between two coaxial cylinders, with fuel fed in through the inner cylinder and heat loss occurring through the outer cylinder. The computations build on and extend analytical results obtained for the problem of a flame stabilized by a line source of fuel. We first describe the analytical results.

We consider the diffusional thermal model for a flame stabilized by a line source of fuel of strength $2\pi\kappa$. This model has been analyzed for the limiting case of infinite activation energy [11, 20]. A stationary axisymmetric solution describing a circular flame front located at $r = \kappa$ has been found for all κ and all Le, where the Lewis number Le is the ratio of thermal diffusivity to mass diffusivity of a deficient component of the fuel mixture. This solution, called the basic solution, is subject to two different classes of instabilities depending on whether $Le > 1$ or $Le < 1$.

There exists a critical value, $Le_{c1} > 1$, such that for $Le > Le_{c1}$, the basic solution is unstable, with two complex conjugate eigenvalues crossing into the right half plane, suggesting the evolution to nonstationary flames, such as axisymmetric pulsating flames, as well as traveling and standing waves along the flame front. Similar behavior has also been found in other geometries. The regime $Le > Le_{c1}$, describing, e.g., lean, heavy hydrocarbon/air mixtures, is referred to as the pulsating regime.

Below a second critical value $Le_{c2} < 1$ there exists a κ_{c2} such that for $\kappa > \kappa_{c2}$ the basic solution is unstable, with a real double eigenvalue crossing into the right half plane, suggestion evolution to stationary cellular flames [11, 20]. This behavior has also been found in other geometries [2, 16, 18, 21, 24]. The regime $Le < Le_{c2}$, describing mixtures where the deficient component is highly diffusive, e.g., lean hydrogen/air mixtures or rich, heavy hydrocarbon/air mixtures, is referred to as the cellular regime.

The analysis in [11, 20] provides an explanation for the onset of cellular flames as a diffusional thermal instability. For $Le > 1$ the analysis also explains the onset of nonstationary behavior. However, the analysis does not describe the subsequent evolution to more complex spatial and temporal patterns and in particular the transition to chaotic behavior. Complex patterns may arise through secondary or higher order bifurcations and transitions, which can be found numerically.

The analysis in [11, 20]) identified the roles of the two fundamental parameters in promoting diffusional thermal instabilities of cylindrical flames. In the cellular regime the stationary circular flame front can be destabilized by either increasing κ or decreasing Le. In the pulsating regime increasing Le is destabilizing. Analytical methods have also identified heat loss as a destabilizing mechanism [3, 15, 25]. In this paper we describe examples of complex spatial and temporal behavior for both the regimes $Le < 1$ and $Le > 1$. In both cases we concentrate on the development of modulated traveling waves (MTWs) and their subsequent transitions to chaotic behavior. For the cellular regime we consider the behavior as Le is decreased for fixed κ. For the pulsating regime we consider the role of heat loss in generating MTWs.

In previous work a variety of higher order instabilities associated with cellular flames have been found numerically. The behavior of stationary cellular flames was considered in [9] where modal transitions between stationary cellular flames of increasing mode number occurred as κ was varied, with Le held fixed. Transitions from axisymmetry to a 3-cell, to a 4-cell, to a 5-cell and to a 6-cell stationary cellular flame were computed. All of the transitions, except for the first, are accompanied by a region of bistability where stationary cellular solutions of adjacent mode numbers coexisted, each with its own domain of attraction. We expect that unstable mixed mode solution branches connect the computed (stable) cellular branches. Since we determine solutions by solving an initial boundary value problem and marching forward in time until a steady state is achieved, we can generally only compute stable solutions.

In [7] transitions to nonstationary behavior were found as Le was reduced. These solutions described very slowly rotating cellular flames. The rotation occurred at a constant angular velocity and corresponded to a traveling wave (TW) along the flame front. This transition from stationary to rotating behavior occurred as an infinite period, symmetry breaking bifurcation, in which the reflection symmetry of the cells was destroyed. We refer to this branch as the TW4 branch. Mixed mode branches, i.e., branches composed of incommensurate spatial modes, appeared after the transition to nonstationary behavior and were identified as MTWs in [10]. These are solutions in which after a certain time interval the solution recovers its structure along the front except for a phase shift. Thus the solution becomes periodic when viewed in an appropriately chosen rotating coordinate system. Both periodically modulated and quasiperiodically modulated waves were found.

Higher order transitions have also been found for $Le > 1$ for both adiabatic and nonadiabatic flames [5, 6]. Transitions from stationary to singly periodic to doubly periodic pulsating flames were found for fixed κ as Le was increased. Pulsating cellular flames were also found. As κ was increased transitions between traveling and standing waves were found with regions of bistability.

In [8] the behavior of nonadiabatic pulsating flames was considered. In this case the problem of a flame established in the region between two coaxial cylinders with fuel fed in through the inner cylinder and heat loss through the outer cylinder was computed. The solution behavior was determined as the heat transfer coefficient η through the outer wall was increased. Transitions from TW solutions to MTW solutions occurred when η exceeded certain critical values. Unlike the MTW solutions in the cellular regime these transitions are due to the introduction of subharmonic or near subharmonic modes in the solution as η increased. Strikingly different behavior was found depending on whether a true subharmonic mode could be excited (even number of cells) or whether only near subharmonics (odd number of cells) could enter. In both cases a subsequent transition to chaos occurred when η became sufficiently large and the flame was near extinction. We illustrate some of the results from [8] here as well as providing new results in this parameter regime.

Our computations are for the case of finite, though large, activation energy. In this case the chemical reaction terms are important only in a narrow reaction zone. In the limit of infinite activation energy the reaction zone shrinks to a surface called the flame front. The analyses in [11, 20] were obtained in this limit. In the case of finite activation energy there is, strictly speaking, no flame front, although we will use this terminology to describe the solution.

The accuracy of the computed solution is very sensitive to the resolution of the reaction zone, the location of which is not known beforehand. As a result, adaptive procedures are needed to locate and resolve the reaction zone as the solution evolves in time. We employ an adaptive pseudo-spectral method which has been used successfully for a variety of other problems in combustion (see, e.g., [5, 6]).

2 Mathematical model.

We first consider the problem of a flame stabilized by a line source of fuel of strength $2\pi\kappa$. We assume that the reaction is limited by a single deficient component and is governed by one-step, irreversible Arrhenius kinetics. We denote dimensional quantities by $\tilde{}$. The unknowns are the temperature $\tilde{T}$ and the concentration $\tilde{C}$ of the deficient component. $\tilde{T}_u$ and $\tilde{T}_b$ are the temperatures of the unburned and burned fuel, respectively, and $\tilde{C}_u$ is the unburned value of $\tilde{C}$. Other dimensional quantities are the coefficient of thermal conductivity $\tilde{\lambda}$, the activation energy $\tilde{E}$, and the gas constant $\tilde{R}$. We introduce the non-dimensional reduced temperature and nondimensional concentration by $\Theta = (\tilde{T} - \tilde{T}_u)/(\tilde{T}_b - \tilde{T}_u)$, $C = \tilde{C}/\tilde{C}_u$. The independent variables are suitably non-dimensionalized. We let t denote non-dimensional time and (r, ψ) polar coordinates. The nondimensional velocity field is $\hat{\mathbf{V}} = (\kappa/\mathbf{r})\,\hat{\mathbf{r}}$, where $\hat{\mathbf{r}}$ is the unit radial vector. The equations of the diffusional thermal model are [20]

$$\begin{aligned} \partial_t\Theta &= \Delta\Theta - \frac{\kappa\Theta_r}{r} + C\Lambda\exp\left(\frac{N(1-\sigma)(\Theta-1)}{\sigma+(1-\sigma)\Theta}\right), \\ \partial_t C &= \frac{\Delta C}{Le} - \frac{\kappa C_r}{r} - C\Lambda\exp\left(\frac{N(1-\sigma)(\Theta-1)}{\sigma+(1-\sigma)\Theta}\right). \end{aligned} \tag{1}$$

Here Δ is the Laplacian, $\sigma = \tilde{T}_u/\tilde{T}_b, N = \tilde{E}/(\tilde{R}\tilde{T}_b)$, Le is the Lewis number, and $\Lambda = M^2/(2Le)$, where $M = N(1-\sigma)$. We note that Λ, which is referred to as the flame speed eigenvalue, depends on the non-dimensionalization. The value employed above arises from the use of the planar, adiabatic flame velocity in the infinite activation energy limit $M \to \infty$. The boundary conditions are $C \to 1(0)$, $\Theta \to 0(1)$ as $r \to 0(\infty)$. In our calculations these boundary points are imposed at finite points sufficiently far from the reaction zone so that there was no effect on the solution.

The model for flames established between two coaxial cylinders is similar, except that at the outer cylinder we allow for heat transfer so that $\partial_r\Theta = -\eta(\Theta - \Theta_w)$, where η is a nondimensional heat transfer coefficient and Θ_w is the nondimensional temperature exterior to the outer wall. We describe transitions obtained by increasing η.

3 Cellular regime.

We first describe results obtained in the cellular regime, $Le < 1$. The following parameters were held fixed, $N = 20, \sigma = .615, \kappa = 14.8$. An overview of the different solution branches is given in Fig. 1 where we plot the maximum norm of the difference between the computed cellular solution and the (unstable) axisymmetric solution as a function of Le. We generally concentrate on determining the nature of the transitions that occur and on the properties of solutions along different solution branches rather than on determining the precise numerical values of Le at which transitions occur. We note that we solve the initial value problem, marching forward in time until a steady state is achieved, so that in general we only compute stable solutions. However unstable solutions can sometimes be computed by modifying the computer program to impose certain symmetries in the angular direction. For example, unstable axisymmetric solutions can be computed by allowing no variation in ψ. Unstable 4-cell solutions can be computed by restricting the computation to the interval $0 \leq \psi \leq 2\pi/4$ and imposing periodicity. Unstable reflection symmetric solutions can be computed by restricting to the interval $0 \leq \psi \leq 2\pi/8$ and imposing reflection symmetry.

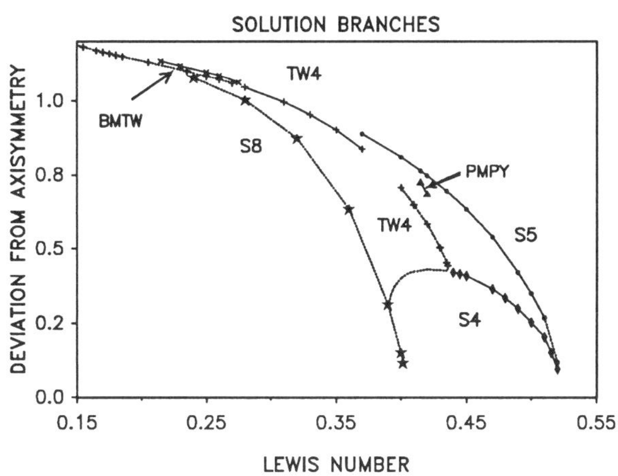

Figure 1: Solution branches as a function of Le.

We find 3 branches of stationary solutions, S4, S5 and S8, which have, 4, 5 and 8 cells respectively. The cells associated with solutions along these branches are reflection symmetric. Only the S4 branch bifurcates stably from the basic axisymmetric state; the S5 and S8 branches are unstable at onset. The S5 branch becomes stable, however, for smaller Le. This behavior is not unexpected since in general only the state closest to the minimum of the neutral curve is stable to sideband instabilities at onset, whereas the other states become stable at the Eckhaus stability boundary. When decreasing Le the S4 branch merges with the S8 branch, i.e. the first harmonic becomes increasingly stronger whereas the fundamental eventually goes to zero. Thus, the S4 branch bifurcates from the S8 branch in a pitchfork bifurcation. We did not follow the S8 branch for values of Le below those indicated in Fig. 1.

Before reaching the S8 branch, the S4 branch becomes unstable to perturbations which break the reflection symmetry (parity-breaking bifurcation) [7]. To calculate solutions on the S4 branch beyond this point we therefore restrict the solutions to be reflection symmetric. At the parity-breaking bifurcation point a branch of traveling 4-cells, TW4, emanates from the S4 branch. These waves are similar to the rotating cells observed in [14] which were also found to be asymmetric while the nonrotating cells were symmetric. In the computations the rotation rate goes to 0 continuously at the bifurcation point where the asymmetry vanishes [7], indicating an infinite period bifurcation. Strikingly, the TW4 solutions appear to be unstable at onset, although the bifurcation corresponds to a supercritical pitchfork bifurcation. Perturbed solutions evolve either to a solution on the S5 branch or, in a small window of Le, to a solution on a MTW branch which we refer to as the "Pushmi-Pullyu" (PMPY) branch, as discussed below. The relevant instability has the character of a sideband instability and can be suppressed by restricting the domain of computation to $0 \le \psi \le 2\pi/4$ and imposing periodicity. In the full domain the TW4 solutions eventually become stable for smaller Le ($Le \le 0.37$). The instability close to onset may explain why the rotation rates of the waves observed in the experiments [14] do not approach 0 continuously as the transition point is approached. Analyses describing both the evolution of the TW4 branch from the S4 branch and the formation of the PMPY branch is given in [10] and briefly sketched below.

When Le is decreased further, the TW4 solutions become unstable again and a branch of modulated waves arises in what appears to be a supercritical Hopf bifurcation. Due to the visual appearance of a breathing of the cells we call this 2-frequency state a breathing MTW (BMTW). For even smaller values of Le two additional, different branches of modulated waves arise; the quasiperiodic MTW (QPMTW), which arises from the BMTW branch and appears to be a quasiperiodically modulated wave with three frequencies, and the half period MTW (HPMTW) branch which bifurcates from the (unstable) TW4 branch. These branches are shown in detail in Fig. 2.

We first consider the PMPY branch. The 4 cells are no longer identical; two of the cells are nearly reflection symmetric, characteristic of the behavior along the branch S4, while two of the cells are not symmetric, characteristic of the behavior along the branch TW4. The dynamics of this solution is shown in Fig. 3 where $\Theta(r_*, \psi, t)$ is shown as a function of ψ for increasing values of t. Here and in the following r_* is chosen so that the circle $r = r_*$ is in the reaction zone. As the forward crest (temperature minimum) of a given cell (cell 3 in the figure if we number the cells from the left) is jerked clockwise (to the left in Fig. 3), the cell expands, becomes asymmetric and moves clockwise. This, in turn, pulls the cell behind it (cell 4) so that it too expands, becomes asymmetric and moves clockwise. At the same time cell 3 pushes the cell ahead of it, compressing it and causing it to become nearly symmetric and nearly stationary. This process then repeats with cells 1 and 2 etc. in turn and can be

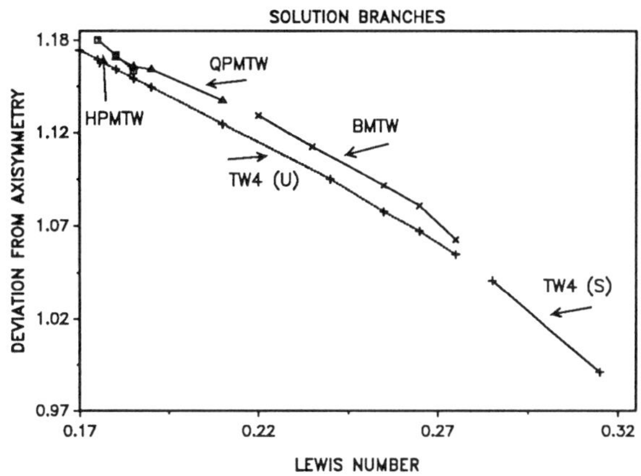

Figure 2: Solution branches over restricted region in Le.

viewed as a *localized* wave of asymmetry propagating counter-clockwise through the stationary symmetric cellular array. This dynamics suggests that the solution is best described as a "Pushmi-Pullyu" [17] solution. Its behavior is analogous to that of the hopping modes observed experimentally in [12].

Alternatively, Fig. 3 shows that the dynamics of the PMPY solution can also be characterized as that of a MTW, since after a discrete time τ the pattern reemerges, though shifted in space,

$$\Theta(\psi - c\tau, t + \tau) = \Theta(\psi, t). \tag{2}$$

Thus, in a coordinate system rotating uniformly with speed c, $\tilde{\psi} = \psi - ct$, the solution would be periodic. In addition, the PMPY solutions possess the more restrictive symmetry

$$\Theta(\tilde{\psi} + 2\pi/4, t + \tau/4) = \Theta(\tilde{\psi}, t), \tag{3}$$

which implies that all 4 cells perform the same dynamics though shifted in time by 1/4 of the period in the rotating coordinate system. In the context of a discrete system with k cells the symmetry

$$\Theta(\tilde{\psi} + 2\pi/k, t + \tau/k) = \Theta(\tilde{\psi}, t), \tag{4}$$

referred to as "ponies on a merry-go-round" (POM), has been observed [4].

The BMTW solutions exhibit the same symmetry (3) as the PMPY solutions. The form of modulation, however, is different. This is shown in Fig. 4 which illustrates the temporal evolution of $\Theta(r_*, \psi, t)$. Analysis of the spectrum of the computed solutions indicates that the BMTW branch arises due to a sideband instability whereby TW4 solutions are modulated by a mode with wave number 1.

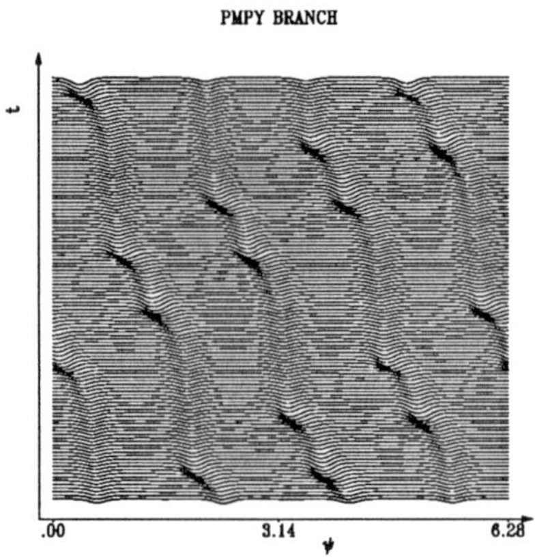

Figure 3: $\Theta(r_*, \psi, t)$ for solution on PMPY branch, $Le = 0.42$, $r_* = 13.5$. Time t (angle ψ) increases along vertical (horizontal) axis.

The following dynamical behavior is observed. Similar to the PMPY solutions the cells undergo an overall rotation with a modulation of cell size and intensity. In contrast to the PMPY solutions, the rotation rate never gets close to 0 and the cells are asymmetric at all times. Thus the asymmetry is global rather than localized as for the PMPY solutions. In order to illustrate the nature of the modulation, we consider the maxima of $\Theta(r_*, \psi, t)$. We find that for each fixed t, Θ attains exactly 4 maxima for $0 \leq \psi < 2\pi$ which we denote by $\hat{\Theta}_i (i = 1, \ldots 4)$. Clearly each such maximum is identified with an individual cell and may be thought of as a measure of the intensity of the cell. Examination of $\hat{\Theta}_i$ for adjacent cells indicates that the modulation is periodic (close to sinusoidal) and is essentially the same for each cell except for a constant phase shift. Similar properties hold for the other two cells as well. The phase shift is the same between all pairs of adjacent cells. In addition to their breathing motion the cells undergo a periodic oscillation in rotation rate.

When Le is reduced to about $Le = 0.21$ the BMTW branch becomes unstable and a transition to the QPMTW branch occurs, as the symmetry (3) is broken. Near the transition point the symmetry breaking is most pronounced in the phase differences between pairs of adjacent cells. Each cell also undergoes a modulation in size, similar to the dynamics of the BMTW solutions. However, adjacent cells now undergo different modulations and the 4 cell array splits into two pairs. Within each pair the modulations are nearly in phase while there is a large phase difference between the two pairs. This behavior has the character of a spatial period doubling. In Fig. 5 we plot $\Theta(r_*, \psi, t)$ for Le=0.180. We see that cells 1 and 2 (numbered from the left) have a small phase difference between them as do cells 3 and 4. In contrast, the phase difference between cells 2 and 3 and between cells 4 and 1 is considerably larger.

While Fig. 5 gives the visual appearance of a periodic modulation, closer exam-

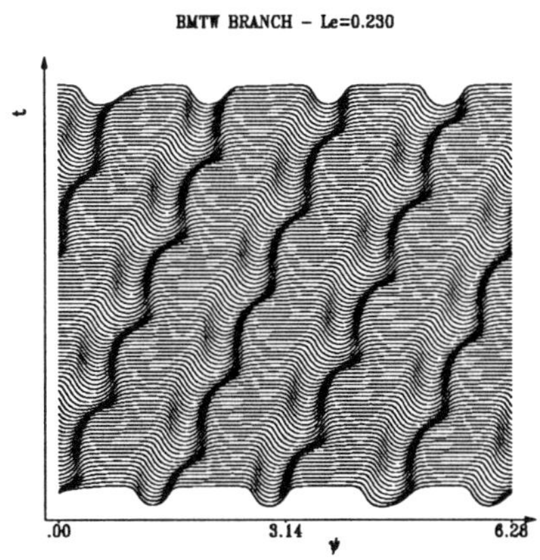

Figure 4: $\Theta(r_*, \psi, t)$ for solution on BMTW branch, $Le = 0.23$, $r_* = 10.4$. Time t (angle ψ) increases along vertical (horizontal) axis.

ination reveals that the amplitude of the modulation is quasiperiodic. This can be seen from the temporal evolution of the $\hat{\Theta}_i$. Examination of this quantity for each individual cell indicates that the modulation is quasiperiodic, with a high frequency oscillation modulated by a low frequency envelope. Alternate cells have the same envelope, and their modulations differ only by a constant phase shift with respect to the envelope. Adjacent cells have distinctly different envelopes. In the laboratory frame the rotation rate will add a third frequency to the spectrum.

Following the QPMTW branch further by decreasing Le, we find a transition to solutions exhibiting seemingly random creation and annihilation of cells through phase slips. For values of Le near the transition point, the chaotic behavior is characterized by intervals where cells are created and annihilated without undergoing any organized motion, alternating, in a seemingly random fashion, with intervals in which there are 4 cells which undergo a roughly organized motion similar to that of QPMTW solutions. This general type of behavior is qualitatively similar to the intermittently ordered states observed in [13]. Note that although a transition to chaos can generally be expected from a branch of quasiperiodic solutions with three independent frequencies, here the transition may not be via the quasiperiodic route as a result of symmetries. This is also supported by the fact that phase slips are very important in the chaotic dynamics so that the qualitative features of the chaotic attractor strongly differ from those of the quasiperiodic attractor.

We also find a narrow window in Le in which a fourth branch of stable modulated waves exists. In contrast to the BMTW and QPMTW solutions, however, alternate cells are *in phase*, i.e. the solution has the discrete translation symmetry $\Theta(\psi + 2\pi/2, t) = \Theta(\psi, t)$, which does not involve any phase shifts. Thus the spatial period is only half the system size, as compared to the BMTW solutions which only have the

QPMTW BRANCH - Le=0.185

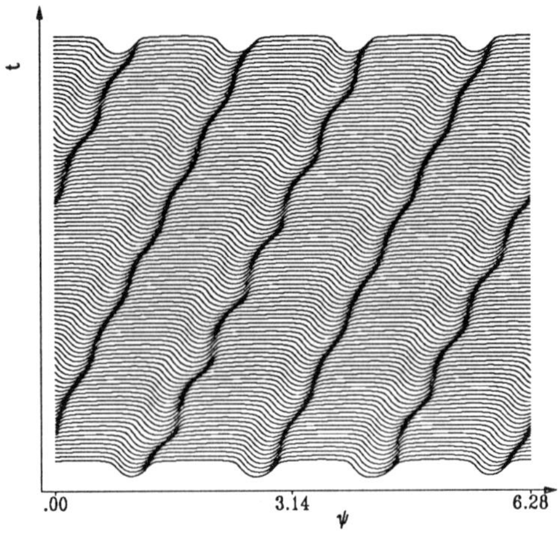

Figure 5: $\Theta(r_*, \psi, t)$ for QPMTW solution , $Le = 0.185$, $r_* = 10.0$. Time t (angle ψ) increases along vertical (horizontal) axis.

period 2π. We therefore call this new branch a HPMTW (Half Period MTW). The HPMTW branch appears to bifurcate from the TW4 branch. Since the TW4 solutions have the spatial symmetry $\Theta(\psi + 2\pi/4, t) = \Theta(\psi, t)$, the bifurcation corresponds to a pure period doubling in space. For comparison, the analogous transition from the TW4 branch to the BMTW branch constitutes a period quadrupling. A solution along the HPMTW branch is illustrated in Fig. 6.

The HPMTW branch can be obtained by following the (unstable) TW4 branch and imposing periodicity over the interval $0 \leq \psi \leq \pi$. From the computation in the restricted domain we have observed that the subharmonic mode (mode 2) appears to grow continuously from 0 as Le is reduced thus indicating that the HPMTW branch bifurcates from the TW4 branch. At onset the HPMTW branch is unstable when considered over the full angular interval $0 \leq \psi \leq 2\pi$. This is expected since it bifurcates from the TW4 branch which is itself unstable for this range of Le. In the full domain the HPMTW branch becomes stable for lower values of Le. Both in the restricted and the full domain the HPMTW branch loses stability to an apparently chaotic attractor for Le approximately 0.165. Thus, its range of stability extends beyond that of the QPMTW branch and at $Le = 0.170$ the HPMTW branch coexists stably with an apparently chaotic solution arising from an instability of the QPMTW solutions for Le sufficiently small.

For appropriate parameter values a new class of solutions can be found in which abrupt changes in the flame front, having a kink-like character, occur. These kinks rotate as time progresses. We have found examples where the kinks rotate at a uniform

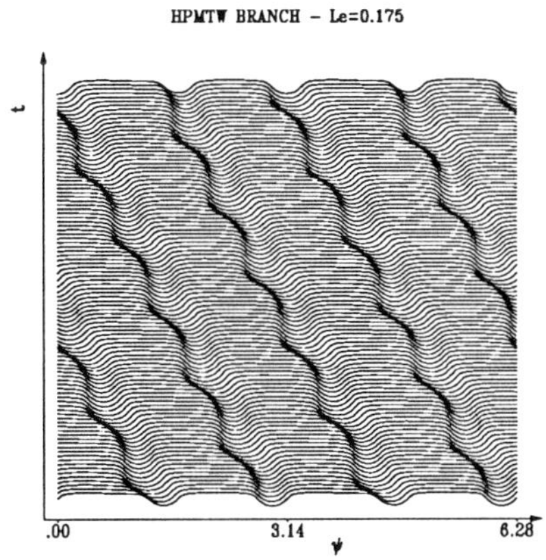

Figure 6: $\Theta(r_*, \psi, t)$ for solution on HPMTW branch, $Le = 0.175$, $r_* = 9.8$.

rate, thus describing traveling waves, and at an oscillatory rate, thus describing modulated traveling waves. Both single and double kinks have been found. An example of a single kink mode is shown in Fig. 7, where the flame location (characterized by the location of the maximum of the reaction rate term for each ray emanating from the origin) is shown at a fixed instant of time. We note the qualitative similarity to modes found in [14].

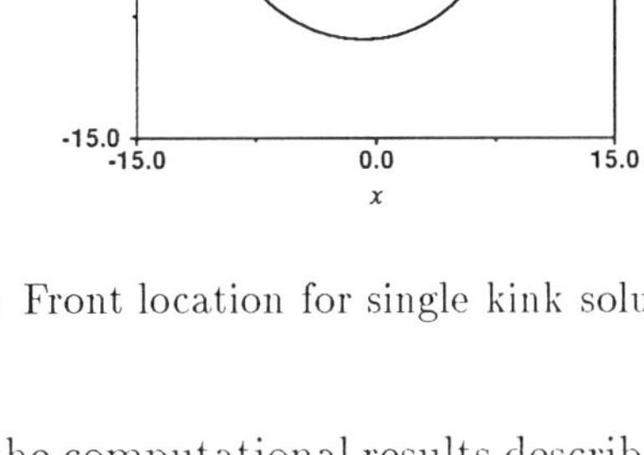

Figure 7: Front location for single kink solution, $Le = 0.23$.

Various features of the computational results described here can also be understood analytically in the framework of simpler equations which can, in principle, be derived from (1) in certain asymptotic limits. The origin of the parity-breaking bifurcation can be traced to a resonant mode interaction of modes with wave numbers q and $2q$, respectively. The stability of solutions on the TW4 branch, as well as the relation of the TW4 branch to the PMPY branch can be explained by considering phase-amplitude equations for perturbations of the S4 branch. The forms of these equations can be derived from general symmetry considerations and scaling arguments. Under certain conditions it can be shown that the resulting phase-amplitude equations will admit traveling wave solutions which are unstable to small perturbations. Further analysis of the system indicates that traveling wave solutions can stabilize further from the bifurcation point as occurs in the computation. The existence of localized, propagating waves waves of asymmetry, as in the PMPY solution branch, can also be shown. We refer to [10] for the details of these analyses.

4 Pulsating regime.

We now consider the behavior of nonadiabatic pulsating cellular flames with $Le > 1$. We describe computations for 8 cell, 7 cell, 6 cell and 5 cell traveling waves. In all cases the control parameter is η and the solutions are pure traveling waves for $\eta = 0$. For each set of computations the following parameters were held fixed, $N = 20, \sigma = 0.5$, $Le = 4.4$,. The numbers of cells was determined by varying κ, Θ_w and δ. For all computations we considered the effect of increasing η until extinction occurs. The predominant effect of increasing η is to promote the development of subharmonic modes along the front. This leads to different behavior for the different cases considered.

We first describe the case where the basic solution is an 8-cell TW. When the subharmonic, mode 4, enters, it breaks the symmetry corresponding to $2\pi/8$ periodicity, leading to a symmetry based on $2\pi/4$ periodicity. There is now an 8-cell array in which cell heights alternate. Correspondingly the period (in time) of the solution at any fixed point in space is doubled. Thus there is a space/time period doubling. Each cell is unaltered as it rotates and the resulting solution is still a TW. Increasing η further leads to additional transitions in which subharmonic modes enter. Mode 2 enters for η between 15 and 18.5 leading to a second space/time period doubling. Mode 1 enters for a value of η between 18.5 and 18.8 leading to a third space/time period doubling. This sequence of transitions cannot continue beyond this point as there are no additional subharmonics. Shortly after the third period doubling there is apparently chaotic temporal dependence [8]. We have not found any other stable state between the third period doubled state and the apparently chaotic state.

We next consider the case where the basic solution is a 7-cell TW. In this case there is no exact subharmonic. Increasing η leads to a breaking of the symmetry of $2\pi/7$ periodicity by the introduction of the near subharmonic modes 3 and 4. While this behavior is qualitatively similar to that observed for the 8-cell TW, the introduction of the two modes leads to radically different behavior. Over short timescales there is a roughly alternating behavior in cell size. However, the pattern cannot be regular because of the odd number of cells. Thus at any fixed instant of time the spatial pattern looks roughly period doubled in ψ, i.e. there is a rough alternating behavior in the size of the cells as ψ increases from 0 to 2π. This alternating behavior is superimposed on a

TW so that the resulting structure rotates. Over short timescales the temperature (or concentration) at a fixed point reflects this rough period doubling in ψ. In particular, as the alternating array of cells rotates past the fixed point, the time evolution of the temperature also exhibits a rough period doubled type of behavior. However, the spatial period doubling is not maintained over long timescales, as an alternating pattern is not compatible with an odd number of cells. Over long timescales each cell undergoes an oscillation in both cell size and maximum temperature. This cell oscillation is such that there is a constant phase difference between successive cells. The result is a MTW.

With the introduction of modes 3 and 4 into the solution there is a pulsation in the temperature of the cells. The modulation is initially sinusoidal but quickly becomes non-sinusoidal as η is increased. In Fig. 8 we plot the temperature of each cell as a function of time for $\eta = 15.3$. This figure is obtained by examining the data at a value of r_* in the reaction zone, in the manner described in [8]. This value of η is near the transition point where the TW is destabilized by the subharmonic modes. For η closer to the transition point a more sinusoidal modulation is obtained [8].

Examination of the solution at a fixed spatial location indicates quasi-periodicity. There are rapid oscillations within an envelope over a longer length scale. Each individual spike or maximum in Θ corresponds to the arrival of a temperature peak (associated with a cell) at the given spatial location. The long envelope corresponds to modulation of the peak temperature within each cell. An important difference between these MTWs and those found for $Le < 1$ is that for $Le < 1$ there is a noticeable modulation in the rotation rate as well as cell size and maximum temperature, while for $Le > 1$ the rotation rate, suitable defined, e.g., the rotation rate of each temperature peak, is approximately constant, though the cell size and maximum temperatures are modulated.

Often, transitions from one pattern to another are described by a modulation of the first pattern. We describe the modulation of the 7-cell traveling wave solution by a function generated by modes 3 and 4. We consider variations of Θ in ψ at a fixed radial location. For purposes of illustration, we describe the 7-cell traveling wave by the functional form

$$\Theta_{TW}(\psi,t) = A + B\cos(\omega_7 t + 7\psi),$$

where A and B are constant. We assume that the modulated traveling wave $\Theta_{MTW}(\psi,t)$ is given by

$$\Theta_{MTW}(\psi,t) = (1 + \epsilon\cos(\omega_3 t + 3\psi) + \epsilon\cos(\omega_4 t + 4\psi))\Theta_{TW}(\psi,t), \tag{5}$$

where ϵ is small near the transition point. In (5) we have taken the amplitudes of modes 3 and 4 to be equal, in accordance with the data in our computations. To leading order the maxima of Θ_{MTW} occur at $\psi^l(t) = (-\omega_7 t + 2\pi l)/7$, for $l = 0, \ldots, 6$. Consider for simplicity one of the maxima, say $l = 0$. The peak value of Θ_{MTW} for $l = 0$ satisfies, to leading order

$$\Theta^0_{MTW}(t) = (1 + \epsilon\cos(\omega_3 t + 3\psi^0(t)) + \epsilon\cos(\omega_4 t + 4\psi^0(t)))(A + B). \tag{6}$$

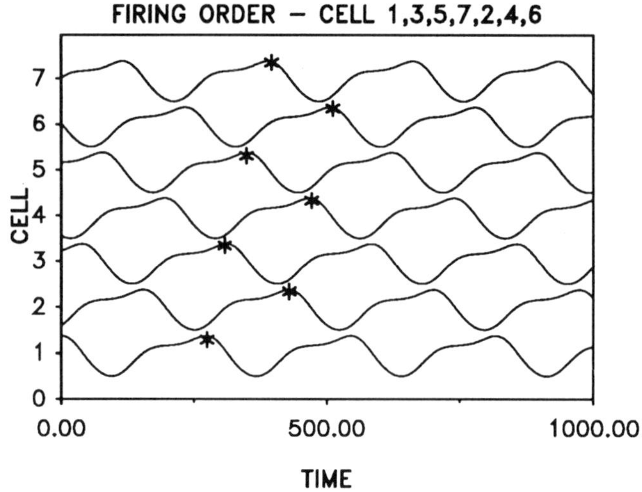

Figure 8: Cell temperatures for all 7 cells, $\eta = 15.3$, $r_* = 32.714$, illustrating the JPOM symmetry. The asterisks illustrate the "firing order", cell 1,3,5,7,2,4,6

Thus (6) describes the oscillation of the cell peak, which is the same quantity that is presented in Fig. 8 for the full system of PDEs. We now consider in greater detail the behavior in time of Θ^0_{MTW}. Suppose first that the frequencies ω_3 and ω_4 are locked to the frequencies $3\omega_7/7$ and $4\omega_7/7$ respectively. In this case Θ^0_{MTW} is independent of time so that $\Theta_{MTW}(\psi, t)$ is periodic in t and describes a pure traveling wave. However, in our problem we find that ω_3 and ω_4 are detuned from $3\omega_7/7$ and $4\omega_7/7$, $\omega_3 - 3\omega_7/7 = 4\omega_7/7 - \omega_4 = 0.0230$. It follows that

$$\Theta^0_{MTW}(t) = 1 + 2\epsilon\cos(\Omega t),$$

where $\Omega = \omega_3 - 3\omega_7/7 = 4\omega_7/7 - \omega_4$ is the degree of detuning.

We next describe the symmetries associated with the MTW described by (5). If the MTW satisfied the symmetry (4) then, ordering the cells sequentially around the circle, cell 1 would achieve a maximum first and then cells 2, 3, 4 ...in turn. Examination of Fig. 8 indicates that this is not the case. Rather, there is a jump of 1 cell, i.e., first cell 1 heats up then cell 3 etc.. Thus, the cells "fire" in the order 1, 3, 5, 7, 2, 4, 6.

This is a manifestation of the more general symmetry

$$\Theta(\tilde{\psi} + l \times 2\pi/k, t + \tau/k) = \Theta(\tilde{\psi}, t),$$

where $l = 2$ and $\tilde{\psi} \equiv \psi - ct$ is the coordinate in the moving frame. Whereas the POM symmetry relates each cell to the cell adjacent to it, this symmetry relates cells according to the sequence $1, l+1, 2l+1, ...(\text{mod } k)$, effectively skipping the $l-1$ cells between successive members of the sequence. Thus, for example, if at a certain time

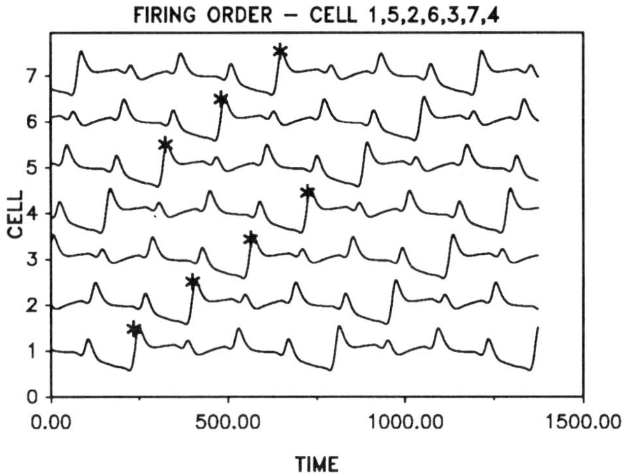

Figure 9: Cell temperatures for all 7 cells, $\eta = 49.5$, $r_* = 32.714$, illustrating the JPOM symmetry. The asterisks illustrate the "firing order", cell 1,5,2,6,3,7,4.

cell 1 attains its maximum temperature then $1/7th$ of a period later in time cell $1 + l$ would attain its maximum. In an experiment a temperature maximum is observed as a spot of maximum brightness. Thus, with $l \neq 1$ the bright spot would be seen as jumping over the $l - 1$ cells between cells 1 and $1 + l$. We refer to this symmetry as "jumping ponies on a merry-go-round" (JPOM). The POM symmetry corresponds to the special case $l = 1$. This general symmetry has been noted in [23] in the context of rotating fluids.

We can determine l from the modes associated with the modulation. Transforming to moving coordinates we find that the temporal period of the modulation $\tau = 2\pi/\Omega$. Upon letting $t \rightarrow t + \tau/7$ and $\tilde{\psi} \rightarrow \tilde{\psi} + l \times 2\pi/7$, the condition for the JPOM symmetry leads to $l = 2$ in agreement with Fig. 8. We note that this analysis is independent of the functional form of the modulation as well as the value of ϵ in (5).

Two period doublings in the cell oscillation occur as η is increased further. Each period doubling is accompanied by a change in symmetry. After one period doubling the solution exhibits the JPOM symmetry with $l = 4$ (Fig. 9). The cells "fire" in the order 1, 5, 2, 6, 3, 7, 4. The solution after the next period doubling is illustrated in Fig. 10. In this case the solution exhibits the JPOM symmetry with $l = 8(\text{mod}7) = 1$, i.e., the POM symmetry (Fig. 10). The cells "fire" in the order 1, 2, 3, 4, 5, 6, 7. Both symmetries can be described by functional forms for the modulation as illustrated above [8].

Upon increasing η further we find that the temperature of each cell no longer exhibits periodic behavior. Each individual cell undergoes an oscillation, however the oscillatory behavior is different for each cell and is not periodic. Upon increasing η

yet further extinction occurs.

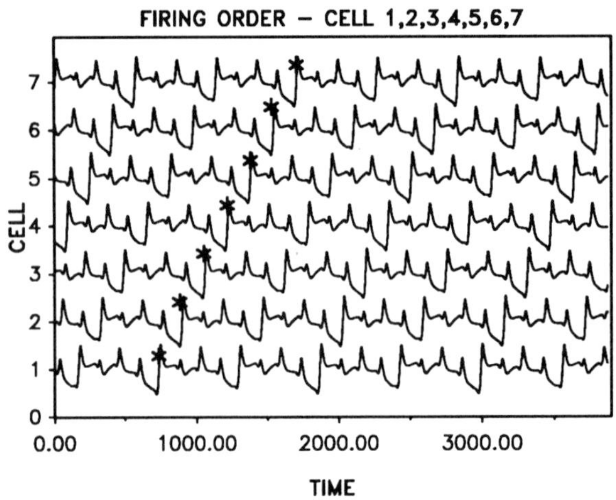

Figure 10: Cell temperatures for all 7 cells, $\eta = 50.0$, $r_* = 32.714$, illustrating the JPOM symmetry. The asterisks illustrate the "firing order", cell 1,2,3,4,5,6,7.

Additional examples of this phenomenon have been found. In particular we considered a 6 cell TW solution and studied the effect of increasing η. This solution exhibited features of the 8 cell and 7 cell branches described above. Upon increasing η the TW6 solution exhibited a spatiotemporal period doubling so that the solution consisted of a rigidly rotating array of 6 cells with an alternating structure so that cells 1-3-5 were identical and cells 2-4-6 were identical. We note that this solution exhibits periodicity over the interval $2\pi/3$ so that it can also be considered a 3 cell TW, where each "cell" is composed of of a pair of alternating cells. Upon further increasing η modulation with a JPOM symmetry was found.

Somewhat different behavior was found by following a branch of 5 cells. For small values of η we found a branch of 5 cells exhibiting uniform rotation, i.e. a TW5 branch. Upon increasing η a periodic (T-periodic) modulation developed, with the cells exhibiting a firing order 1-3-5-2-4 (i.e., a JPOM symmetry with $l = 2$). An example of this type of solution is shown in Fig. 11a. Upon further increasing η the cell modulation exhibited a period tripling, i.e., a 3T periodic modulation, while the spatial symmetry remained unchanged. This solution is illustrated in Fig. 11b. A further increase in η leads to a 4T periodic modulation as shown in Fig. 11c. Further increasing η led to chaotic behavior.

Acknowledgement: This work was supported by N.S.F. grants DMS 93-01635 and MSS 91-02981 and D.O.E. grant DEFG02ER25027.

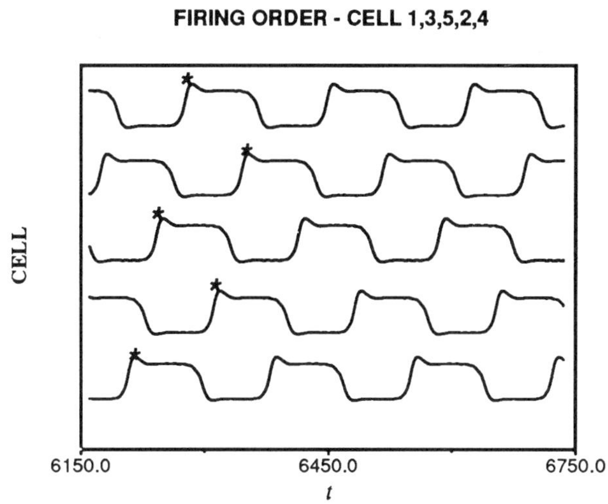

Figure 11a. Cell temperatures for 5 cell MTW, $\eta = 3.0$, $r_* = 30.495$, illustrating the JPOM symmetry. The asterisks illustrate the "firing order", cell 1,3,5,2,4.

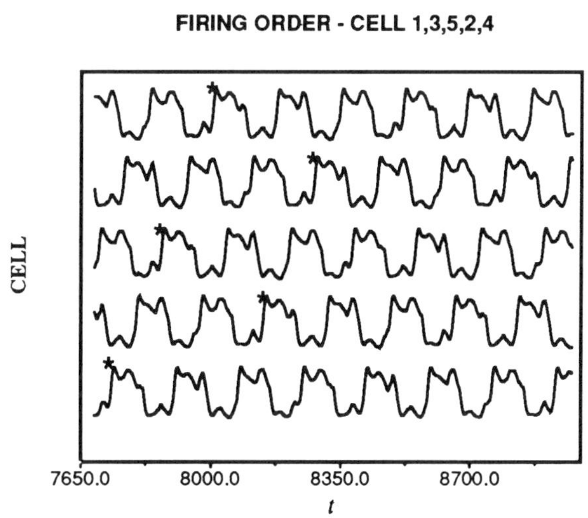

Figure 11b. Cell temperatures for 5 cell MTW, $\eta = 4.0$, $r_* = 30.495$, illustrating the JPOM symmetry with period tripling. The asterisks illustrate the "firing order", cell 1,3,5,2,4.

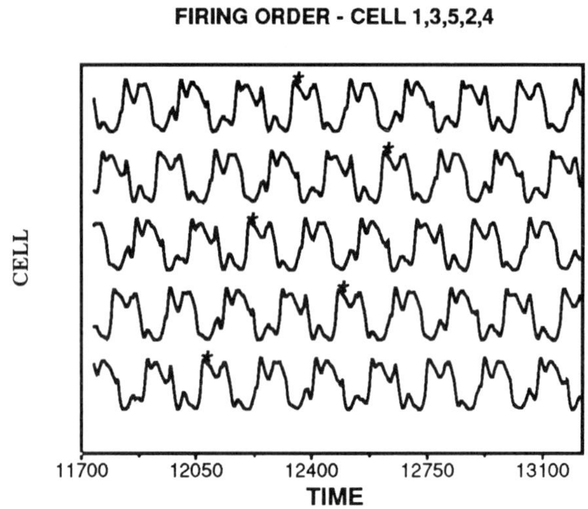

Figure 11c. Cell temperatures for 5 cell MTW, $\eta = 4.5$, $r_* = 30.495$, illustrating the JPOM symmetry with period quadrupling. The asterisks illustrate the "firing order", cell 1,3,5,2,4.

References

[1] Aldushin, A.P., and Kasparyan, S.G.: Soviet Physics - Doklady 1979, 24, 29.

[2] Aldushin, A.P., Kasparyan, S.G. and Shkadinsky, K.G.: Soviet Physics - Doklady 1978, 24, 636.

[3] Aldushin, A.P., and Kasparyan, S.G.: Comb. Expl. Shock Waves 1981, 17, 177.

[4] Aronson D. G., Golubitsky M. and Mallet-Paret J.: Nonlinearity 1991, 4, 903.

[5] Bayliss A., Gottlieb D., Matkowsky B. J., and Minkoff M.: J. Comput. Phys. 1989, 81, 421.

[6] Bayliss A. and Matkowsky B. J.: in "Dynamical Issues in Combustion Theory", Fife, Linan and Williams eds., Springer-Verlag 1991 1.

[7] Bayliss A. and Matkowsky B. J.: SIAM J. Appl. Math., 1992, 52, 396.

[8] Bayliss A. and Matkowsky B. J.: SIAM J. Appl. Math., 1994, 54, 147.

[9] Bayliss A., Matkowsky B. J. and Minkoff M.: SIAM J. Appl. Math. 1989, 49, 1421.

[10] Bayliss A., Matkowsky B. J. and Riecke H.: Physica D, 1994, 74, 1.

[11] Garbey M., Kaper H. A., Leaf G. K. and Matkowsky B. J.: Quart. Appl. Math. 1989, 47, 691.

[12] Gorman M., el-Hamdi M. and Robbins K. A.: Combust. Sci. and Tech., 1994, 98, 71.

[13] Gorman M., el-Hamdi M. and Robbins K. A.: Combust. Sci. and Tech., 1994, 98, 79.

[14] Gorman M., Hamill C. F., el-Hamdi M. and Robbins K. A.: Combust. Sci. and Tech., 1994, 98, 25.

[15] Joulin G. and Clavin P.: Comb. and Flame 1979, 35, 139.

[16] Kuske R. and Matkowsky B. J.: *Two Dimensional Cellular Burner Stabilized Flames,* to appear.
[17] H. Lofting: The Voyages of Doctor Dolittle. Philadelphia, J.B. Lippincott, 1922.
[18] Margolis S. B. and Matkowsky B. J.: Comb. Sci. and Tech. 1983, 34,45.
[19] Markstein G. H.: Nonsteady Flame Propagation. NY, Pergamon Press, 1967.
[20] Matkowsky B. J., Putnick L. J. and Sivashinsky G. I.: SIAM J. Appl. Math. 1980, 38,489.
[21] Olagunju D. O. and Matkowsky B. J.: Quart. J. Appl. Math. 1990, 48, 645.
[22] Olagunju D. O. and Matkowsky B. J.: SIAM J. Appl. Math. 1991, 51, 73.
[23] Rand D.: Arch. Rat. Mech. Anal. 1982, 79, 1.
[24] Sivashinsky G. I.: Ann. Rev. Fluid Mech. 1983, 15, 179.
[25] Sivashinsky, G.I. and Matkowsky, B.J.: SIAM J. Appl. Math. 1981, 40, 255.
[26] Zeldovich, Ya. B.: The Theory of Combustion and Detonation of Gases (in Russian)., Moscow, Izd-vo, Akad. Nauk. SSSR, 1944.
[27] Zeldovich, Ya. B., Barenblatt, G. I., Librovich, V. B., and Makhviladze, G. M.: The Mathematical Theory of Combustion and Explosions. Consultants Bureau, New York, 1985.
[28] Zeldovich, Ya. B. and Frank-Kamenetskii, D. A.: Zhi. Fiz. Khim. 1938, 12, 100.
[29] Zeldovich, Ya. B. and Voevodskii, V. V.: Thermal Explosions and Flame Propagation in Gases (in Russian). Lecture Notes Moscow Mechanical Inst., 1947.

On the Justification of the Ginzburg-Landau approximation

P. BOLLERMAN[1], A. van HARTEN[1], G. SCHNEIDER[2]

[1] *Department of Mathematics, University of Utrecht, P.O. Box 80.010, 3508 TA Utrecht, The Netherlands*

[2] *Institut für Angewandte Mathematik, Universität Hannover, Welfengarten 1, 30167 Hannover, Germany*

Abstract

The Ginzburg-Landau equation arises as an approximate model for modulated wave phenomena in various nonlinear distributed systems. It is derived from a (set of) nonlinear PDE by a formally consistent expansion with respect to a small parameter $\varepsilon > 0$, which measures the distance to the instability threshold for a certain control parameter. Though the derivation is formally consistent, an important and highly nontrivial question is whether G-L's equation really captures the dynamics of certain solutions of the original problem in a reliable sense. In this paper we discuss several mathematical aspects of this question, namely: validity of G-L's equation, attractivity of the G-L manifold and long-term stability of the G-L manifold. A survey of recent results is given and the background of the mathematical complexity of the various facets of this question is investigated. Also, some new results are presented concerning (i) the process of the peak formation involved in attraction to G-L's manifold and (ii) the proof of validity of G-L's equation for problems of Poiseuille flow.

1 Introduction

Modulation phenomena are an intriguing and frequently occurring aspect of oscillatory pattern formation in various contexts. Under certain conditions such long space, long time variations in waves can be described by rather simple looking modulation equations, which contain a wealth of interesting behaviour of solutions. One of the most famous modulation equations is the Ginzburg-Landau equation:

$$A_T = (1 + ia)A_{XX} + A \pm (1 + ib) \mid A \mid^2 A \tag{1}$$

This equation arises typically in the following situation. Consider a model consisting of a system of nonlinear evolution equations on a cylindrical domain $\mathbb{R} \times \Omega$ with boundary conditions on $\partial\Omega$. This system is translation invariant in the unbounded direction and it admits a basic solution with the same property. Now, consider perturbations p of this basic solution and reformulate the model in terms of p. Suppose, that there is a control parameter R of the system such that for $R < R_c$ we have stability. At $R = R_c$ there is one critical mode with a separable structure: a harmonic wave with wavenumber k_c and wave speed ω in the unbounded direction and, say, $\phi(y)$ in the transversal direction(s) $y \in \Omega$. Just above the critical level, say at $R = R_c + \varepsilon^2$

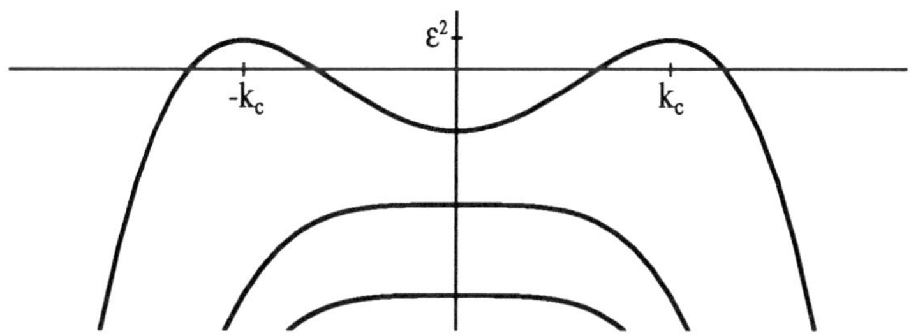

Figure 1: Eigenvalue curves

with ε a small parameter, a small band of modes of width $\mathcal{O}(\varepsilon)$ around k_c becomes unstable.

In a natural way this gives rise to solutions where the amplitude of the critical mode is modulated:

$$\varepsilon g = \varepsilon A(X,T)\phi(y)e^{ik_c(x-\omega t)} + c.c. + \dots \tag{2}$$

Here we denote $X = \varepsilon(x - \nu t)$, a long space scale in a frame moving with group velocity of the waves, and $T = \varepsilon^2 t$, a slow time variable.

It can be shown that under rather generic conditions A has to satisfy a version of the Ginzburg-Landau equation. After some rescaling and renaming of the variables this equation reduces to (1). Note that by its derivation G-L's equation is likely to be a rather universal way to describe modulations. Classical examples, where G-L's equation arises in this way are Bénard convection, Taylor-Couette flow and Poiseuille flow [NW69], [dPES71], [HS72]. However, the scope of applicability of G-L's equation is much wider. It includes problems in reaction diffusion theory, seabed- and river morphology, solid state physics, etc., discussed elsewhere in this volume.

Observations in nature confirm qualitatively a wide range of behaviour as predicted by G-L's equation [CH93]. This illustrates the practical usefulness of G-L's equation as a reduced model. From a theoretical point of view insight in the merits of G-L's equation can be obtained by comparing its solutions with solutions of the original problem from which it is derived. This can be done numerically, as well as analytically. We restrict ourselves to the latter approach. Let us first introduce a few concepts which are crucial for a discussion of an analytic justification of G-L's equation: validity, attractivity and long term stability.

The first concept refers to an existence property of nearby exact solutions.

Validity: Given an $\mathcal{O}(1)$ solution A of G-L's equation for $0 \leq \varepsilon^2 t \leq T_1$ with $T_1 = \mathcal{O}(1)$ and its corresponding formal approximation as defined in (2) then there exists a solution p of the original problem defined on the same time interval, which satisfies:

$$\|p(\cdot,t) - \varepsilon g(\cdot,t)\| \le C\varepsilon^2 \tag{3}$$

In this, definition $\|\cdot\|$ is some norm for functions on the whole cylinder and the constant C is independent of ε. It should be emphasized, that in this concept the choice of the norm is crucial. This choice is intimately related to the Banach space $\mathcal{B}$, in which the evolution equation is solved. In the sequel we shall refer to various choices for this Banach spaces and its norm. A natural choice is a space of bounded continuous functions with the supremum norm. Sometimes we will require stronger regularity, e.g. local integrability of derivatives up to a certain order or even analyticity. We shall specify the spaces in some detail, where we need it. Note that in this validity concept one asks for the existence of only one nearby exact solution p. A somewhat stronger concept of validity is also very natural:

Strong Validity: Given an $\mathcal{O}(1)$ solution A of G-L's equation for $0 \le \varepsilon^2 t \le T_1$ with corresponding εg as in (2) then for all initial perturbations $\varepsilon^2 p_0$ such that $\|p_0\| \le C_0$ with some constant $C_0 > 0, C_0 = \mathcal{O}(1)$, the solution p of the original problem starting at $\varepsilon g(\cdot,0) + \varepsilon^2 p_0$ is defined on the same time-interval and it satisfies the same estimate as in (3).

In the existing literature the former validity result is proved for several interesting classes of problems using a supremum like norm for the error estimate, [CE90], [vH91], [Bo93], [Sch94a], [Sch94b]. Without severe difficulties these results can be extended to the stronger validity result choosing a stronger norm, which includes supremum estimates of the derivatives up to a certain degree. For examples of such results on strong validity we refer to [KSM92]. In section 3 we shall demonstrate the idea of the proof of validity for a famous example: the Kuramoto-Shivashinsky equation. Moreover, we shall discuss some of the functional analytic difficulties arising in dealing with certain realistic problems compared with prototype examples.

However, from a practical point of view validity is only a partial answer to the question of justification of G-L's equation. It says roughly speaking that in well-prepared experiments for which the initial conditions are set sufficiently accurately, observations of phenomena as predicted by G-L's equation should be possible. The next question is what happens if such initial conditions are perturbed in a more rigorous way. The concept of local attractivity refers to a decay property of more general initial perturbations of a solution of G-L's equation. Let G be the G-L manifold:

$$G = \{\varepsilon g \text{ as in (2)} | A \text{ satisfies G-L's equation with i.c. in the set } B_0 \text{ with norm } \le R\}$$

Note that this definition depends on the space of initial amplitudes B_0. Now a definition of local attractivity is as follows:

Local Attractivity: Let A be an $\mathcal{O}(1)$ solution of G-L's equation for $0 \le \varepsilon^2 t \le T_1 = \mathcal{O}(1)$ and let εg corresponding to A by (2). Then for each initial perturbation p_0 satisfying $\|p_0\|_0 \le \delta_0(\varepsilon)$ for some order function $\delta_0(\varepsilon) = o(1)$ the solution $\tilde{\mathrm{p}}$ of the original problem with initial condition $\varepsilon g(\cdot,0)+p_0$ is well-defined on the same time interval and after some transient behaviour for $0 \le \varepsilon^2 t \le T_0$ with $T_0 > 0$ and $T_0 = \mathcal{O}(1)$ but further arbitrarily small, this solution behaves like $\varepsilon\tilde{g}$ corresponding with a G-L solution $\tilde{A}$ through (2), i.e.

$$\|\tilde{p}(\cdot,t) - \varepsilon\tilde{g}(\cdot,t)\| \le C\varepsilon^2 \text{ for } T_0 \le \varepsilon^2 t \le \hat{T}_1$$

with $T_0 < \hat{T}_1 < T_1, T_1 - T_0 = \mathcal{O}(1)$ while $g = \mathcal{O}(1)$-bounded on that interval.

Note that the norm to measure initial perturbations $\|\cdot\|_0$ is allowed to be different from the norm $\|\cdot\|$ in which the error with the solution $\tilde{g}$ corresponding to G-L's equation is measured. This freedom is mainly exploited to study initial perturbations decaying at infinity of bounded, but non-decaying solutions corresponding with G-L's equation. In the definition of local attractivity $\delta_0(\varepsilon)$ characterizes the size of the domain of attraction. Note, that for a choice $\|\cdot\|_0 = \|\cdot\|$ and $\delta_0 = \mathcal{O}(\varepsilon^2)$ the definition of local attractivity would coincide with strong validity. Interesting choices occur for $\delta_0 \gg \varepsilon^2$, say $\delta_0 = \mathcal{O}(\varepsilon)$ (i.e. the same order as εg) or even larger. In case strong validity has already been established it suffices to check the condition:

$$\|\tilde{p}(\cdot, T_0/\varepsilon^2) - \varepsilon g(\cdot, T_0/\varepsilon^2)\| \le C\varepsilon^2$$

for an arbitrary, but $\mathcal{O}(1)$ $T_0 > 0$.

Recently several local attractivity results have been demonstrated for decaying perturbations; in [Eck93] with an $\mathcal{O}(\varepsilon)$ domain of attraction, in [ES94], [MS94] with an even larger domain of attraction. All these proofs work without condition on the sign of the real part of the coefficient of the nonlinearity in G-L's equation. The decay mechanism is induced by the properties of the linearized equation, for the original problem, which leads to peak formation in the spectrum of the solution near integer multiples of the critical wavenumber. This is sufficient to show the local attractivity. A rather simple, short proof is sketched in section 2 and the process of peak formation is analysed in some depth showing some new facets compared with existing literature. For instance, at the end of section 2, an argument is given to show, that in case of a positive sign in front of the nonlinearity one cannot expect a domain of local attraction of size $\mathcal{O}(\varepsilon^\nu)$ with $0 < \nu < 1$ in the sense of the sup-norm.

Finally, we want to conclude this introduction with some remarks on even stronger justification properties which do not only take into account a finite time interval on the intrinsic modulation time scale, but which refer to global properties for all positive times.

Global Long-term Stability: There are balls S_i, i=1, 2, 3 with radii $R_i > 0, R_i = \mathcal{O}(1), R_1 < R_2 < R_3$, such that

i) all solutions of G-L's equation with initial conditions in S_2 stay within S_3 for all time and they stay inside S_1 for $\varepsilon^2 t > T_1$

ii) the corresponding εg defined as in (2) have a local attraction and a strong validity property with a uniformly valid error estimate.

As a consequence of the contraction property (i) solutions of G-L's equation starting in S_2 exist for all time. Due to (ii) all solutions of the original problem with initial conditions $\varepsilon g(\cdot, 0) + p_0$ with g corresponding with $A \in S_2$ and $\|p_0\|_0 \leq \delta_0$ exist for all $t > 0$. Moreover, on each compact time interval $T' \leq \varepsilon^2 t \leq T' + \Delta T, \Delta T > 0, \Delta T = \mathcal{O}(1)$ they can be approximated with some solution $\varepsilon \tilde{g}_{(T', T'+\Delta T)}$ corresponding with G-L's equation and the error is $\mathcal{O}(\varepsilon^2)$ in the sense of $\|\cdot\|$. Hence, this conglomerate of solutions of G-L's equation starting in S_2 contains all the information on the possible behaviour of solutions of the original equation starting as indicated. It is of course not true in general, that one solution of the original problem with some particular initial conditions can be approximated after some transient behaviour by one particular εg corresponding with one solution of G-L's equation. The reason is that in general in the long run $\mathcal{O}(\varepsilon^2)$ errors for this particular approximation will build up and the approximation becomes meaningless. However, the solution of the original problem cannot escape from the set induced by S_3 and its behaviour can be described by a sequence of solutions induced by G-L's equation. In this paper we shall not elaborate on this global long-term stability property. It is clear that it requires a correct sign (i.e. negative) of the real part of the coefficient in front of the nonlinearity in G-L's equation in order to get the described contraction property. For more details we refer to [Sch94c]. In the next section we focus on a more detailed discussion of the local attraction property in case of the Kuramoto-Shivashinsky equation.

2 Attractivity

To explain the results let us consider as example the so called Kuramoto-Shivashinsky equation

$$\partial_t v = -(1 + \partial_x^2)^2 v + \varepsilon^2 v + v \partial_x v \tag{4}$$

for $t > 0$ and $x \in \mathbb{R}$. As already said, we are interested in the case when the trivial solution $v \equiv 0$ is weakly unstable. Therefore we consider ε^2 to be small. Linearizing at $v \equiv 0$ we find solutions of the form $v(x,t) = e^{\lambda t + ikx}$, where $\lambda(k, \varepsilon^2) = -(1-k^2)^2 + \varepsilon^2$ is positive for k close to $\pm k_c = \pm 1$. Plotting the eigenvalue curve in the (k, ε^2)-plane one gets the stability picture from the introduction (fig.1). Thus this equation can serve as a prototype model for our purposes.

As in the case of compact domains, where center manifold theory or inertial manifold theory is available, also here one expects that the weakly nonlinear dynamics is driven by the unstable modes. In the introduction it was stated that the equation which the critical modes fulfill is given by a Ginzburg-Landau equation. But what sort of process allows us to say that this equation dominates the dynamics of the full problem? Such an attractivity property was first stated in [dPES71] and first proved by W. Eckhaus in [Eck93].

Starting computer simulations of the dynamics of the above equation on a large, but finite interval with periodic boundary conditions one observes for the modulus of the Fourier modes a picture which looks more or less the same for all small initial

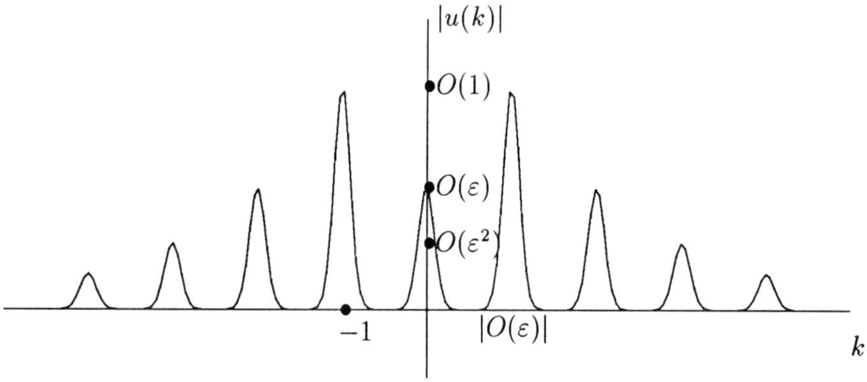

Figure 2: Peaks in the power spectrum

conditions. Some strongly concentrated peaks appear at integer multiples of the critical wave number k_c.

Since numerical experiments on the infinite line are not possible we do the following experiment in mind. Starting with initial conditions which are in Fourier space of order $\mathcal{O}(\varepsilon)$ in L^1 and of order $\mathcal{O}(1)$ in L^∞ we are interested in the evolution of the Fourier modes. What one observes and what W. Eckhaus proved in [Eck93] is that for $t = \mathcal{O}(1/\varepsilon^2)$ peaks as in figure 2 appear. Such a picture corresponds with the structure of a Ginzburg-Landau solution.

$$v = \varepsilon\{A_1 e^{ix} + c.c.\} + \varepsilon^2\{A_0 + (A_2 e^{2ix} + c.c.)\} + \ldots \tag{5}$$

where A_1,A_0,A_2 depend on $T = \varepsilon^2 t$ and $X = \varepsilon x$ only.

Let us now consider an initial perturbation of such a solution, for which the Fourier-transform is of $\mathcal{O}(\varepsilon)$ in both L^1- and L^∞-sense. The essence of local attractivity is a decay property of the spectral density, which is initially present in between the peaks. In accordance with the definition of local attractivity in the introduction, it should be demonstrated, that the original peak structure is restored at some time T_0/ε^2 with $T_0 > 0$ and $T_0 = \mathcal{O}(1)$.

First we demonstrate such a result by using rather elementary tools. Next, we shall discuss some improvements. To do so we execute a Fourier-transform $u(k) = \frac{1}{2\pi}\int v(x)e^{-ikx}dx$ of (4) and get

$$\partial_t u = \lambda(k,\varepsilon^2)\; u - \rho(k)\; u * u, \tag{6}$$

where $\rho(k) = (ik)/2$ and $u * u$ denotes the convolution. The solutions of this equation will live in weighted L^1-spaces. The appearing peaks are described by the choice of appropriate weights. For $n \geq 1$ we define the family of weights $f_n(k)$ by

$$1/f_n(k) = \varepsilon^{(n+3)/2} \max_{j=\pm1,\pm2;l=\pm3,\dots,\pm n} \{2, 1/(\varepsilon+|k-j|)^{n/2-|j|+3/2}, 1/(\varepsilon+|k|)^{n/2-1/2}, 1/(\varepsilon+|k-l|)^{n/2-|l|/2+1/2}\}.$$

We see that a plot of $k \mapsto 1/f_n(k)$ looks similar to figure 2. Note that $u(\cdot,t)f_n$ in L^1 with $\mathcal{O}(\varepsilon)$ norm allows for peaks near $k = \pm1$ and $k = 0, \pm2$ as shown. In between these peaks the spectrum vanishes up to any polynomial order of ε.

For the solutions of (6) we show

Theorem 1 *Let $u(\cdot,t)$ be an initial condition of (6) with $\|u(\cdot,t)\|_1 \leq C\varepsilon$ for an ε independent constant C. Then for all $n > 0$ there exists a $C_n, \tilde{T}_n, \varepsilon_0 > 0$ such that for all $0 < \varepsilon < \varepsilon_0$ the associated solution $u(\cdot,t)$ fulfill $\|u(\cdot,\tilde{T}_n/\varepsilon^2,)f_n(\ \cdot\ )\|_1 \leq C_n$.*

Proof: In a first step we prove that there exists a $\tilde{T}_1$ such that $\|u(\cdot,\tilde{T}_1/\varepsilon^2)f_1\|_1 \leq C_1$. Then we show that from $\|u(\cdot,\tilde{T}_n/\varepsilon^2)f_n\|_1 \leq C_n$ the estimate $\|u(\cdot,\tilde{T}_{n+1}/\varepsilon^2)f_{n+1}\|_1 \leq C_{n+1}$ for a $\tilde{T}_{n+1} > \tilde{T}_n$ follows. Note that the spaces in which the equation is solved are defined in Fourier space with weighted L^1 norms. In physical space this corresponds with bounded continuous solutions. Questions of extra regularity will be addressed later on. With this bootstrap method the assertion follows. In the following we need the set $I_c = [-5/4, -3/4] \cup [3/4, 5/4]$. We denote by E_c the characteristic function on I_c and by E_s the one on the complement.

I) We start with the second step and define a second family of weights

$$1/f_n^*(k) = \varepsilon^{(n+5)/2} \max_{l=\pm1,j=\pm2;m=\pm3,\dots,\pm n} \{2, 1/(\varepsilon+|k-j|)^{n/2+1/2}, 1/(\varepsilon+|k|)^{n/2+1/2}, 1/(\varepsilon+|k-l|)^{n/2-1/2}, 1/(\varepsilon+|k-m|)^{n/2-|m|/2+1}\}$$

to control the convolution. Moreover, we need $1/f_n^c = E_c/f_n$, $1/f_n^s = E_s/f_n$, $1/f_n^{c,*} = E_c/f_n^*$, and $1/f_n^{s,*} = E_s/f_n^*$. I.e. the so defined weights have values in $\mathbb{R} \cup \{\infty\}$. A simple calculation shows

$$f_n^{c,*}(k+l) \leq f_n^s(k)f_n^c(l), \qquad f_n^{c,*}(k+l) \leq \varepsilon f_n^s(k)f_n^s(l), \qquad f_n^{s,*}(k+l) \leq f_n^c(k)f_n^c(l),$$
$$f_n^{s,*}(k+l) \leq \varepsilon^{1/2}f_n^s(k)f_n^c(l), \quad \text{and } f_n^{s,*}(k+l) \leq \varepsilon f_n^s(k)f_n^s(l)$$

As a consequence of this, one finds e.g. $\|(u*v)f_n^{s,*}\|_1 \leq \varepsilon\|uf_n^s\|_1\|vf_n^s\|_1$. These estimates are a consequence of one of the main points of this theory on which the attractivity and the approximation property heavily depends, namely the fact that the convolution of critical modes (modes with positive growth rates) gives noncritical modes.

i) We set $u_n(0) = u(\tilde{T}_n/\varepsilon^2)$ and show first that there exist $\tilde{C}_n, T_n, \varepsilon_0 > 0$ such that $\sup_{0\le t\le T_n/\varepsilon^2} \|u_n(t)f_n\|_1 \le \tilde{C}_n$ for all $0<\varepsilon<\varepsilon_0$. To do so we set $S_i(t) = \sup_{0\le s\le t} \|E_i u_n(s)f_n\|_1$ with $i=c,s.$ and define $\|u_n(0)f_n\|_1 =: C_n/4$. Then with $T = \varepsilon^2 t$

$$
\begin{aligned}
S_c(t) &\le e^{\varepsilon^2 t}\|u_n(0)f_n\|_1 + \textstyle\int_0^t \sup_{k\in I_c} |e^{\lambda(t-\tau)} f_n/f_n^*|(S_c(\tau)S_s(\tau) + \varepsilon S_s(\tau)^2)d\tau \\
&\le C_n/2 + C(\sqrt{T}+T)(C_nC_s+1)
\end{aligned}
$$

as long as $S_c(t) \le C_n$, $S_s(t) \le C_s$ and if we choose ε such that $\varepsilon C_s^2 < 1$. There exists $T_n > 0$ so that $e^{T_n} < 2$ and $C(\sqrt{T_n}+T_n)(C_nC_s+1) \le C_n/2$. Thus for a $\sigma > 0$

$$
S_s(t) \le e^{-\sigma t}\|u_n(0)f_n\|_1 + C(S_c^2(t) + \varepsilon^{1/2}S_cS_s + \varepsilon S_s^2) \;\le\; C_n/2 + CC_n^2 + 1 =: C_s,
$$

where we have chosen ε such that $\varepsilon^{1/2}C_nC_s + \varepsilon C_s^2 < 1$. Thus $\tilde{C}_n = C_n + C_s(n)$ and $\tilde{T}_{n+1} = T_n + \tilde{T}_n$.

ii) Now we show that there exists $\varepsilon_0, \hat{C}_n > 0$ such that for all $0<\varepsilon<\varepsilon_0$ the estimate $\|u_n(T_n/\varepsilon^2)f_{n+1}\|_1 \le \hat{C}_n$ holds. We obtain

$$
\begin{aligned}
\|E_c u_n(T_n/\varepsilon^2)f_{n+1}\|_1 &\le C\sup_{|k|<1/2} |e^{-k^2T_n/\varepsilon^2}(1+|k/\varepsilon|)^{1/2}|\|u_n(0)f_n\|_1 \\
&\quad + \textstyle\int_0^t \sup_{k\in I_c} |e^{\lambda(t-\tau)} f_{n+1}/f_n^*| d\tau (C_nC_s+1) \\
&\le C(1+T_n^{-1/4})C_n + C(T_n + T_n^{1/4})(C_nC_s+1) = \mathcal{O}(1)
\end{aligned}
$$

and

$$
\begin{aligned}
\|E_s u_n(T_n/\varepsilon^2)f_{n+1}\|_1 &\le C_n e^{-\sigma T_n/\varepsilon^2}/\varepsilon + C(C_n^2 + C_nC_s + C_s^2) \\
&\le C(C_n+C_s)^2 + 1 = \mathcal{O}(1)
\end{aligned}
$$

where we made use of $f_{n+1}^s \le f_n^{s,*}$ and choose ε so that $C_n e^{-\sigma T_n/\varepsilon^2}/\varepsilon < 1$.

II) The first step, i.e. to come to f_1, consists of two parts. In the first part we go with the same arguments as above from $1/\varepsilon$ to $f_0 = 1/(\varepsilon E_c + \varepsilon^2 E_s)$. For this we only need the time $\tilde{T}_0 = \varepsilon^{7/4}$. Therefore $\|u(1/\varepsilon^{1/4})f_0\|_1 = \mathcal{O}(1)$. Note that with the weight $1/\varepsilon$ a priori solutions only exist up to a time $\mathcal{O}(1/\varepsilon)$. The next step works exactly as in I) with $u_0(0) = u(1/\varepsilon^{1/4})$, this f_0, and $f_0^* = (\varepsilon^3 E_c + \varepsilon^2 E_s)$. ■

Now some remarks on the proof. The proof follows the ideas of [Eck93]. For the above mentioned initial conditions in $L^1 \cap L^\infty$, it was shown in [Eck93] that the attractivity happens on a much faster time scale, namely $\mathcal{O}(1/\varepsilon^\nu)$ with $\nu < 2$ which is much smaller than the Ginzburg-Landau time-scale $\mathcal{O}(1/\varepsilon^2)$. Moreover, we only show that the peaks for $|m| \ge 3$ are of order $\varepsilon^{1+|m|/2}$ instead of $\varepsilon^{|m|}$ as in [Eck93].

The chosen spaces contain the right orders of solutions since the later on chosen approximations are covered. But the space L^1 in Fourier space is much too small to contain spatial periodic or front-solutions. As a corollary to the above theorem it follows that the functions in Fourier space can be written for a $T > 0$ as

$$u(T/\varepsilon^2, k) = \sum_{m=0}^{N} \varepsilon^{\beta(m)}(\varepsilon^{-1}\hat{A}_m((k-m)/\varepsilon, \varepsilon^2 t)) + \mathcal{O}(\varepsilon^{(N+3)/2}),$$

where the function β is defined by $\beta(\pm 2) = \beta(0) = 2$, $\beta(\pm 1) = 1$, and $\beta(m) = 1+|m|/2$ for $|m| \geq 3$. The functions $\hat{A}_m$ fulfill $\|(1+k^2)^{N/2}\hat{A}_m\|_1 \leq C_0$ for a constant C_0 independent of ε. Since the above spaces are too small we will cite a formulation in physical space which is applicable to a greater class of functions, namely all bounded sufficiently smooth functions. Since

$$\sup_{x \in \mathbb{R}, r=0,\ldots N} |\partial_x^r \int u(k) e^{ikx} dk| = \sup_{r=0,\ldots N} |\int u(k)(ik)^r dk| \leq C\|(1+|\cdot|)^N u(\cdot)\|_1 < \infty$$

the inverse Fourier-transform $v(x,t) = \mathcal{F}^{-1}u(k,t)$ of such a function is N times differentiable. Therefore one can expect that the attractivity result can be stated in the following way (compare [Sch94d] Theorem 1):

For $C > 0$, all $N \in \mathbb{N}$ and all initial conditions v_0 of (4) with $\|v_0\|_{C_b^n} \leq C\varepsilon$ exist $\varepsilon_0, T_0 > 0$ such that for all $0 < \varepsilon < \varepsilon_0$ there is a representation of the associated solution v at a time T_0/ε^2 through

$$v(x, T_0/\varepsilon^2, \varepsilon) = \sum_{m=0}^{N} \varepsilon^{\beta(m)} A_m(\varepsilon x) e^{imx} + \mathcal{O}(\varepsilon^{(N+3)/2}).$$

The functions $A_m \in C_b^N$ fulfill $\|A_m\|_{C_b^N} \leq C_0$ for a constant C_0 independent of ε.

Generalizations of this result are possible. In [Sch94d] the exponential instead of the polynomial decay of the peaks is shown, i.e. analyticity of the functions A_m. In [vH92] it is shown that for analytic initial perturbations the perturbed peak w near $k = 1$ shrinks in the following way

$$|w(m,t)| \leq C\varepsilon e^{-\alpha|m|\sqrt{1+t}} \tag{7}$$

where $m = k - 1$.

With the so called method of pseudo-orbits it is possible to extend the attractivity result to an $\mathcal{O}(1)$-neighborhood of the origin instead of only $\mathcal{O}(\varepsilon)$-neighborhoods ([MS94]). Extensions of the local attractivity domain can also be found in [ES94]. However, one should be careful. Looking, in a general situation, at initial perturbations $\varepsilon^\nu e^{ik_c x}$ with $0 < \nu < 1$ it is found, that such perturbations explode in $\mathcal{O}(\varepsilon^{-2\nu})$ time due to nonlinear interactions, if the sign of the nonlinear term in the Ginzburg-Landau equation is wrong (positive). Hence, extension of the domain of attractivity for non-decaying initial perturbations is impossible in that case.

3 Validity

3.1 Validity in the Kuramoto-Shivashinsky case

To start with we discuss the validity of the Ginzburg-Landau approximation for our prototype example, the Kuramoto-Shivashinsky equation. The essence of the proof of validity will be discussed. Next, we consider how to deal with some complications, which occur in showing the validity of the Ginzburg-Landau equation in the context of well-known classical examples such as the two-dimensional Poiseuille flow. In the previous section it was shown, that the main dynamics of the problem is found by the ansatz

$$v(x,t,\varepsilon) = \sum_{m\in Z}\sum_{n=0}^{\infty} \varepsilon^{1+||m|-1|+n} A_m^n(X,T)e^{imx}, \tag{8}$$

where $T = \varepsilon^2 t$ and $X = \varepsilon x$. Inserting this ansatz in (4) gives after equating the coefficients in front of all $\varepsilon^l e^{imx}$ to zero, equations which the A_m^n have to fulfill. After eliminating some algebraic equations one obtains in lowest order the Ginzburg-Landau equation

$$\partial_T A_1^0 = A_1^0 + 4\partial_X^2 A_1^0 - (1/9)A_1^0|A_1^0|^2. \tag{9}$$

This is the equation for the critical modes and it drives the nonlinear dynamics. Note that all A_m^n exist and can be computed from A_1^0 as long as the solution A_1^0 of the Ginzburg-Landau equation exists. Therefore it is possible to define a m-order approximation ϕ_m by taking the finite sum in (8) consisting of all terms up to order ε^m (see [vH91]).

Since this equation is only an approximation, it cannot be expected that it describes the problem for all times. But on $\mathcal{O}(1)$ time-scales of the Ginzburg-Landau equation it should be possible to find solutions of the original problem which are nearby the associated approximation. So in several papers ([CE90], [vH91], [KSM92], [Bo93], [Sch94a], [Sch94b]) versions of the following validity property as discussed in the introduction, are shown:

Let $A = A(X,T) \in C([0,T_1], C_b^4(\mathbb{R},\mathbb{C}))$ be a solution of the Ginzburg-Landau equation (9). Then there exist $\varepsilon_0, C > 0$, such that for all $0 < \varepsilon \leq \varepsilon_0$ there are solutions $v(x,t,\varepsilon)$ of (4) with

$$\sup_{0\leq t\leq T_1/\varepsilon^2} \|v(x,t,\varepsilon) - (\varepsilon A(\varepsilon^2 t, \varepsilon x)e^{ix} + c.c.)\|_{C_b^0(\mathbb{R},\mathbb{R})} < C\varepsilon^2.$$

Again, for the sake of simplicity, we will restrict the proof to initial conditions which are in L^1 in Fourier space. This induces the result in a physical space of bounded, continuous functions, as mentioned above. Let $\hat{A}_1^0$ be a solution of the Fourier transform of the Ginzburg-Landau equation (9) with $\sup_{0\leq T\leq T_0} \|\hat{A}_0^1(K,T)(1+|K|)^{n_0}\|_1 \leq C$, with n_0 chosen later on sufficiently large. The error will be estimated by the differential equation it fulfills. This differential equation should have $\mathcal{O}(\varepsilon^2)$-bounded solutions on a time interval of length $\mathcal{O}(1/\varepsilon^2)$. As said before we have to take the space L^1 with weight f_0 instead of $1/\varepsilon$ to assure this. For arbitrary n we can choose n_0 and n_1 sufficiently large such that the norm of the residual $\|Res(\phi_{n_1})f_0\|_1 = \|(-\partial_t\phi_{n_1}+\lambda\phi_{n_1}-\rho\phi_{n_1}*\phi_{n_1})f_0\|_1$ is of order $\mathcal{O}(\varepsilon^n)$ and that $\|\phi_{n_1}f_0\|_1 = \mathcal{O}(1)$ uniformly on the required time interval.

We are done once we have shown that a solution u can be written as sum $u = \phi_{n_1} + \varepsilon^2 R$ and that the error function R fulfills $\sup_{0\leq t\leq T_1/\varepsilon^2}\|R(t)f_0\|_1 \leq C = \mathcal{O}(1)$. Equivalent to this is to show that $u = \varepsilon\phi_c + \varepsilon^2\phi_s + \varepsilon^2 R_c + \varepsilon^3 R_s$ with $\phi_{n_1} = \varepsilon\phi_c + \varepsilon^2\phi_s$, $\phi_c = E_c\phi_c$, $\phi_s = E_s\phi_s$, $R_c = E_c R_c$, and $R_s = E_s R_s$ and $\sup_{0\leq t\leq T_0/\varepsilon^2}(\|R_c\|_1 + \|R_s\|_1) = \mathcal{O}(1)$. We have $E_c(R_c * R_c) = E_c(\phi_c * R_c) = 0$ and so

$$\begin{aligned}
\partial_t R_c &= \lambda R_c + E_c(2\varepsilon^2\rho(R_c*\phi_s + R_s*\phi_c) \\
&\quad +2\varepsilon^3\rho(R_c*R_s+\phi_s*R_s) + \varepsilon^4\rho(R_s*R_s) + \mathrm{Res}(\phi_{n_1})/\varepsilon^2) \\
\partial_t R_s &= \lambda R_s + E_s(\rho(\phi_c*R_c) + 2\varepsilon\rho(R_c*\phi_s + R_s*\phi_c + R_c*R_c/2) \\
&\quad +2\varepsilon^2\rho(R_c*R_s+\phi_s*R_s) + \varepsilon^3\rho(R_s*R_s) + \mathrm{Res}(\phi_{n_1})/\varepsilon^3)
\end{aligned} \tag{10}$$

The local existence is no problem, and so it remains to bound the solutions on the long time interval $[0, T_1/\varepsilon^2]$. Constants coming from the residual are denoted in the following by $C_{Res} = \mathcal{O}(1)$. Applying the variation of constant formula to these equations and making the abbreviations $S_i(s) := \sup_{t\leq s}\|R_i(t)\|_1$ allows to obtain the estimates

$$\begin{aligned}
S_c(t) &\leq \varepsilon^2 C \textstyle\int_0^t (S_c(\tau) + S_s(\tau) + \varepsilon S_c(\tau)S_s(\tau) + \varepsilon^2 S_s(\tau)^2)d\tau + C_{Res} \\
S_s(t) &\leq C(S_c(t) + \varepsilon S_s(t) + \varepsilon S_c(t)^2 + \varepsilon^2 S_c(t)S_s(t) + \varepsilon^3 S_s(t)^2) + C_{Res}.
\end{aligned}$$

Inserting the second equation in the first one gives as long as $S_c \leq C_c$ and $S_s \leq C_s$ the estimate

$$S_c(t) \leq CC_{Res} + \varepsilon^2\int_0^t (CS_c(\tau)+1)\,d\tau,$$

where $\varepsilon CC_s + \varepsilon CC_c^2 + \varepsilon^2 C_c C_s + \varepsilon^3 C_s^2 + \varepsilon C_c C_s + \varepsilon^2 C_s^2 < 1$. Applying Gronwall's inequality to this gives $S_c(t) \leq C(C_{Res} + T_0)e^{CT_0}) =: C_c$. In the second equation we obtain by this $S_s(t) \leq C(C_c + 1 + C_{Res}) =: C_s$, where $C(\varepsilon C_s + \varepsilon C_c^2 + \varepsilon^2 C_c C_s + \varepsilon^3 C_s^2) < 1$. Choosing now ε_0 in such a way that the restrictions on ε are fulfilled we are done. ■

The proof works for different weights. All existing papers work in physical space. Transferring these proofs into Fourier space one can see that they differ in the choice of weight in Fourier space. If quadratic terms are absent it is possible to use $1/\varepsilon$ as weight ([CE90], [KSM92]). For problems with quadratic terms in [Bo93], [Sch94a], [Sch94b] the weight f_0 and in [vH91] a weight with exponentially concentrated peaks is taken.

Hence, the conclusion is that validity in this sort of prototype problems is well-understood nowadays.

However, in reality the system of equations that constitute a physical model is usually more complex and more difficult to analyze than the Kuramoto-Shivashinsky equation. There are several complications that can arise.

First of all the physical models are usually posed in dimensions higher than one. Either one has to add another unbounded spatial direction or one has to add a bounded transverse direction (or do both). Each extension of the model has its own implication to the characteristics of the spectrum of the linear operator involved. The first case results in a continuous spectral surface. As a degeneracy it is now possible to have an infinite number of wavevectors becoming unstable at the same time, e.g. in models with rotational symmetry (cf. [Sch94e]). In the latter case of bounded transversal directions the spectrum will be a discrete set of spectral curves. The possibility of intersecting spectral curves might present a serious complication when it happens near a critical wave number. It is also possible to have several spectral curves producing instabilities at the same time, due to different instability mechanisms. Such situations as sketched here above give rise to variations in the Ginzburg-Landau equation, which we don't want to discuss here. In the case of Poiseuille flow other complications are appearing. An extra difficulty arises from working with vectorvalued functions; another problem is the presence of a pressure gradient, which is not satisfying an evolution equation itself. Altogether the transversal direction, the vector character and the elimination of the pressure make it necessary to develop some highly non-trivial functional analytic tools to make the proof of validity work.

The Navier-Stokes Equations for an incompressible fluid form the basis of the classical hydrodynamical models. One complication which is not immediately obvious for these problems is to make them well-posed for unbounded domains. This is because unbounded domains leave a lot of freedom in the balance between flow field and pressure gradient. To overcome this problem one has to pose a sensible constraint on the flow or the pressure. Examples such as Poiseuille flow between two plates, Bénard convection and Taylor-Couette flow between two rotating cylinders have been studied extensively in the past and modulation equations have been derived for them. Of course it turns out that these modulation equations depend on the choice of the constraint.

The case of two-dimensional Poiseuille flow contains all complications described above.

3.2 Validity in the case of Poiseuille flow

Let $\Omega = I\!R \times [0,1]$. The equations describing two-dimensional Poiseuille flow are

$$\begin{cases} \partial_t \mathbf{u} = -\nabla p + \frac{1}{R}\Delta \mathbf{u} - (\mathbf{u}\cdot\nabla)\mathbf{u} \\ \nabla\cdot\mathbf{u} = 0 \\ \mathbf{u}|_{y=0,1} = \mathbf{0} \end{cases} \tag{11}$$

where $\mathbf{u} = (u,v)$, $\nabla = (\partial_x, \partial_y)$, $\Delta = \partial_x^2 + \partial_x^2$ and R the Reynolds number. We shall assume that there is a basic flow $\mathbf{u}_0 = (U,0) = (y-y^2,0)$, $p_0 = -\frac{2}{R}x$. Perturbing the basic flow by setting $\mathbf{u} = (U + u', v')$, $p = p_0 + p'$ gives us

$$\begin{cases} \partial_t \mathbf{u}' = -\nabla p' + \frac{1}{R}\Delta \mathbf{u}' - (\mathbf{u}' \cdot \nabla)\mathbf{u}_0 - (\mathbf{u}_0 \cdot \nabla)\mathbf{u}' - (\mathbf{u}' \cdot \nabla)\mathbf{u}' \\ \nabla \cdot \mathbf{u}' = 0 \\ \mathbf{u}'|_{y=0,1} = \mathbf{0} \end{cases} \tag{12}$$

It is well known that a two-dimensional divergence free flow can be described by a streamfunction $\mathbf{\Psi}$ such that $u' = \partial_y \mathbf{\Psi}$, $v' = -\partial_x \mathbf{\Psi}$. This streamfunction then satisfies the vorticity equation:

$$\begin{cases} \partial_t \Delta \mathbf{\Psi} = \frac{1}{R}\Delta^2 \mathbf{\Psi} - U\partial_x \Delta \mathbf{\Psi} + \partial_y^2 U \partial_x \mathbf{\Psi} - \partial_y \mathbf{\Psi} \partial_x \Delta \mathbf{\Psi} + \partial_x \mathbf{\Psi} \partial_y \Delta \mathbf{\Psi} \\ \partial_x \mathbf{\Psi}|_{y=0,1} = \partial_y \mathbf{\Psi}|_{y=0,1} = 0 \end{cases} \tag{13}$$

It is important to note that this problem is not yet well-posed. Because we know that when we add a constant to $\mathbf{\Psi}$ we still get the same solution (u', v') of (12), we are allowed to replace $\partial_x \mathbf{\Psi}|_{y=0} = 0$ by $\mathbf{\Psi}|_{y=0} = 0$ We cannot replace the boundary condition $\partial_x \mathbf{\Psi}|_{y=1} = 0$ by an acceptable and consistent boundary condition without posing a constraint on the flow. E.g. we can assume a constant flow though a cross-section of the channel, i.e. $\int_0^1 u\, dy = \int_0^1 (U + u')\, dy = C_F$. It follows that

$$\mathbf{\Psi}|_{y=1} = 0.$$

Another sort of constraint is keeping the pressure difference along the channel constant. One can achieve this by posing $\lim_{L\to\infty} \frac{1}{2L}\int_{-L}^{L} \partial_x p\, dx = \lim_{L\to\infty} \frac{1}{2L}\int_{-L}^{L} \partial_x (p_0 + p')\, dx = C_P$. We now split $\mathbf{\Psi}$ in two parts: $\mathbf{\Psi} = \mathbf{\Psi}' + \Psi_0$ where Ψ_0 depends only on the time and the space variable y. Now $(\mathbf{\Psi}', \Psi_0)$ should satisfy

$$\begin{cases} \partial_t \Delta \mathbf{\Psi}' = \frac{1}{R}\Delta^2 \mathbf{\Psi}' - U\partial_x \Delta \mathbf{\Psi}' + \partial_y^2 U \partial_x \mathbf{\Psi}' - \partial_y \Psi_0 \partial_x \Delta \mathbf{\Psi}' + \partial_x \mathbf{\Psi}' \partial_y^2 \partial_y \Psi_0 \\ \qquad\qquad -\partial_y \mathbf{\Psi}' \partial_x \Delta \mathbf{\Psi}' + \partial_x \mathbf{\Psi}' \partial_y \Delta \mathbf{\Psi}' - \lim_{L\to\infty} \int_{-L}^{L} \partial_y (\partial_x \mathbf{\Psi}' \partial_y^2 \mathbf{\Psi}')\, dx \\ \partial_t \partial_y \Psi_0 = \frac{1}{R}\partial_y^2 \partial_y \Psi_0 + \lim_{L\to\infty} \frac{1}{2L}\int_{-L}^{L} \partial_x \mathbf{\Psi}' \partial_y^2 \mathbf{\Psi}'\, dx \\ \mathbf{\Psi}'|_{y=0,1} = \partial_y \mathbf{\Psi}'|_{y=0,1} = 0 \\ \partial_y \Psi_0|_{y=0,1} = 0 \end{cases} \tag{14}$$

We now have a well posed problem for this case also.

For the sake of simplicity we shall explore the first case. Linear stability analysis of the model leads us to the so called Orr-Sommerfeld equation (cf. [DR81]):

$$\begin{cases} s(-k^2+\partial_y^2)f = \dfrac{1}{R}(-k^2+\partial_y^2)^2 f - Uik(-k^2+\partial_y^2)f + \partial_y^2 Uikf \\ \\ f|_{y=0,1} = \partial_y f|_{y=0,1} = 0 \end{cases} \tag{15}$$

This analysis provides us with values k_c, R_c at which instability of the basic flow sets in, and a critical eigenfunction $f(k,R)$ with its companion critical eigenvalue $\lambda(k,R)$. As a first approximation of a bifurcating solution we now take

$$\mathbf{\Psi}(x,y,t) = \varepsilon(A_1(\varepsilon x+\varepsilon\nu t,\varepsilon^2 t)e^{i(k_c x+\omega t)}f(k_c,R_c;y)+c.c.)+\mathcal{O}(\varepsilon^2) \tag{16}$$

where $\varepsilon = \sqrt{R-R_c}$ is a small parameter, and $\lambda(k,R) = i(\omega+\varepsilon\nu(\frac{k-k_c}{\varepsilon}))+\mathcal{O}(\varepsilon^2)$ near $k=k_c$ and $\varepsilon=0$. Since $f(k_c,R_c)$ is just an $\mathcal{O}(1)$ approximation of $f(k,R)$ it makes sense to incorporate the higher order terms in $f(k,R)$. In going to higher order approximations one should also consider the appearance of slaved modes at integer multiples of the critical wavenumber k_c. Summarizing all this for a second order approximation this means

$$\begin{cases} \mathbf{\Psi}(x,y,t) &= \varepsilon(A_1^0(\varepsilon x+\varepsilon\nu t,\varepsilon^2 t)e^{i(k_c x+\omega t)}f(k_c,R_c;y)+c.c.)+ \\ & \quad \varepsilon^2(\partial_X A_1^0(\varepsilon x+\varepsilon\nu t,\varepsilon^2 t)e^{i(k_c x+\omega t)}\partial_k f(k_c,R_c;y)+c.c.)+ \\ & \quad \varepsilon^2 A_0^0(\varepsilon x+\varepsilon\nu t,\varepsilon^2 t)g_0(y)+ \\ & \quad \varepsilon^2(A_2^0(\varepsilon x+\varepsilon\nu t,\varepsilon^2 t)e^{2i(k_c x+\omega t)}g_2(y)+c.c.)+ \\ & \quad \varepsilon^2(A_1^0(\varepsilon x+\varepsilon\nu t,\varepsilon^2 t)e^{i(k_c x+\omega t)}f(k_c,R_c;y)+c.c.)+\mathcal{O}(\varepsilon^3) \\ &= \varepsilon\mathbf{\Psi}_0^0(A_1^0,f)+\varepsilon^2\mathbf{\Psi}_1^0(A_2^0,A_0^0,g_0,g_2)+\varepsilon^2\mathbf{\Psi}_1^1(A_1^1,f)+\mathcal{O}(\varepsilon^3) \end{cases} \tag{17}$$

Here $A_0^0 = |A_1^0|^2$, $A_2^0 = (A_1^0)^2$, and g_0 and g_2 are solutions of

$$\begin{cases} 0 &= L(0,R_c)g_0 + \partial_y f ik_c(\partial_y^2-k_c^2)\bar f + ik_c f\partial_y(\partial_y^2-k_c^2)\bar f \\ & \quad - \partial_y \bar f ik_c(\partial_y^2-k_c^2)f - ik_c\bar f\partial_y(\partial_y^2-k_c^2)f \\ 2i\omega(\partial_y^2-4k_c^2)g_2 &= L(2k_c,R_c)g_2 + ik_c f\partial_y(\partial_y^2-k_c^2)f - \partial_y f ik_c(\partial_y^2-k_c^2)f \end{cases} \tag{18}$$

with $L(k,R)g = \frac{1}{R}(\partial_y^2-k^2)^2 g - Uik(\partial_y^2-k^2)g + \partial_y^2 Uikg$. The A_1^1 term represents a perturbation of a smaller magnitude and is only present for completeness. By inserting the expansion in (13) we can formally conclude that A_1^0 satisfies the Ginzburg-Landau equation.

For a mathematical justification we set the solution

$$\mathbf{\Psi} = \varepsilon\mathbf{\Psi}_0^0(A_1^0,f)+\varepsilon^2\mathbf{\Psi}_1^0(A_2^0,A_0^0,g_0,g_2)+\varepsilon^2 w_c(f)+\varepsilon^3 w_s \tag{19}$$

Here $w_c(f)$ describes the critical error made in the $\lambda(k, R)$ eigenspace for an $\mathcal{O}(1)$ range of wave numbers k around $\pm k_c$, and w_s describes the noncritical error. The equations for w_c and w_s are then of the form

$$\begin{cases} \partial_t \Delta w_c(f) &= Lw_c(f) + \varepsilon^2 \{\mathcal{L}_c(w_c(f), w_s) + \varepsilon \mathcal{N}_c(w_c(f), w_s) + \mathcal{R}_c\} \\ \partial_t \Delta w_s &= Lw_s + \mathcal{L}_s(w_c(f)) + \varepsilon \mathcal{N}_s(w_c(f), w_s) + \mathcal{R}_s \end{cases} \tag{20}$$

The longtime existence of $\mathcal{O}(1)$ functions $w_c(f)$, w_s satisfying the above equations (20) would prove validity of the G-L approximation for this case.

Let us now discuss briefly the mathematical apparatus to show this. If we want to find solutions of (10) or (20) we must decide on a function space in which to look for them.

We desire results in physical space for a wide class of relevant solutions. Hence, estimates made in Fourier space must be transferred to physical space. However, the results in the previous section are not strong enough to consider periodic solutions. To include them would require a Fourier space of distributions which rules out the use of weighted L^1-spaces. It turns out to be a good idea to consider a physical space of L^∞ functions with some degree of smoothness.

In [vH91] this smoothness property is taken to the extreme, considering a space of analytic functions on a strip in the complex plane with the sup-norm. The Fourier transform of such a function decays exponentially as $|k| \to \infty$ in distributional sense. It therefore certainly has the decay imposed by the weights in the previous function space, but it now allows for periodic solutions as well. A drawback is that it only allows for solutions with analytic initial conditions.

This restriction was overcome in [Tea94]. It was shown that solutions of semilinear parabolic systems with essentially bounded initial conditions become analytic instantaneously on a strip with width that initially grows as $\sqrt[2m]{t}$ where $2m$ is the order of the linear operator. When using this property in our proof we still must assume some extra smoothness of the initial condition in order to be able to split critical and stable modes.

Now we face the problem of choosing an appropriate space in which we can solve the problem (20). We choose to work in physical space bounded, analytic functions on a strip in the x-domain, with values in the space of continuous functions in y on $[0, 1]$. For details we refer to [Bo93], [Bo95]. In choosing such a space of solutions we run into a technical problem. The characteristic functions E_c and E_s defined in the proof of theorem 1 of section 2 can no longer be used. Instead we define E_c to be a C^∞-function with support in $[\pm 1 - 1/4, \pm 1 + 1/4]$ and identical to 1 on $[\pm 1 - 1/5, \pm 1 + 1/5]$, and $E_s = 1 - E_c$. This does however in no way affect the proof.

Generally, incorporating a bounded transverse dimension does not require building function spaces more complicated than the one described above. This is because the bifurcating solution is driven through only one eigenfunction. It is sufficient to consider a space of functions $u(x, y, t)$ on $\mathbb{R} \times \Omega$ (with Ω a bounded domain) such that $u(\cdot, y, t)$

can be analytically extended to a strip in the complex plane $\forall y \in \Omega$, $t > 0$, and $u(x,\cdot,t) \in C(\Omega)$ $\forall x \in I\!R$, $t \geq 0$.

Returning to Poiseuille flow we make two remarks. First we recall that in fact we are interested in bounded solutions of the equations for $\mathbf{u} = (u, v)$. This implies a boundedness of $\boldsymbol{\Psi}$ due to the zero boundary conditions on $I\!R \times \partial\Omega$. Thus the function space is suitable to accommodate for solutions $\boldsymbol{\Psi}$. Second we note that the presence of the Laplacian in the time derivative part of the equations implies that we need initial conditions which are twice differentiable. This is of course not a restriction since we already imposed some smoothness on the initial data.

The unconventional form of the time derivative term prompts us to develop special functional analytic tools for the following type of inhomogeneous abstract Cauchy problem.

$$\begin{cases} \partial_t M u = L u + f(t) \\ u|_{t=0} = u_0 \end{cases} \tag{21}$$

The general idea is the following: Suppose that the evolution of $\partial_t M u = Lu$ is given by the operator $W(t)$ in the following way: $u(t) = W(t)Mu_0$, then the solution of (21) is given by

$$u(t) = W(t)Mu_0 + \int_0^t W(s-t)f(s)\,ds \tag{22}$$

The one parameter family $W(t)$ is not an ordinary semigroup which can be estimated with the help of knowledge of the resolvent of L. It is related to the so-called Existence- and Uniqueness-families which are described in [dL94]. It is possible though to give an analogous characterization of $W(t)$ in terms of the linear operators M and L. Let the resolvent set $\rho(L, M)$ be given by $\rho(L, M) := \{\lambda \in \mathbb{C} : (\lambda M - L)^{-1} \text{ exists}\}$. Let Γ be a path in $\rho(L, M)$ connecting $\infty e^{-i(\frac{1}{2}\pi+\delta)}$ and $\infty e^{i(\frac{1}{2}\pi+\delta)}$ $(0 < \delta < \frac{1}{2}\pi)$ such that the spectrum $\sigma(L, M) := \mathbb{C}/\rho(L, M)$ lies to the left of Γ. Then we have the following representation of $W(t)$:

$$W(t) = \frac{1}{2\pi i}\int_\Gamma e^{\lambda t}(\lambda M - L)^{-1}\,d\lambda \tag{23}$$

By this representation of $W(t)$ we are able to prove that the integral equations which follow by (22) from (20), define a contractive mapping. More details on this analysis can be found in [Bo95].

References

[Bo93] **P. Bollerman.** Validity Results for Ginzburg-Landau's equation. Preprint 800. (1993) University of Utrecht.

[Bo95] **P. Bollerman.** Validity of the Ginzburg-Landau Equation in Two-dimensional Poiseuille Flow: A Streamfunction Approach. Paper in preparation.

[CE90] **P. Collet, J.-P. Eckmann.** The time dependent amplitude equation for the Swift-Hohenberg problem. Comm. Math. Phys., 132 (1990) 139-153.

[CH93] **M.C. Cross, P.C. Hohenberg.** Pattern formation outside of equilibrium. Rev. Mod. Phys., 65 (1993) 851-1090.

[DR81] **P.G. Drazin, W.H. Reid.** Hydrodynamic Stability. Cambridge University Press (1981).

[Eck93] **W. Eckhaus.** The Ginzburg-Landau equation is an attractor. J. Nonlinear Science, 3 (1993) 329-348.

[ES94] **W. Eckhaus, A. Shepeleva.** Attraction properties of the Ginzburg-Landau manifold. Preprint 878. (1994) University of Utrecht.

[vH91] **A. van Harten.** On the validity of Ginzburg-Landau's equation. J. Nonlinear Science, 1 (1991) 397-422.

[vH92] **A. van Harten.** Decay to the Ginzburg-Landau manifold. Lecture held in Stuttgart, 1992. Internal Communication, University of Utrecht

[HS72] **L.M. Hocking, K. Stewartson** On the nonlinear response of a marginally unstable plane parallel flow to a two-dimensional disturbance. Proc. R. Soc. Lond. A, 326 (1972) 289-313.

[IMD89] **G. Iooss , A. Mielke et Y. Demay.** Theory of steady Ginzburg-Landau equation, in hydrodynamic stability problems. Eur. J. Mech. , B/Fluids, 8, nr.3, (1989), 229-268.

[KSM92] **P. Kirrmann, G. Schneider, A. Mielke** The validity of modulation equations for extended systems with cubic nonlinearities. Proceedings of the Royal Society of Edinburgh, 122A (1992) 85-91.

[dL94] **R. deLaubenfels.** Existence Families, Functional Calculi and Evolution Equations. LNiM 1570. Springer-Verlag (1994).

[MS94] **A. Mielke, G. Schneider** Attractors for modulation equations on unbounded domains. Preprint IfAM, Universität Hannover 1994.

[NW69] **A. Newell, J. Whitehead.** Finite bandwidth, finite amplitude convection. J. Fluid Mech., 38 (1969) 279-303.

[dPES71] **R.C. di Prima, W. Eckhaus & L.A. Segel.** Nonlinear wave-number interaction in near-critical two-dimensional flows. J.Fluid.Mech. 49 (1971), 705-744.

[Sch94a] **G. Schneider.** A new estimate for the Ginzburg-Landau approximation on the real axis. J. Nonlinear Science 4 (1994) 23-34.

[Sch94b] **G. Schneider.** Error estimates for the Ginzburg-Landau approximation. J. Appl. Math. Physics (ZAMP), 45 (1994) 433-457.

[Sch94c] **G. Schneider.** Global existence via Ginzburg-Landau formalism and pseudo-orbits of Ginzburg-Landau approximations. Comm. Math. Phys., 164 (1994) 157-179.

[Sch94d] **G. Schneider.** Analyticity of Ginzburg-Landau modes. J. Diff. Eqns., 1994. To appear.

[Sch94e] **G. Schneider.** Validity and Limitation of the Newell-Whitehead equation, Preprint IfAM, Universität Hannover, 1994.

[Tea94] **P. Takáč, P. Bollerman, A. Doelman, A.v. Harten, E.S. Titi.** Analyticity of Essentially Bounded Solutions to Semilinear Parabolic Systems and Validity of the Ginzburg-Landau Equation. To appear in SIAM J. Math. An.

Bifurcations far from criticality in fluid systems

F.H. BUSSE[1] and R.M. CLEVER[2]
[1] *Institute of Physics, University of Bayreuth, D-95440 Bayreuth*
[2] *Institute of Geophysics and Planetary Physics, University of California at Los Angeles, U.S.A.*

Abstract

Many aspects of the evolution from simple to complex fluid flow with increasing control parameter in classical problems of fluid dynamics can be understood through the analysis of subsequent symmetry breaking bifurcations. The sequence-of-bifurcations approach is based on the Galerkin method for the numerical analysis of two- or three-dimensional solutions. Steady solutions or solutions in the form of traveling waves are obtained through a Newtôn-Raphson iteration and the stability of these solutions is analyzed in the form of a linear eigenvalue problem. Typical features introduced by the bifurcations from spatially periodic two-dimensional flows are often observed as coherent structures in the turbulent flows at higher values of the control parameter.

1 Introduction

Most of the theoretical research on bifurcating states of flow in fluid mechanics has focussed on the first bifurcation where a basic state reflecting all the symmetries of the external conditions becomes first unstable. The bifurcating secondary solution can often be derived analytically and numerous tools for the analysis of weakly nonlinear problems can be applied (Cross & Hohenberg, 1993). Even some secondary and tertiary bifurcations can be analyzed in this way through the unfolding of a suitably chosen codimension-2 or -3 bifurcation point in the parameter space of the problem (Golubitsky and Schaeffer, 1985). Other secondary bifurcations can not be captured in this way, especially if they correspond to new mechanisms of instability which depend on a sufficiently high value of the amplitude of the secondary solution. Numerical methods must be used in this case in order to describe first the secondary state of flow in the fully nonlinear regime and to analyze its stability in a second step. The physical phenomena that can be described in this way rival those found at primary bifurcation points and many features of turbulent flows can be understood on the basis of the physical processes giving rise to secondary and tertiary bifurcations. It is the purpose of this article to give an introduction to this approach and to illustrate it with a few examples from the theory of convection which will include some new results.

In order to maximize the symmetries that can be broken in subsequent bifurcations we shall restrict the attention to fluid systems that are homogeneous in two spatial dimensions and in time. The remaining inhomogeneous dimension is required by the energy flux which must be applied in order to keep the system away from thermodynamic equilibrium. The general structure of the problem is sketched in figure 1a and

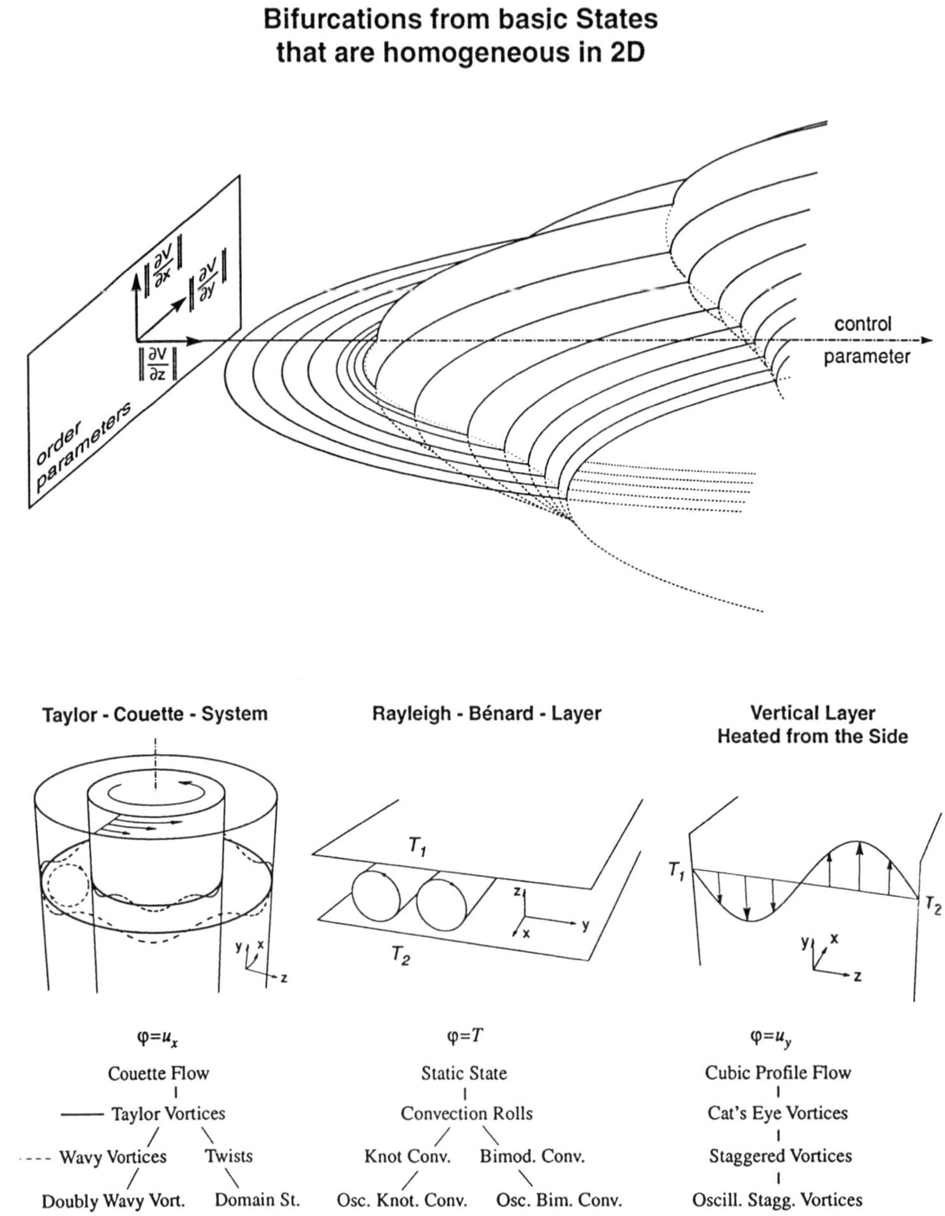

Figure 1: a) Sketch of the bifurcation structure in the order parameter space as a function of the control parameter R. b) Examples of fluid systems that are homogeneous in two dimensions and exhibit supercritical primary, secondary and tertiary bifurcations.

the principal examples are visualized in figure 1b. From our general point of view it does not matter whether a plane layer or a cylindrical annulus is considered, although the latter configuration would be preferrable for experimental realisations because of its periodic boundary condition in the azimuthal direction. End walls cause unavoidable inhomogeneities which often turn bifurcations into imperfect bifurcations. But through variations of the aspect ratio in fluid layers the effect of side walls can be minimized. From a theoretical point of view the reasons for focussing on the configurations that are homogeneous in two dimensions are overwhelming:

(i) The physical mechanisms of instability can be described in their simplest form

(ii) The transitions to more complex flows occur through exact bifurcations

(iii) Transitions can be readily identified through broken symmetries

(iv) The high degree of symmetry permits a maximum number of subsequent bifurcations

(v) Last not least, the simplicity of the formulation of the problem and its symmetries facilitate the analytical and numerical studies considerably.

In the following we shall first outline the analysis of higher bifurcations for spatially periodic solutions and then present some new results on convection in the presence of a constant shear and on oscillatory knot convection.

2 Rolls as Generic Secondary Solutions

As proposed above we consider a fluid system which is homogeneous with respect to two spatial dimensions as in the case of the examples of figure 1b. The primary solution or basic state of the system exhibits the same symmetry as the external conditions. Solutions bifurcating from this basic state will be described in general by nonlinear equations of the type

$$Lv - RBv = \frac{\partial}{\partial t} Vv + N(v) \tag{1}$$

where L, B, V are linear operators, R is the control parameter, and N is a nonlinear operator which typically will be bilinear in v. The variable v is vector valued in most problems of interest. But for simplicity's sake we may think of it as the velocity component in the z-direction normal to the walls of the examples of figure 1b. Since v vanishes for the basic state the point of bifurcation is determined by equation (1) without the term $N(v)$. Without losing generality the solution can be assumed in the form

$$v = \exp\{i\boldsymbol{l} \cdot \boldsymbol{r} + \sigma t\} g(z) \tag{2}$$

where $\boldsymbol{l}$ is an arbitrary wave vector tangential to surfaces of spatial homogeneity. In expression (2) we have used a notation appropriate for planar surfaces of homogeneity and we shall use the Cartesian geometry for simplicity in the following. But it is obvious, that a suitably modified notation applies in the case of a cylindrical or even spherical geometry. The minimal value R_c of R for which non-decaying solutions of the form (2) exist corresponds to a particular wave vector $\boldsymbol{l}_c$. In problems with additional symmetries several or even an infinite manifold of vectors $\boldsymbol{l}_c$ may exist as in the case of the Rayleigh-Bénard problem with its property of isotropy. In the following we shall identify the y-direction with the direction of $\boldsymbol{l}_c$. We then can write the two-dimensional roll solution evolving from the preferred disturbance (2) in the form

$$v = \sum_{m,n} a_{mn} \exp\{im\alpha y\} g_n(z) \text{ with } a_{-mn} - a^*_{mn} \tag{3}$$

where the star indicates the complex conjugate. For simplicity we have assumed the case of a monotonous bifurcation leading to steady rolls. But the case of an oscillatory time dependence can be treated analogously. We have introduced a Galerkin representation in which the real functions $g_n(z)$ represent a complete system of functions which individually satisfy the appropriate boundary conditions. Since we are primarily interested in problems of high symmetry we may restrict the attention to problems with a symmetry plane $z = 0$ in which case the property

$$g_n(-z) = (-1)^{n+1} g_n(z) \tag{4}$$

can be assumed.

Solutions of the form (3) describing steady rolls exhibit the five symmetries listed in table 1

Table 1: Symmetry properties of two-dimensional rolls

translation in time:	$\frac{\partial}{\partial t} v = 0$
translation along roll axis:	$\frac{\partial}{\partial x} v = 0$
transverse periodicity:	$v(y + 2\pi/\alpha_1 z) = v(y, z)$
transverse reflection:	$v(-y, z) = v(y, z)$ or $a_{-mn} = a_{mn}$
inversion about roll axis:	$v(y + \frac{\pi}{\alpha}, z) = -v(y, -z)$ or $a_{mn} = 0$ for odd $m + n$

of which the first three characterize all roll type solutions while the fourth property is lost, for instance, in the presence of a mean shear with a y-component and the fifth property requires the symmetry plane $z = 0$ as expressed by relationship (4). We like to emphasize that even Taylor vortices satisfy all properties of table 1 in the limit of the small gap approximation and when the two cylinders are rotating with nearly the same speed.

The stability of the steady roll solution (3) can be analyzed through the superposition of infinitesimal disturbances. According to Floquet theory the general disturbances can be written in the form

$$\tilde{v} = \exp\{ibx + idy + \sigma t\} \sum_{m,n} \tilde{a}_{mn} \exp\{im\alpha y\} g_n(z) \tag{5}$$

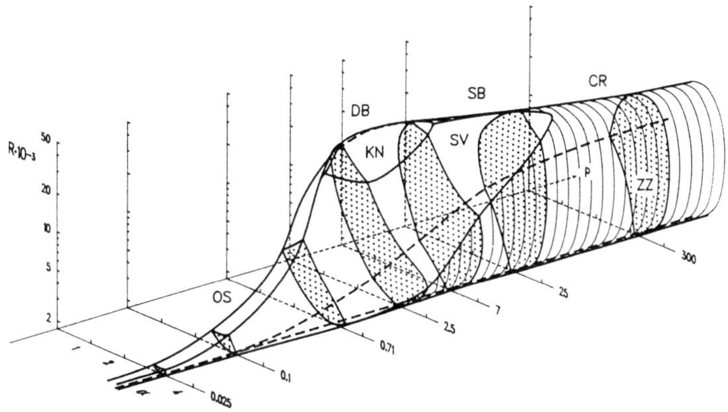

Figure 2: Stability boundaries of convection rolls in a Rayleigh-Bénard layer as a function of the wavenumber α of the rolls, the Rayleigh number R and the Prandtl number P. The stability boundaries corresponding to the oscillatoey (OS), the skewed varicose (SV), the knot (KN), the single blob (SB), the double blob (DB), the zig-zag (ZZ), and the cross-roll (CR) instability are shown.

where the growthrate σ is to be determined as the eigenvalue of the system of linear homogeneous equations for the unknown coefficients $\tilde{a}_{mn}$. As in the case of the steady solution (3) the actual implementation of the numerical analysis on a computer requires a truncation of the infinite summations in expressions (3), (5). Usually fairly rapid convergence of the numerical results with N_T is achieved if coefficients $a_{mn}, \tilde{a}_{mn}$ with

$$m + n > N_T \tag{6}$$

are neglected and a N_T of the order 10 is chosen. Whenever there exists as a function of b and d an eigenvalue σ with positive real part σ_r for given parameters $R, \alpha, \ldots$ of the steady solution (3) the latter is unstable. In this way regions of stability of steady rolls can be determined in their parameter space. An example of such a stability region is shown in figure 2 for Rayleigh-Bénard convection. The stability boundaries correspond to a variety of instabilities differing in their symmetry properties as listed in table 2 where σ_i refers to the imaginary part of σ. For further details on these instabilities we refer to the original publications by Busse (1967), Clever and Busse (1974, 1978), Busse and Clever (1979), Busse and Bolton (1984), and Bolton et al. (1987).

3 Tertiary Solutions and Their Instabilities

The growing disturbances of the form (5) evolve in general into three-dimensional solutions that are periodic in x and in y. The solutions generated by the 2. through 7. instabilities of table 2 have been analyzed in a number of papers (Frick et al. 1982;

Clever and Busse, 1987, 1989a,b, 1991, 1992, 1994a,b; Busse and Auer, 1994).

Table 2: Symmetries Broken by Bifurcations from Convection Rolls

Broken Symmetries	translation in time	longitud. translation	transverse periodicity	transverse reflection	inversion about axis
Properties of disturbance	$\sigma_i \neq 0$	$b \neq 0$	$d \neq 0$	$\tilde{a}_{mn} \neq \tilde{a}_{-mn}$	$\tilde{a}_{mn} \neq 0$ for $m+n=$ odd
Eckhaus Instab.			X	X	
Crossroll Inst.		X			X
Knot-Instab.		X			X
Even Blob-Inst.	X	X			
Odd Blob-Inst.	X	X			X
Oscillatory Inst.	X	X		X	
Zig-Zag-Instab.		X		X	
Skewed Var. Inst.		X	X	X	
Osc.Skewed Var.I.	X	X	X	X	

The solutions can be represented in the form

$$v = \sum_{l,m,n} a_{lmn} \exp\{i\alpha_x x + i\alpha_y y\} g_n(z) \text{ with } a_{-l-mn} = a^*_{lmn} \tag{1}$$

where the star indicates the complex conjugate and where the possibility of convection flows in the form of waves traveling along the axis of the rolls is included if x is replaced by $\hat{x} = x - ct$. In the latter case $-\alpha_x c$ corresponds to σ_i at the point of bifurcation and varies along the bifurcating branch. For the determination of c the time invariance of the basic problem must be used. By fixing the phase of the traveling wave at the time $t = 0$, i.e. by requiring for instance $a_{-111} = a_{111}$, we gain just the extra equation that is needed to determine c. Besides the traveling wave solution a standing wave solution is also possible which can be described by the representation (7) only if time dependent coefficients, $a_{lmn} = a_{lmn}(t)$, are assumed. In table 3 a variety of solutions of the form (7) with constant coefficients a_{lmn} are listed.

Table 3: Tertiary Solutions and Their Symmetries

Tertiary Solution	Reflection Symmetries	Inversion Symmetry
knot-/bimodal convection	$a_{-lmn} = a_{lmn} = a_{l-mn}$	$a_{lmn} = 0$ for $l+m+n =$ odd
ondulating rolls	$a_{-lmn} = a_{lmn}, a_{l-mn} = (-1)^l a_{lmn}$	$a_{lmn} = 0$ for $m+n =$ odd
symmetric traveling wave convection or wavy rolls with Poiseuille flow	$a_{l-mn} = (-1)^l a_{lmn}$	$a_{lmn} = 0$ for $m+n =$ odd
Wavy rolls with Couette flow or wavy Taylor vortices in small gap limit	$a_{l-mn} = (-1)^l a_{lmn}$	$a_{lmn} = (-1)^{m+n} a_{-lmn}$
Traveling blob convection	$a_{l-mn} = a_{lmn}$	$a_{lmn} = 0$ for $l+m+n =$ odd

The stability of the stationary solutions of the form (7) can be analyzed in the same way as in the case of the roll solution (3) through the superposition of infinitesimal disturbances

$$\tilde{v} = \sum_{l,m,n} \tilde{a}_{lmn} \exp\{i(l\alpha_x + d)x + i(m\alpha_y + b)y + \sigma t\} g_n(z) \tag{2}$$

where the growthrate σ becomes again the eigenvalue of a system of linear homogeneous equations for the unknowns $\tilde{a}_{lmn}$. Because in this system the symmetry properties of the stationary solutions listed in table 3 can not be used in general, the number of equations far exceeds the number of different coefficients in the stationary solution corresponding to the same truncation parameter N_T. The latter is usually defined in that coefficients with subscripts l, m, n satisfying

$$l + m + n > N_T \tag{3}$$

are neglected. Fortunately some of the most dangerous disturbances correspond to $d = b = 0$ in which case the general set (8) of disturbances can be separated into subclasses each of which contains approximately the same number of coefficients as the stationary solution for a given value of the truncation parameter N_T. For example, stationary solutions of the form (7) satisfy three symmetry properties according to table 3. In the case $d = b = 0$ the equations for disturbances (8) separate into equations for the coefficients which have either the same symmetry or the opposite one as the stationary coefficients a_{lmn}. Since there are 3 symmetries, 2^3 subclasses of disturbances are obtained which we list in the form

$$ECC,\ ESC,\ ECS,\ ESS,\ OCC,\ OSC,\ OCS,\ OSS \tag{4}$$

depending on whether $l+m+n$ is even (E) or odd (O) and on whether x, y-dependence is described by cosines (CC), or by sines (SS), or by sines in the x- and by cosines in the y-direction (SC) or vice versa (CS). Similar separations into subclasses can be obtained in other cases of three-dimensional stationary solutions with a corresponding simplification of the stability analysis.

4 Quarternary Solutions

It is evident from table 3 that the symmetries that have not been broken by the instabilities of table 2 give rise to symmetries of the three-dimensional stationary solution. Additional symmetry breaking instabilities must thus be expected. The most commonly encountered instabilities of the form (7) with $d = b = 0$ involve a finite imaginary part σ_i of the growthrate σ. The quarternary solutions bifurcating from the tertiary solutions thus correspond to time dependent coefficients $a_{lmn}(t)$. Some examples of quarternary solutions are listed in table 4.

For details on oscillatory knot convection and oscillatory bimodal convection we refer to the papers of Clever and Busse (1989a, 1994a). The asymmetric traveling wave convection has been treated by Clever and Busse (1992). In the following we shall discuss first an example of a tertiary solution of convection in the presence of a mean shear and in section 6 some results on steady and oscillatory knot convection will be presented.

Table 4: Quarternary Solutions and Their Symmetries

	Reflection Symmetry	Inversion Symmetry	Remarks
Oscillatory knot convection ($ECC + ESC$) or wavy oscillatory bimodal convection	$a_{l-mn} = a_{lmn}$	a_{lmn} for $l + m + n =$ odd	a_{lmn} periodic in time
Oscillatory knot convection ($ECC + OSC$)	$a_{l-mn} = a_{lmn}$ $a_{-lmn} = -a_{lmn}$ for odd $l + m + n$ $a_{-lmn} = a_{lmn}$ for even $l + m + n$		a_{lmn} periodic in time
Symmetric oscillatory bimodal convection	$a_{-lmn} = a_{l-mn} = a_{lmn}$	$a_{lmn} = 0$ for $l + m + n$ odd	
Asymmetric traveling wave convection	$a_{l-mn} = -a_{lmn}$ for odd $l + m + n$ $a_{l-mn} = a_{lmn}$ for even $l + m + n$		a_{lmn} time independent in moving frame $\hat{x} = c - ct$

5 Convection with a Mean Shear

In the presence of a plane Couette flow generated by a relative motion of the upper and lower boundaries of a horizontal fluid layer heated from below, the onset of convection assumes the form of longitudinal rolls. These rolls are unstable with respect to the zig-zag or wavy instability as soon as the Rayleigh number Ra and the Reynolds number Re of the Couette flow exceed threshold values depending on the Prandtl number of the fluid. This bifurcation is supercritical for Prandtl numbers of the order unity or higher and for moderate Reynolds numbers. When the Prandtl number is less than unity and the Reynolds number is sufficiently large, the bifurcations become subcritical. For Reynolds numbers of the order 500 it is found that wavy roll solutions exist even for vanishing Rayleigh numbers. This situation is of special interest since it has long been known that there exists no solution bifurcating from plane Couette flow. This situation is illustrated in the sketch of figure 3. The existence of separated steady solutions of the Couette system was first demonstrated by Nagata (1990). In context to the convection problem with shear it has been analyzed by Clever and Busse (1992) who also demonstrated that the steady wavy roll solutions are unstable in the isothermal case, $Ra = 0$. Nevertheless, the steady tertiary solutions in the form of wavy rolls may describe some of the properties that are found in experimentally realized turbulent Couette flow. In figures 4 and 5 the structure of the wavy roll solution is illustrated. For this purpose it is convenient to use the general representation of the solenoidal velocity field $\boldsymbol{u}$ in terms of poloidal and toroidal components, which are

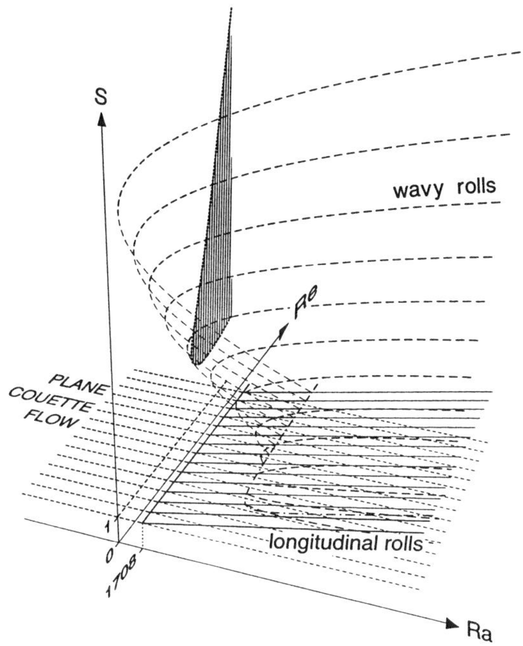

Figure 3: Sketch of the bifurcation structure of a layer heated from below with moving boundaries. The solutions of plane Couette flow (short dashes), longitudinal convection rolls (solid lines) and steady wavy rolls (long dashes) are indicated by the space spanned by the Rayleigh number Ra, the Reynolds number Re and the shear Nusselt number S.

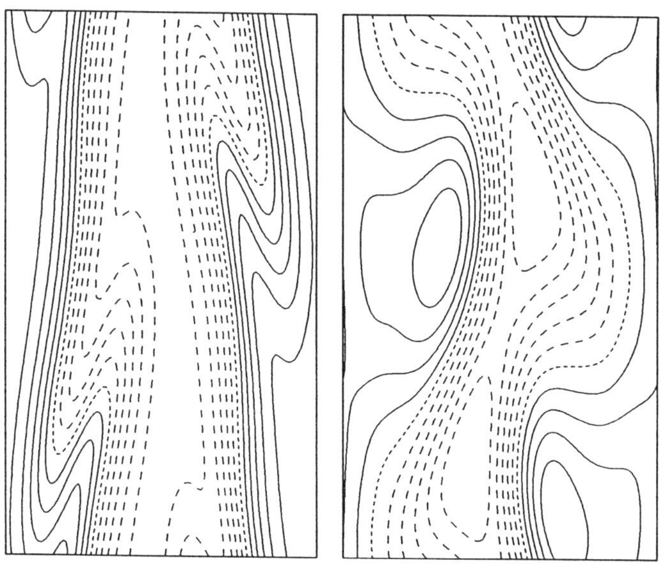

Figure 4: Lines of constant z-component of the velocity in the planes $z = 0$ (left) and $z = -0.4$ (right) for steady wavy rolls in the case $Ra = 0, Re = 1000$ with $\alpha_x = 1.5, \alpha_y = 2.5$.

described by the scalar functions φ and ψ, respectively

$$\boldsymbol{u} = \nabla \times (\nabla \times \boldsymbol{k}\varphi) + \nabla \times \boldsymbol{k}\psi \tag{1}$$

where $\boldsymbol{k}$ is the unit vector normal to the layer. Rather similar structures are found in numerical simulations of turbulent plane Couette flow (Hamilton et al., 1993). Characteristic features are the inclined eddies in the planes y = const. which are described by the stream function $\partial\varphi/\partial x$ in figure 5 and which accomplish a strong momentum transport near the upper and lower boundaries.

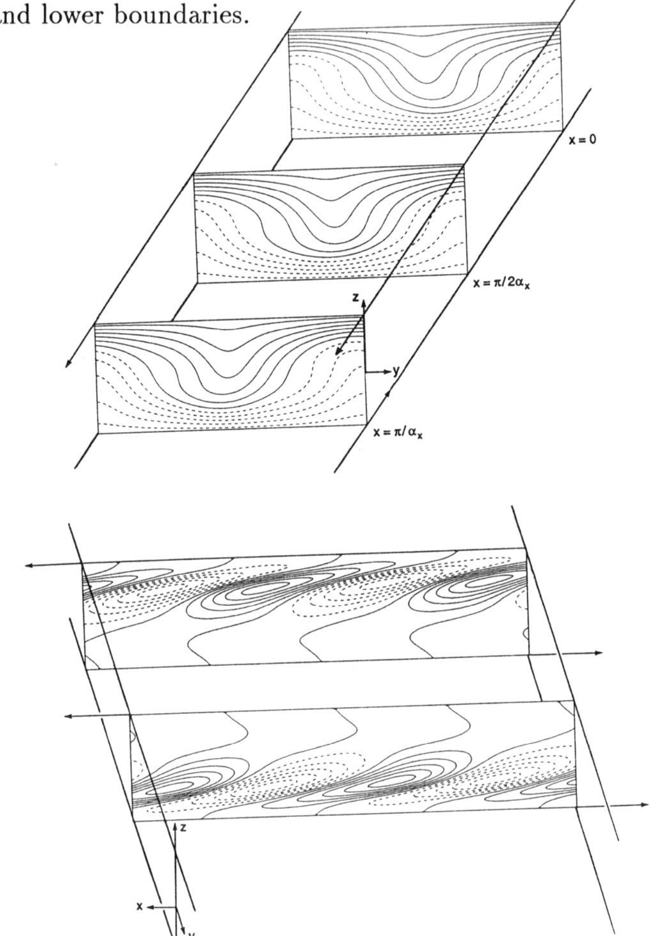

Figure 5: Lines of constant x-component of the velocity in the planes $x = n\pi/2\alpha_x$ with $n = 0, 1, 2$ (upper plot) and lines of constant $\partial\varphi/\partial y$ in the planes $y = 0, \pi/\alpha_y$ (lower plot) for the same case as in figure 4.

6 Steady and Oscillatory Knot Convection

The transition from two-dimensional rolls to the tertiary convection solution occurs typically in a monotonous fashion for Prandtl numbers of the order 10 or larger. Among the steady three-dimensional convection flows, knot convection is of special interest because it accomplishes the metamorphosis from the sheet like convection in rolls to convection in the form of thermal plumes which persists as a typical structure even in highly turbulent convection. Figure 6 shows the narrow plumes in the isotherms of the midplane, $z = 0$, and also demonstrates the effective heat transport as the plume meets the opposite boundary. In the case of figure 6 the cold plume occupies only a small part of the thermal boundary layer while a much larger part of it is used for the collection of hot fluid into the rising plume.

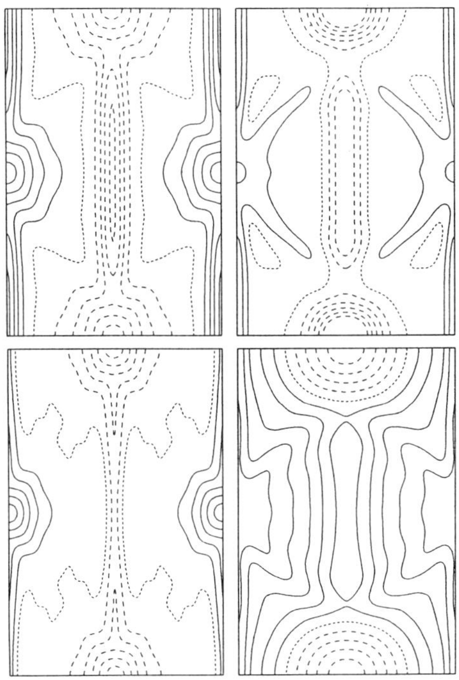

Figure 6: Lines of constant vertical velocity in the planes $z = 0$ (upper left) and $z = -0.4$ (upper right) and isotherms in the planes $z = 0$ (lower left) and $z = -0.4$ (lower right) for steady knot convection with $P = 2.5, R = 6 \cdot 10^4, \alpha_x = 1.5, \alpha_y = 2.2$.

Since the spatial degrees of freedom are already occupied, the third bifurcation breaks the translation symmetry in time. This time dependence assumes the form of oscillating blobs similar to the thermal blobs of the blob instabilities listed in table 1. As a result we obtain oscillatory knot convection as quarternary solutions. Blobs of fluids that are hotter or colder than the average fluid at the respective position circulate with the basic convection roll. They represent an instability of the thermal boundary layers in that the periodic breakups of these layers occurs in resonance with the circulation period of the roll velocity field. The time-dependence visualized in the 5 plots of the lower right part of figure 7 clearly exhibits this process. The shift during each period in the position of the erupting fluid from the thermal boundary layer is

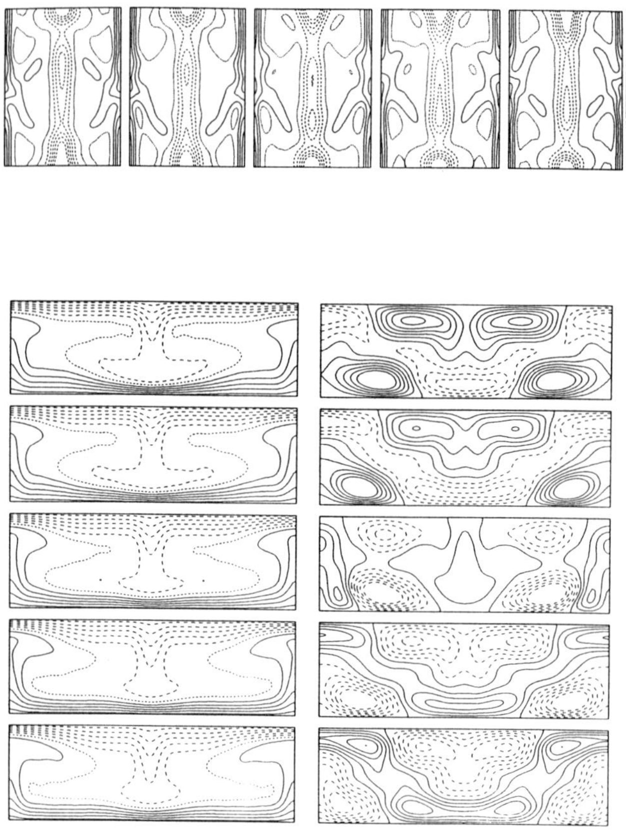

Figure 7: Lines of constant vertical velocity in the plane $z = -0.4$ (upper row of 5 plots) and isotherms in the plane $x = \pi/\alpha_x$ (left column of 5 plots) for oscillatory knot convection of the $ECC + ESC$ type with $P = 4, R = 35 \cdot 10^3 \cdot 10^3, \alpha_x = 1.5, \alpha_y = 2.0$. The plots correspond approximately to the times $t = n\pi/4\omega$ with $n = 0, 1, 2, 3, 4$ from left to right or from top to bottom, respectively. The frequency of oscillation is given by $\omega = 157$. The right column of 5 plots shows the deviations of the isotherms from the time average at the same times as on the left side.

associated with an oscillating mean flow which tends to move the convection cell back and forth along the axis of the original rolls. Both, the monotonous transition to plume like convection and the oscillatory onset of thermal blobs contribute to an increase of the heat transport beyond that carried by the preceding mode of convection at the same Rayleigh number. But the maximisation of the heat transport is certainly not the principal cause for transition since there are many transitions that are actually connected with a degradation of the transport properties.

7 Conclusions

In summarizing some of the main results of the analysis of tertiary and quarternary solutions we may conclude

(i) The spatially periodic tertiary and quarternary solutions often have only small basins of attraction when they are stable and thus they will not be realized in experiments unless special controlled initial conditions are used. Even when the tertiary and quarternary solutions are unstable they have similar properties as the turbulent realisations of the fluid flow for the same external parameters.

(ii) In particular, typical coherent structures of highly turbulent states such as the thermal plumes and the oscillatory emergence of thermal blobs from the boundary layers in the case of convection have their first and simplest manifestations in the tertiary and quarternary states of fluid flow.

(iii) The mechanisms of instability that are responsible for secondary and tertiary bifurcations may interchange their sequential order. Since they typically break different symmetries they are orthogonal to each other. The instabilities of two-dimensional rolls can thus often be found with minor modifications as instabilities of three-dimensional convection flows if another instability has preceded them.

(iv) In the extended parameter spaces in which bifurcation problems can be formulated for systems that are homogeneous in two spatial dimensions and in time it is possible to extend tertiary solutions into regions of stable basic states such as plane Couette flow which do not possess bifurcating solutions. The separated tertiary states appear to resemble turbulent states of the respective fluid system.

With respect to point (i) we may add that the theoretical assumptions of strict periodicity have long been regarded as idealisations that are not reflected by the imperfect patterns of physical reality. The careful experiments of Bodenschatz et al. (1991) have demonstrated, however, that rather perfect lattices of hexagonal convection cells can be obtained without controlled initial conditions. The more complex spatially periodic structures corresponding to higher bifurcations may not be realizable in this way. Nevertheless, they represent natural solutions for systems with spatial homogeneity and they are thus an obvious subject for the study of the transport mechanisms in moderately turbulent fluid systems.

The research reported in this paper has been supported by the Atmospheric Sciences Section of the U.S. National Science Foundation and by a NATO travel grant.

References

Bodenschatz, E., de Bruyn, J.R., Ahlers, G., and Cannell, D.S., (1991) Transitions between Patterns in Thermal Convection, *Phys. Rev. Letts.* 67, 3078-3081

Bolton, E.W., Busse, F.H., and Clever, R.M., (1985) Oscillatory instabilities of convection rolls at intermediate Prandtl numbers, *J. Fluid Mech.* 164, 469-485

Busse, F.H., (1967) On the stability of two-dimensional convection in a layer heated from below, *J. Math. Phys.* 46, 140-150

Busse, F.H., and Auer, M., (1994) Undulating Rolls and Domain Instability, *Phys. Rev. Lett.* 72, 3178-3181

Busse, F.H., and Bolton, E.W., (1984) Instabilities of convection rolls with stress-free boundaries near threshold, *J. Fluid Mech.* 146, 115-125

Busse, F.H., and Clever, R.M., (1979) Instabilities of convection rolls in a fluid of moderate Prandtl number, *J. Fluid Mech.* 91, 319-335

Clever, R.M., and Busse, F.H., (1974) Transition to time-dependent convection, *J. Fluid Mech.* 65, 625-645

Clever, R.M., and Busse, F.H., (1978) Large wavelength convection rolls in low Prandtl number fluids, *J. Appl. Math. Phys. (ZAMP)* 29, 711-714

Clever, R.M., and Busse, F.H., (1987) Nonlinear oscillatory convection, *J. Fluid Mech.* 176, 403-417

Clever, R.M., and Busse, F.H., (1989a) Nonlinear oscillatory convection in the presence of a vertical magnetic field, *J. Fluid Mech.* 201, 507-523

Clever, R.M., and Busse, F.H., (1989b) Three-dimensional knot convection in a layer heated from below, *J. Fluid Mech.* 198, 345-363

Clever, R.M., and Busse, F.H., (1990) Convection at very low Prandtl numbers, *Phys. Fluids A* 2, 334-339

Clever, R.M., and Busse, F.H., (1991) Instabilities of longitudinal rolls in the presence of Poiseuille flow, *J. Fluid Mech.* 229, 517-529

Clever, R.M., and Busse, F.H., (1992) Three-dimensional convection in a horizontal fluid layer subjected to a constant shear, *J. Fluid Mech.* 234, 511-527

Clever, R.M., and Busse, F.H., (1994a) Steady and Oscillatory Bimodal Convection, *J. Fluid Mech.* 271, 103-118

Clever, R.M., and Busse, F.H., (1994b) Standing and Traveling Oscillatory Blob Convection, submitted to *J. Fluid Mech.*

Cross, M.C., and Hohenberg, P.C., (1993) Pattern formation outside of equilibrium, *Rev. Mod. Phys.* 65, 851-1112

Frick, H., Busse, F.H., and Clever, R.M., (1983) Steady three-dimensional convection at high Prandtl number, *J. Fluid Mech.* 127, 141-153

Golubitsky, M., and Schaeffer, D.G., (1985) "Singularities and Groups in Bifurcation Theory, Vol. I", Springer: New York, Berlin, Heidelberg, Tokyo

Hamilton, J.M., Kim, J., and Waleffe, F., (1993) Regeneration of near-wall turbulence structures, pp. 11-5-1 - 11-5-6 in Proc. 9th Symposium on "Turbulent Shear Flows", Kyoto

Nagata, M., (1990) Three-dimensional finite-amplitude solutions in plane Couette flow: bifurcation from infinity, *J. Fluid Mech.* 217, 519-527

Self-organization and chaotic advection in quasi-2D confined flows

H.J.H. CLERCX and G.J.F. van HEIJST
Fluid Dynamics Laboratory, Department of Technical Physics, Eindhoven University of Technology, P.O. Box 513, 5600 MB Eindhoven, The Netherlands.

Abstract

Nonlinear dynamics is responsible for two important properties of two-dimensional incompressible Euler flows. Due to the inverse energy cascade kinetic energy is transferred from the smaller to the larger scales. This process results in the formation of large coherent structures. Due to incompressibility the Lagrangian trajectories of so-called passive tracers in an unsteady flow can be calculated from a perturbed Hamiltonian system, and result in chaotic trajectories of the tracers when particle trajectories and streamlines do not coincide. We will discuss self-organization in 2D fluids, the two-dimensionalization mechanism in rotating and stratified fluids, and present an overview of some relevant experimental results on self-organization and chaotic advection in these fluids. We will also briefly discuss some analytical techniques, based on statistical-mechanical theories of self-organization in inviscid or slightly viscous flows, which are relevant in order to extract information concerning the evolution (decay) of the self-organized state found in the experiments in rotating and stratified fluids.

1 Introduction

The study of two-dimensional (2D) flows has received an important impetus from increased interest in the dynamics and advection properties of geophysical flow systems. With the advent of satellite imagery it became clear that long-lived coherent vortex structures are abundant in the Earth's atmosphere and oceans. The emergence and persistence of these vortex structures are the result of a remarkable property of 2D flows: in contrast to their three-dimensional counterparts, these flows are characterized by the so-called inverse energy cascade, according to which kinetic energy shows a spectral flux to the larger scales of motion. Phenomenologically, this intriguing property - now commonly referred to as self-organization - results in the formation of larger, organized coherent flow structures.

In the oceans and the planetary atmospheres the background rotation and density stratification, associated with both temperature and salinity gradients, provide the required dynamical constraints to make the large-scale geophysical flows to good approximation two-dimensional. In addition, the geometrical constraint (atmospheres and oceans are only thin fluid shells covering a rotating globe) provides another two-dimensionalization mechanism. Well-known examples of rather persistent, coherent oceanic vortex structures are the Gulf Stream rings and Agulhas rings that are pinched off from the unstable, meandering currents they are named after. Other examples are the subsurface eddies that are formed by outflow of Mediterranean Sea water over

the sill in the Strait of Gibraltar: invisible for an observer at the surface, this overflow results in the formation of so-called Meddies, which take on the appearance of pancake-shaped lens-like eddies with a diameter approximately 100 km and a vertical size of $\sim$ 400 m. At a depth of approximately 1 km, these vortex structures slowly drift in western or southern direction. Even larger vortices have been observed to exist in Nature, viz. in the atmospheres of other planets: the best known examples are probably Jupiter's Great Red Spot (first observed by Hooke, more than 300 years ago) and the Great Dark Spot on Neptune. Dynamically similar vortices have been created on much smaller scales, namely in laboratory experiments on rotating and stratified fluids, see e.g. Hopfinger & van Heijst (1993) and van Heijst & Flór (1989).
The presence of a magnetic field may affect the motion of a magnetohydrodynamic fluid in the same way as background rotation does on an ordinary fluid. This mechanism was exploited by Nguyen Duc & Sommeria (1988) in their experiments on vortices in a shallow layer of mercury. As in large-scale geophysical flows, where the horizontal scale of motion of the dominant flow phenomena is much larger than the vertical scale, the motion in a soap film is also close to two-dimensional. The nice laboratory experiments by Couder & Basdevant (1986) clearly demonstrated the self-organization property of such flows: by towing a thin cylinder through the soap film they revealed that the wake flow had a remarkably ordered structure consisting of a multitude of dipolar vortices. A better understanding of the self-organization process in 2D flows is not only relevant to the fields of fluid dynamics mentioned before, but also to plasma physics and astrophysics where similar 2D flow phenomena are encountered.
The aim of this paper is to give an overview of some aspects of self-organization of 2D or quasi-2D flows. Some basic properties of such flows, including their inverse energy cascade, are discussed in section 2. In order to appreciate the experiments to be discussed later, the basic dynamics of rotating and stratified flows are briefly reviewed in section 3. In section 4 a concise review is presented of some analytical approaches that are used to investigate the dynamics of inviscid, incompressible 2D flows. Next, a few different laboratory experiments on self-organization in rotating and stratified fluids are described in section 5; here a distinction is made between local forcing on a large domain, and global forcing of the flow in a confined geometry. In section 6, the important aspect of chaotic advection in organized 2D flow systems is discussed. Finally, some conclusions are formulated in section 7.

2 Self-organization of quasi-2D flows

Self-organization of 2D flows is in sharp contrast with our daily experience with turbulence in 3D flows. In 3D turbulence, kinetic energy is transported to the smaller length scales, at which it is finally dissipated, even for vanishingly small viscosity. In remarkable contrast, 2D turbulence is characterized by an *inverse* energy cascade, resulting in self-organization of the flow. The principal difference between 3D and 2D fluid dynamics is the absence of vortex stretching (and tilting) in the latter. This can be explained by considering the Navier-Stokes equation describing the flow in a

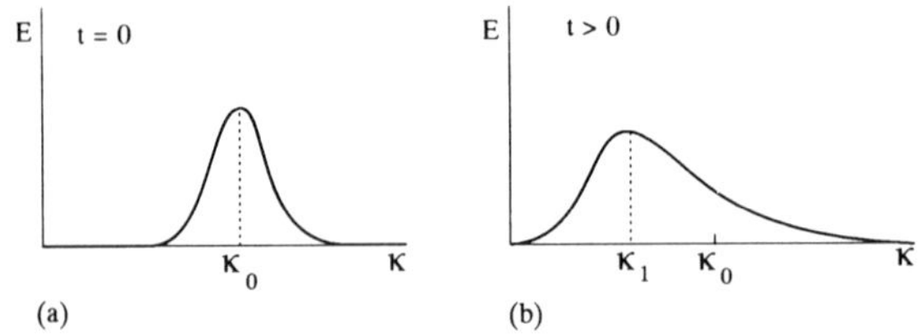

Figure 1: The evolution of the spectral-distribution of the kinetic energy in a 2D flow: the peak which initially is at a wavenumber κ_0 (a) moves to smaller κ-values (b).

barotropic fluid in the velocity-vorticity $(\mathbf{u}, \boldsymbol{\omega})$ formulation, with the vorticity defined as $\boldsymbol{\omega} = \nabla \times \mathbf{u}$,

$$\frac{D\boldsymbol{\omega}}{Dt} \equiv \frac{\partial \boldsymbol{\omega}}{\partial t} + (\mathbf{u} \cdot \nabla)\boldsymbol{\omega} = (\boldsymbol{\omega} \cdot \nabla)\mathbf{u} + \nu \nabla^2 \boldsymbol{\omega} \ . \tag{1}$$

The first term on the R.H.S. represents the stretching and tilting of vortex tubes.[1] When a vortex tube is stretched the distance of a fluid particle from the axis of the tube will diminish. Conservation of angular momentum requires the fluid in the tube to spin up. Effectively this means that fluid motion becomes concentrated at smaller and smaller scales. Although it is assumed that viscosity is small, it cannot be neglected at the smallest scales, and the kinetic energy of the flow will finally be dissipated. Consider now an incompressible 2D flow in the xy-plane with the velocity field represented by $\mathbf{u} = (u, v, 0)$. For such a flow the vorticity vector $\boldsymbol{\omega} = \nabla \times \mathbf{u} = (0, 0, \omega)$ points in a direction perpendicular to the plane of flow. As a consequence the stretching term in eq. (1) is zero. The stream function ψ can be defined according to $\mathbf{u} = \nabla \times (\hat{\mathbf{e}}_z \psi)$, and by definition it follows that

$$\nabla^2 \psi = -\omega \ . \tag{2}$$

Owing to the absence of stretching or tilting of vortex tubes, barotropic inviscid 2D flows are characterized by conservation of vorticity ω, which is expressed as

$$\frac{D\omega}{Dt} = \frac{\partial \omega}{\partial t} + (\mathbf{u} \cdot \nabla)\omega = 0 \ . \tag{3}$$

It can be shown straightforwardly that such flows possess an infinite set of other conserved quantities, the most important ones being the kinetic energy $E = \frac{1}{2}(\nabla\psi)^2$ and the enstrophy $V = \frac{1}{2}\omega^2 = \frac{1}{2}(\nabla^2\psi)^2$, which is a measure of total vorticity in the flow. Conservation of kinetic energy per unit mass and unit surface can be expressed as

$$E = \frac{1}{2}(\nabla\psi)^2 \sim \int_0^\infty \epsilon(\kappa, t) d\kappa = \text{constant} \ , \tag{4}$$

[1] A vortex tube is a tube whose surface is composed of vortex lines. The vorticity at a point on a vortex line is always tangential to that line.

and conservation of enstrophy as

$$V = \frac{1}{2}(\nabla^2\psi)^2 \sim \int_0^\infty \kappa^2\epsilon(\kappa,t)d\kappa = \text{constant} \ , \qquad (5)$$

with $\epsilon(\kappa,t)$ the spectral energy density in the wave number interval $(\kappa, \kappa + d\kappa)$ at time t. With these two conservation laws it is possible to understand qualitatively the inverse energy cascade. Consider some symmetric energy spectrum initially ($t = 0$) peaked at κ_0 (see fig. 1). As a result of nonlinear interactions of the flow, which we assume to be the basic mechanism of energy transfer to other wave numbers, a redistribution of kinetic energy will take place. During this process the total amount of kinetic energy remains constant, i.e. broadening of the spectral distribution must result in a lowering of the peak value. However, symmetric broadening of the spectral distribution cannot take place because this would imply an increase of enstrophy, in violation with (5). As a consequence there must be a larger spectral flux of energy to smaller wavelengths than to larger wavelengths, or, equivalently: energy is transferred to the larger scales of motion. In Nature and in experiments we can observe this inverse energy cascade in the emergence of large coherent vortex structures, a process referred to as the *self-organization* of 2D flows. The conservation laws are strictly spoken derived for the inviscid limit. However, the presence of a moderate viscosity does not alter the physical processes significantly, because viscous effects are only important on the smaller scales of the flow and thus hardly affect the larger, organized structures. For this reason, 2D flows are weakly dissipative, and the coherent vortices resulting from the self-organization process are very persistent. This rather intricate phenomenon of ordering in 2D flows has been demonstrated by e.g. van Heijst & Flór (1989) and Flór (1994) in experiments in stratified fluids, by Nguyen Duc & Sommeria (1988) in experiments in magnetohydrodynamic flows, by Couder & Basdevant (1986) in soap-film experiments, and also by direct numerical simulations of the evolution of initially random 2D vorticity distributions (see e.g. McWilliams, 1984; Legras, Santangelo & Benzi, 1988; Matthaeus et al., 1991). The evolution of a random vorticity field into an organized state consisting of several isolated vortices as observed in the numerical simulations by McWilliams (1984) is shown in fig. 2. These simulations were carried out with a pseudo-spectral code with periodic boundary conditions. In both the experiments and the simulations a number of different types of coherent vortices have been found in the final stages of the flow evolution. The most common types are the monopolar and dipolar vortices, but also a tripolar vortex has been found as a stable structure (van Heijst & Kloosterziel, 1989). These vortex structures were also observed to emerge in the direct numerical simulations of Legras et al. (1988). Monopolar and dipolar vortices have been observed in many parts of the world's oceans, and recent satellite observations have also provided evidence of the occurrence of a tripolar structure in the ocean (Pingree & LeCann, 1992).

When they are sufficiently isolated, the vortex structures can be considered as quasi-

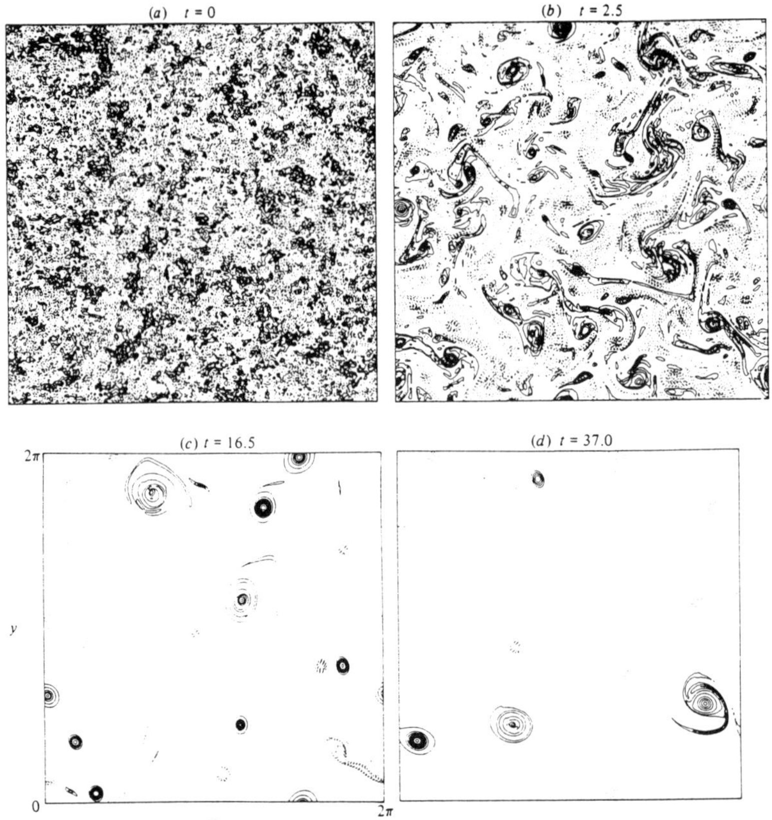

Figure 2: The evolution of the vorticity distribution in 2D turbulence (reproduced from McWilliams, 1984). From initially randomly distributed vorticity (a) gradually isolated vortex structures arise (b,c,d).

stationary in a co-moving frame of reference. In that case $(\mathbf{u}\cdot\nabla)\omega = 0$, or, equivalently

$$J(\omega,\psi) = \frac{\partial\psi}{\partial y}\frac{\partial\omega}{\partial x} - \frac{\partial\psi}{\partial x}\frac{\partial\omega}{\partial y} = 0\ , \tag{6}$$

with J the Jacobian operator. Note that ω and ψ are the vorticity and the stream function, respectively, of the flow relative to the co-moving xy frame. Obviously, any functional relationship between the vorticity and the stream function, $\omega = F(\psi)$, satisfies this relation (6). Once the $\omega - \psi$ relation is known, one is able, at least in principle, to determine the complete flow field by solving the Poisson equation for the stream function, viz. $\nabla^2\psi = -\omega = -F(\psi)$, in which the source term is possibly nonlinear. This seemingly simple approach has a serious drawback: in general we do not know the exact $\omega - \psi$ relation necessary to solve the Poisson equation. Experimental measurements of the $\omega - \psi$ relation may thus provide a very important tool for a better understanding of the flow evolution. It appears experimentally that this tool is even useful for studying the evolution of slightly viscous flows. As will be discussed below, several theoretical approaches predict different $\omega - \psi$ relations and at this stage it is not clear at all which of them is the correct approach.

3 Basic dynamics of rotating and stratified flows

The appearance of coherent vortices in experiments in rotating and stratified fluids suggests quasi-2D fluid dynamics. Rotating and stratified fluids both have their specific two-dimensionalization mechanisms. Consider first a *rotating fluid* and define the motion in the fluid with respect to a coordinate system co-rotating with the fluid. The Navier-Stokes equation, including the Coriolis acceleration, then reads

$$\frac{\partial \mathbf{u}}{\partial t} + (\mathbf{u} \cdot \nabla)\mathbf{u} + 2\boldsymbol{\Omega} \times \mathbf{u} = -\frac{1}{\rho}\nabla p + \nu \nabla^2 \mathbf{u} \ . \tag{7}$$

The centrifugal acceleration can be written in the form of a gradient and is here included in the pressure p. The relative importance of the respective terms can be expressed with the help of dimensionless numbers. Introduce a characteristic velocity U and length scale L over which the velocity varies appreciably. Other characteristic quantities are the angular velocity Ω and the kinematic viscosity ν of the fluid. The Rossby number, $Ro = U/(2\Omega L)$, provides a measure of the relative importance of the inertial and Coriolis accelerations, whereas the Ekman number, $Ek = \nu/(2\Omega L^2)$, is a measure of the relative importance of viscous forces and Coriolis accelerations. When both the Rossby number and the Ekman number are small, which is to good approximation the case for large-scale geophysical flows and for most of the experiments in rotating fluids to be discussed here [2], quasi-steady flows (as observed in the rotating coordinate system) are governed by a balance between Coriolis and pressure gradient forces, i.e.

$$2\boldsymbol{\Omega} \times \mathbf{u} = -\frac{1}{\rho}\nabla p \ , \tag{8}$$

also known as the geostrophic balance equation. A remarkable property can be derived by taking the curl of eq. (8). Using conservation of mass and assuming that the flow is barotropic one obtains $(\boldsymbol{\Omega} \cdot \nabla)\mathbf{u} = 0$, which expresses that the motion is independent of the coordinate parallel to the rotation axis. This result, known as the Proudman theorem, was first verified experimentally by Taylor (1923). His celebrated experiments with moving bodies in rotating fluids clearly demonstrated the existence of a column of more or less stagnant fluid moving with the body. This feature is now generally referred to as a Taylor-column.

The two-dimensionalization mechanism active in *stratified fluids* can be easily understood from the restoring gravitational force. Consider a fluid with a stable linear stratification, i.e. the density decreases linearly with height z in the fluid. When a

[2] An exception should be made for spin-up flows where the very thin Ekman layers generate secondary flows.

fluid particle with density ρ moves in a direction parallel with the z axis it enters an area with different density $\rho + \Delta\rho$ and buoyancy forces the fluid particle back to its original level in the fluid. Consequently the fluid motion is forced to be planar. This can be shown experimentally by considering the evolution of an isolated turbulent region in a stratified fluid (Fonseka, Fernando & van Heijst, 1994; Flór, Fernando & van Heijst, 1994). Three stages in the evolution of the vertical size of the turbulent patch are identified: the vertical growth of the turbulent patch, the stage of maximum height, and finally the collapse of the turbulent region due to gravity. The final stage of the decaying stratified turbulence is characterized by the appearance of coherent structures, a manifestation of 2D turbulence (some typical examples were reported by e.g. Lin & Pao (1979) and van Heijst & Flór (1989)). These intuitive and experimental observations can be founded more thoroughly by a deeper analysis of the fluid dynamics processes in a density-stratified fluid. We assume small density variations ρ' ($\rho' << \rho$), so that the flow is governed by the Navier-Stokes equation in the so-called Boussinesq approximation. In this approximation it is assumed that the reference state is governed by a hydrostatic balance

$$\frac{\partial p}{\partial z} = -\rho g \ , \tag{9}$$

with g the gravity acceleration. Linearization of the Navier-Stokes equation with respect to density variations ρ' and use of eq. (9) results in the following set of equations

$$\frac{\partial \mathbf{u}}{\partial \mathbf{t}} + (\mathbf{u} \cdot \nabla)\mathbf{u} = -\frac{1}{\rho}\nabla p' - \frac{\rho'}{\rho} g \hat{\mathbf{e}}_z + \nu \nabla^2 \mathbf{u} \ , \tag{10}$$

$$\nabla \cdot \mathbf{u} = 0 \ , \tag{11}$$

$$\frac{\partial \rho'}{\partial t} + (\mathbf{u} \cdot \nabla)\rho' + w\frac{d\rho}{dz} = 0 \ , \tag{12}$$

with p' the deviation from the hydrostatic pressure and w the z-component of the velocity $\mathbf{u}$. Furthermore it is assumed that mass diffusion is negligible. Riley, Metcalfe & Weissman (1981) and Lilly (1983) have performed scaling analyses of these equations. Riley et al. studied the decay of initially 3D isotropic turbulence in density stratified fluids and for the case of strongly buoyancy-dominated flow they found that to lowest order the motion can be decomposed in a vertical component (parallel to the z axis), induced by internal gravity waves, and a horizontal component representing the 2D (planar) turbulent motion. We do not repeat their steps but give the final result of the scaling analysis, relevant for understanding of the appearance of horizontal flow structures, only. When viscous effects are neglected these equations can be written in the following non-dimensionalized form:

$$\frac{\partial}{\partial t}\mathbf{u}_H + (\mathbf{u}_H \cdot \nabla_H)\mathbf{u}_H + F^2 w \frac{\partial}{\partial z}\mathbf{u}_H = -\nabla_H p' \ , \tag{13}$$

$$(\frac{L_v}{L_h}F)^2 \left(\frac{\partial w}{\partial t} + (\mathbf{u}_H \cdot \nabla_H) w + F^2 w \frac{\partial w}{\partial z} \right) = -\frac{\partial p'}{\partial z} - \rho' \ , \tag{14}$$

$$\nabla_H \cdot \mathbf{u}_H + F^2 \frac{\partial w}{\partial z} = 0 \ , \tag{15}$$

$$\frac{\partial \rho'}{\partial t} + (\mathbf{u}_H \cdot \nabla_H)\rho' + F^2 w \frac{\partial \rho'}{\partial z} + w = 0 \ , \tag{16}$$

with L_h and L_v horizontal and vertical length scales ($L_v << L_h$), $\mathbf{u}_H$ and w the horizontal and vertical components of the fluid velocity, respectively, and $\nabla_H = \left(\frac{\partial}{\partial x}, \frac{\partial}{\partial y}\right)$. For small values of the Froude number F (based on the vertical length scale), which measures the relative importance of inertial effects over stratification effects, it can be shown that the horizontal motions are decoupled from the vertical motion (except for viscous and diffusive effects), resulting in the following equations for the horizontal motion

$$\frac{\partial}{\partial t}\mathbf{u}_H + (\mathbf{u}_H \cdot \nabla_H)\mathbf{u}_H = -\nabla_H p' \ , \tag{17}$$

$$\nabla \cdot \mathbf{u}_H = 0 \ , \tag{18}$$

Riley et al. (1981) suggested that a possible mode of motion resulting from eqs. (13)-(16) is horizontal flow, but with a vertical variation (the vertical motion can be obtained by going to the next order and results from pressure gradients induced by horizontal motions). The fluid motions in the different layers can thus be assumed to evolve to first order nearly independently of each other, resulting in a quasi-2D flow in the respective layers. Although Riley et al. and Lilly (1983), who also included Coriolis forces in the equations of motion, gave a general overview of scaling studies of the Navier-Stokes equation in the Boussinesq approximation, a similar theory for large scale atmospheric flows in the presence of strong stable stratification was already presented by Drazin (1961).

4 Analytical approaches

Analytical studies of coherent vortex structures in 2D flows are abundant. These include stability studies of vortices, studies of vortex interactions, studies of chaotic advection properties, etc. One particular approach in investigating the self-organization of 2D flows is based on the application of analytical techniques originating from statistical mechanics. In this approach the major goal is to find analytical expressions for the $\omega - \psi$ relations apt to the different 2D flows, which in turn can be compared with measured $\omega - \psi$ relations from experiments or extracted from data obtained by numerical simulations. Here we restrict ourselves to these analytical studies.

Statistical-mechanical theories are used in many fields of physics. In fluid dynamics this approach was first adopted in 1949 by Onsager in a paper about statistical hydrodynamics (Onsager, 1949). The basic idea was to approximate the Euler system by a large number of point vortices, which form a finite-dimensional Hamiltonian system. Methods from statistical mechanics can then be used to study the evolution and equilibrium states of such a large set of point vortices. Since that time many others have used the statistics of point vortices to study 2D flows (Joyce & Montgomery,

1973; Pointin & Lundgren, 1976). Joyce and Montgomery (1973) were able to derive a relationship between the locally averaged vorticity $< \omega >$ and the stream function ψ by a mean field approximation, which turned out to be

$$< \omega > \simeq sinh(\beta\psi) \ . \tag{19}$$

Some of the experimentally measured $\omega - \psi$ plots can be fitted quite well with this relation, but often other nonlinear $\omega - \psi$ relations satisfy just as well (Pasmanter, 1994). Montgomery and coworkers recently presented results about the evolution and self-organization of an initially random vorticity field by solving the full 2D nonlinear Navier-Stokes equation numerically (Montgomery et al., 1992). With a high-resolution pseudo-spectral code they performed long-time integrations and found $\omega - \psi$ relations which resemble the sinh-relation remarkably well, despite the differences between this direct numerical simulation and the point-vortex methods.

The modelling of a continuous vorticity distribution by a finite set of point vortices may sometimes lead to inconsistencies because the maximum vorticity is not always bounded by the initial maximum, and also, as Onsager (1949) already pointed out, there is the problem of different point-vortex distributions describing the same continuous vorticity distribution resulting in different statistical equilibrium states. Some years ago several groups developed alternative statistical-mechanical theories for Euler flows taking into account the infinite number of conservation laws to be obeyed by the flow (Robert & Sommeria, 1991; Miller, Weichman & Cross, 1992). The organized structure appears to correspond to a state of maximum entropy, taking into account all the constants of the motion, and determination of the maximum entropy state thus in principle provides the $\omega - \psi$ relation. However, an explicit expression for the $\omega - \psi$ relation can only be obtained by assuming relatively simple initial vorticity distributions or in the "dilute" vorticity approximation, in which agreement is found between these more advanced theories and mean field approaches like those of Joyce & Montgomery (1973). With their method Sommeria and coworkers studied the appearance of coherent structures in a 2D shear layer (Sommeria, Staquet & Robert, 1991) and the inertial organization of a jet in two dimensions (Thess, Sommeria & Jüttner, 1994). An interesting aspect of these theories is the possibility to extend it from exact 2D flows to multilayer systems (Robert & Sommeria, 1991). As an interesting extension of their work to multilayer systems, as those discussed in section 3, can serve the study of the evolution of pancake-like vortices in stratified fluids.

Point-vortex methods and maximum entropy theories are basically theories applicable to inviscid flows. It is obvious that it is advantageous to follow this approach, but its weaknesses are apparent. The quasi-2D flows are always (weakly) dissipative, destroying the concept of conserved quantities. Another weakness is the application of maximum entropy theories for flows in bounded domains and how to take care of the boundary layers and possible flow separation in these layers. Few analytical approaches exist to study these more complicated flow phenomena. Two approaches are the theory of minimum enstrophy vortices introduced by Leith (1984) and a method to obtain $\omega - \psi$ relations for flows in bounded domains presented recently by Suh (1994).

Two-dimensional flows are characterized by a regular cascade of the enstrophy to the smaller scales of motion, at which it is effectively dissipated even for the slightest amount of viscosity. Apparently, in weakly viscous flows the enstrophy V dissipates at a faster rate than the kinetic energy E. This so-called selective decay mechanism has inspired theoreticians to seek for solutions that are characterized by a minimum of enstrophy with the constraint of a fixed kinetic energy. Leith has shown that these minimum enstrophy vortices for simple flow configurations are represented by a linear $\omega - \psi$ relation. In the recently reported study of two-dimensional spin-up of a fluid in a rectangular container by Suh (1994) it is shown analytically, based on ideas from Batchelor (1956) to study quasisteady flow with closed streamlines at high Reynolds numbers, that the $\omega - \psi$ relation of the vortices in the array should be linear. This analytical result has been confirmed by the numerical simulations of spin-up (Suh, 1994; van de Konijnenberg et al., 1994), but also by experiments in rotating and stratified fluids (van de Konijnenberg et al., 1994; Flór, 1994). Finally we mention here a new analytical/numerical framework proposed by Montgomery, Shan & Matthaeus (1993) in which numerical simulations of the 2D Navier-Stokes equations are performed including weak dissipation. Their method is based on a two-fluid formulation involving interpenetrating positive and negative vorticity fluxes which, as they show, are intimately linked with the decay problem of passive scalars. It seems that such a 2D Navier-Stokes fluid relaxes to a maximum-entropy state.

5 Laboratory experiments and numerical simulations

Some characteristics of the two-dimensionalization process of rotating and stratified fluids, and the subsequent self-organization of the flow, have been examined in the laboratory. In rotating fluids experiments were performed to study the characteristics of organized structures like monopoles, dipoles and tripoles and their stability properties (for a review, see Hopfinger & van Heijst, 1993). These vortices can be created in several ways. Monopolar vortices, for example, can be generated by applying the so-called collapse technique (Kloosterziel & van Heijst, 1992). A rotating tank is filled with a homogeneous fluid and in the centre a bottomless cylinder is placed with the same fluid but with a different height. After vertical lifting of the cylinder a gravity-driven flow in radial direction arises which is quickly deflected by the Coriolis force, finally resulting in a purely azimuthal flow. When the height difference is of the same order as the fluid height in the tank itself the ensuing motion will be 3D turbulent. However, the vertical flow component will be suppressed within a few rotation periods and the two-dimensionalization process is thus quickly completed. Isolated monopolar vortices can be generated by putting a cylinder in a rotating fluid, stirring the fluid inside the cylinder either in the cyclonic or in the anticyclonic direction and then quickly withdrawing the cylinder vertically. The vortices thus created may be stable or unstable, depending on the steepness of the initial radial vorticity gradient. When cyclonic, the unstable monopolar vortex usually transforms into a tripole (see van Heijst, Kloosterziel & Williams, 1991), or into two dipoles when the initial vortex is

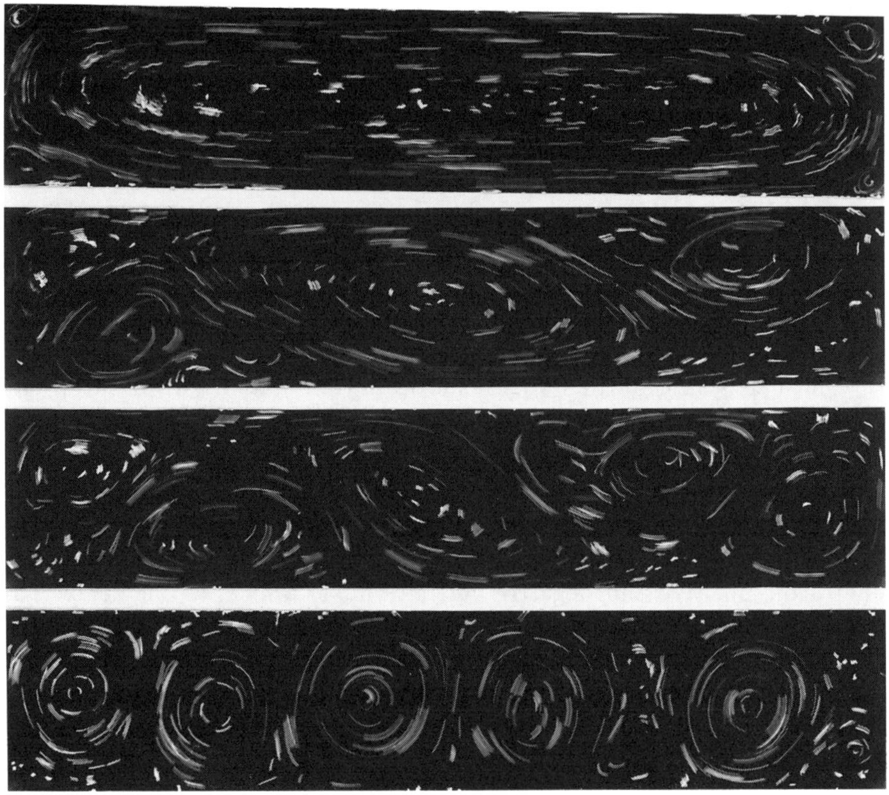

Figure 3: Sequence of plan-view photographs taken during spin-up from rest of a free-surface homogeneous fluid in a rectangular container for $\delta = 5$. The flow is visualized by small tracer particles floating on the fluid surface.

anticyclonic (Kloosterziel & van Heijst, 1991). Higher-order modes of instability have also been found (see Kloosterziel & van Heijst, 1991; Carnevale & Kloosterziel, 1994; Morel & Carton, 1994), but these are rather unstable, as can be concluded from their rapid disintegration.

Spin-up experiments on fluid in a rectangular container, initially rotating with angular velocity Ω, and forced instantaneously to rotate with a higher angular velocity $\Omega + \Delta\Omega$, were conducted by van Heijst, Davies & Davis (1990). These experiments revealed the self-organization of the flow in an array of quasi-2D cells, with the number of cells roughly scaling with the aspect ratio of the container. The self-organization of the flow during the adjustment process can be understood by considering the fluid motion in the container immediately after increasing the rotation rate from Ω to $\Omega + \Delta\Omega$ of the container at $t = 0$. In a coordinate system corotating with the container one observes a large anticyclonic cell that fills the domain entirely; this motion arises because of the inability of the fluid to follow the change in rotation of the boundaries.

At $t = 0^+$ boundary layers are set up and cyclonic vorticity, generated in the boundary layers, will be convected by the primary anticyclonic flow along the sides of the container. In the subsequent stage it is observed that the advected cyclonic vorticity gets accumulated in cyclonic cells in the corners of the domain, with the cells at the downstream ends of the longer side walls being larger and more intense than those in the corners downstream of the shorter walls. The larger cells are then seen to grow and to migrate into the interior, upon which complicated, irregular flow patterns may be observed; the details of the flow evolution in this transition stage depend on the rotation parameters Ω and $\Delta\Omega$, and on the horizontal aspect ratio $\delta = L/B$, with L and B the length and the width of the tank, respectively. During these stages the flow is to a good approximation two-dimensional. Via a complicated process of interactions between cells (new cells being formed by accumulation of viscously-generated wall vorticity) the flow is finally observed to organize into an array of counter-rotating cells which completely fill the domain. A quasi-stationary state sets in because the decay of the cells by the Ekman circulation mechanism, induced by bottom friction in the Ekman boundary layer, occurs on a timescale considerably larger than the eddy turnover time scale. An example of the experimentally observed cellular flow pattern in a rectangular tank with aspect ratio $\delta = 5$ is given in fig. 3.

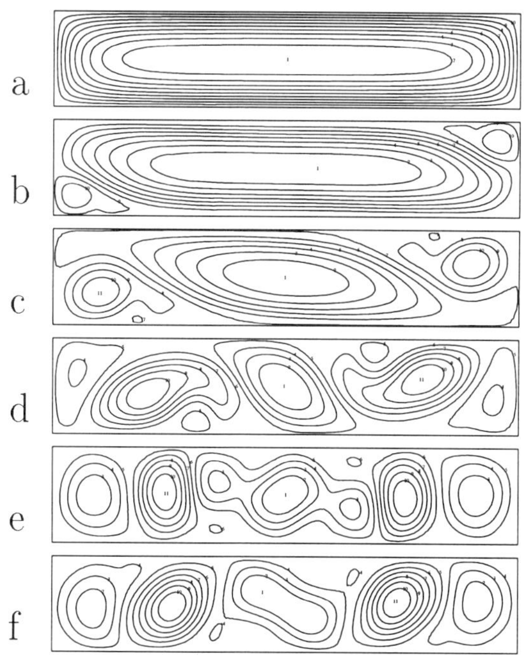

Figure 4: Streamlines for 2D spinup on a rectangular domain ($\delta = 5$). $Re = 5000$; (a) $t = 0$ *sec*; (b) $t = 10$ *sec*; (c) t=20 sec; (d) $t = 30$ *sec*; (e) $t = 40$ *sec*; (f) $t = 50$ *sec*.

Direct numerical simulations of 2D spin-up, based on a spectral element method developed by Timmermans, van de Vosse & Minev (1994), serve as a good example to show the evolution to a cellular pattern in a tank with aspect ratio $\delta = 5$. The results are plotted in fig. 4 and resemble the experimental evolution very well. The experiments and numerical simulations by van de Konijnenberg et al. (1994) and Suh (1994) provide evidence for a linear $\omega - \psi$ relation, although accurate measurements and 3D numerical simulations of the spin-up process suggest that the $\omega - \psi$ relation becomes slightly nonlinear at later stages of the adjustment process to solid body rotation. This effect is due to nonlinear suction effects by the Ekman boundary layer at the bottom of the tank.

In stratified fluids other techniques are used to generate coherent vortex structures. We describe here some frequently used methods. Localized forcing can be realized by (pulsed) horizontal injection of fluid in a linearly stratified fluid. The density of the injected fluid is matched with the layer in which it is injected. The initially 3D-turbulent blob quickly collapses under gravity, and the flow becomes quasi-2D: then the inverse energy cascade sets in, resulting in the formation of a dipolar vortex as shown in fig. 5 (van Heijst & Flór, 1989). Examples of global forcing are the use of multiple sources and sinks to generate turbulence in stratified flows (Boubnov, Dalziel & Linden, 1994), and the use of a stirring rake (a linear array of vertical rods) to generate quasi-2D turbulence on the entire (bounded) domain. Experiments on the self-organization of stratified flows in rectangular containers are described by Flór (1994). In his experiments the flow was generated by either using two horizontal jets in juxtaposition at the longer sidewalls or by stirring the fluid in the entire domain with a rake. In both types of experiments the flow becomes organized in a cellular pattern similar to that observed in the spin-up experiments. Both flow measurements and numerical simulations revealed that the organized, decaying state is to a good approximation characterized by a linear $\omega - \psi$ relation. Figure 6 shows a typical result of the measured stream function and vorticity in a rectangular tank with aspect ratio $\delta = 4$. The cellular structure is clearly visible. Measurements of the vorticity and stream function in the two central cells indicates again a linear $\omega - \psi$ relation, although initially this relation seems to be nonlinear.

6 Chaotic advection

Chaotic advection properties of 2D flows are relevant to many branches of fluid dynamics. Examples are enhancement of stirring efficiency by chaotic advection (see e.g. Aref (1984)), the mixing properties of geophysical flows (see Behringer, Meyers & Swinney (1991), Pierrehumbert (1991) and Pierrehumbert & Yang (1993)), chaotic advection and anomalous diffusion in Rayleigh-Bénard convection (see Solomon & Gollub (1988)), and many others, including industrial flow problems. The chaotic transport properties of these flows are generally studied in experiments by following added or naturally available passive tracers, i.e. tracers which have no dynamical influence on the advecting velocity field. In experiments small solid particles or dye can serve as passive tracers. In Nature temperature, salt, but also pollutants in the atmosphere

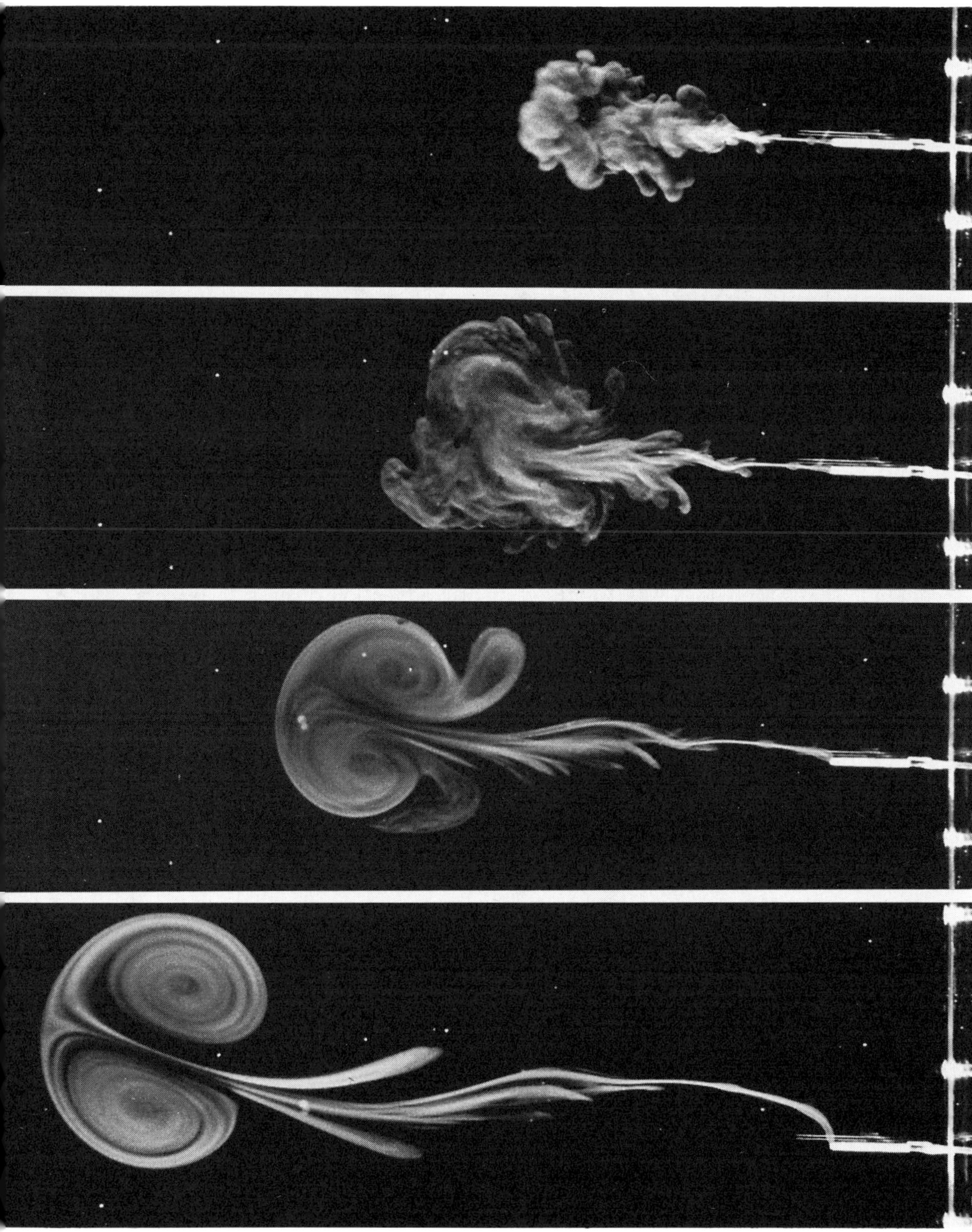

Figure 5: Top view photograph made during a laboratory experiment, which illustrates how from a turbulent "cloud" in a stratified fluid ultimately an organized dipolar vortex appears.

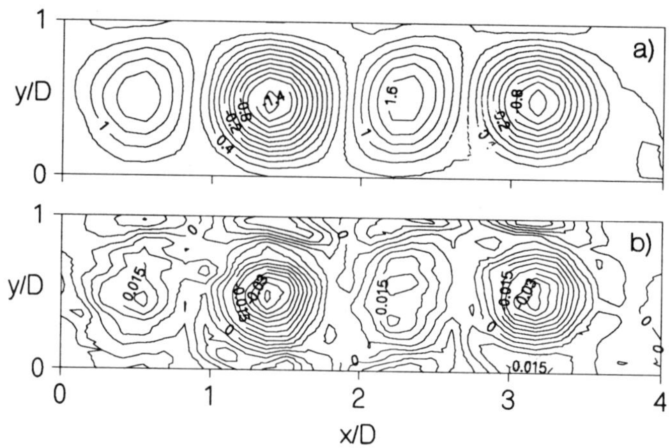

Figure 6: Contour plots of the stream function (a) and the vorticity (b) of a cell pattern for $\delta = 4$ at $t = 45$ *min* after forcing by the rake (reproduced from Flór, 1994).

and the oceans may act as passive tracers. From the above examples we may directly conclude that an obvious way to study chaotic advection of passive tracers is by exploiting the Lagrangian approach, i.e. following the trajectories of the individual fluid particles in time. Supposing that we know the Eulerian velocity field $\mathbf{u}(\mathbf{r}, t)$, either by using some specified simple analytical model of the flow, or by using a numerically obtained velocity distribution, we can follow the trajectory of the fluid particles by solving the following equation of motion

$$\frac{d\mathbf{r}}{dt} = \mathbf{u}(\mathbf{r}, t) , \tag{20}$$

with initial condition $\mathbf{r}(t = 0) = \mathbf{r}_0$. For the case of incompressible 2D flow one can introduce the stream function ψ as before and eq. (20) can be rewritten in the following form

$$\frac{dx}{dt} = \frac{\partial\psi(x, y, t)}{\partial y} \quad , \quad \frac{dy}{dt} = -\frac{\partial\psi(x, y, t)}{\partial x} . \tag{21}$$

This equation defines a Hamiltonian system with two degrees of freedom when ψ is time-dependent and reduces to a system with one degree of freedom when ψ is constant in time. For a more elaborate discussion about these systems we refer to the seminal study by Aref (1984) and the review paper by Ottino (1990). Here we restrict ourselves to some more or less simple flows for which a relatively simple physical interpretation is possible. Let us first briefly consider the time-independent case. In steady flows the streamlines and particle trajectories are time-independent and coincide, thus the particle trajectories are integrable. Mathematically this conclusion was already inferred by the fact that eq. (21) defines a Hamiltonian system with one degree of freedom, which is integrable indeed. For time-dependent flows the Hamiltonian system might be non-integrable. Somewhat surprisingly, it appears that the stream function need not be very complicated to enter the regime of chaotic particle paths, because particles move along different streamlines at different times. We can illustrate this with an interesting example with direct physical relevance presented by Solomon & Gollub (1988) in their study of mixing processes in Rayleigh-Bénard convection cells. Solomon & Gollub observed chaotic transport of passive tracers between the cells during their experiments and they constructed a model by which the transport processes could be studied numerically. They assumed a cellular flow, periodic in the x direction when perturbations are absent and with solid boundaries at $y = 0$ and $y = D$, and proposed the following stream function

$$\psi(x, y, t) = \sin\left(\frac{\pi}{D}\left\{x + \epsilon \sin \omega t\right\}\right) F(y) \ . \tag{22}$$

The frequency and amplitude of the perturbation are denoted by ω and ϵ, respectively. The function $F(y)$ satisfies the rigid (free-slip) boundary conditions. The perturbation of the cellular flow pattern causes a deformation of the separatrices between the cells. This deformation is largest near the stagnation points at the upper and lower boundaries ($y = 0$ and $y = D$), resulting in chaotic advection and mixing of the fluid. With this approach Solomon and Gollub were able to analyse numerically the chaotic advection properties of this particular cellular flow field by varying the oscillation amplitude. Quite surprisingly such a simple model gives results that agree very well with experimental results on advection and mixing of the tracers. Recently, Velasco Fuentes (1994) presented another example of how simple flows can induce chaotic advection of tracers. He modelled the unsteady behaviour of a tripolar vortex with three modulated point vortices only, and the advection of passive tracers according to this model is in remarkable good agreement with experimentally obtained results.

In his experiments on cellular flow patterns in a stratified fluid confined in a rectangular container, Flór (1994) observed a very interesting feature of small, random oscillations of the vortex positions. A qualitative comparison between these experiments and the analytical/numerical modelling described above is worthwile. Three cases were considered more precisely, namely cellular structures in rectangular containers with aspect ratios $\delta = 3, 5$ and 8. Here we will only discuss some qualitative results of experiments in containers with the smaller aspect ratios. A more detailed description of the experimental arrangement and the experimental results can be found in Flór (1994). The linearly stratified fluid was initially forced by two jets at opposing sidewalls, acting during several minutes. During and after the forcing the flow starts to organize itself, eventually taking on the appearance of an ordered array of cells. The Reynolds number, based on the maximum velocity and radius of the cellular structures, is of the order $O(10^3)$, and decreases by viscous effects to order $O(10^2)$ during the experiment. The number of cells was found to depend on the aspect ratio of the container: for instance, three cells were observed in the tank with $\delta = 3$, five in the tank with $\delta = 5$, and seven in the tank with $\delta = 8$. Accurate measurements

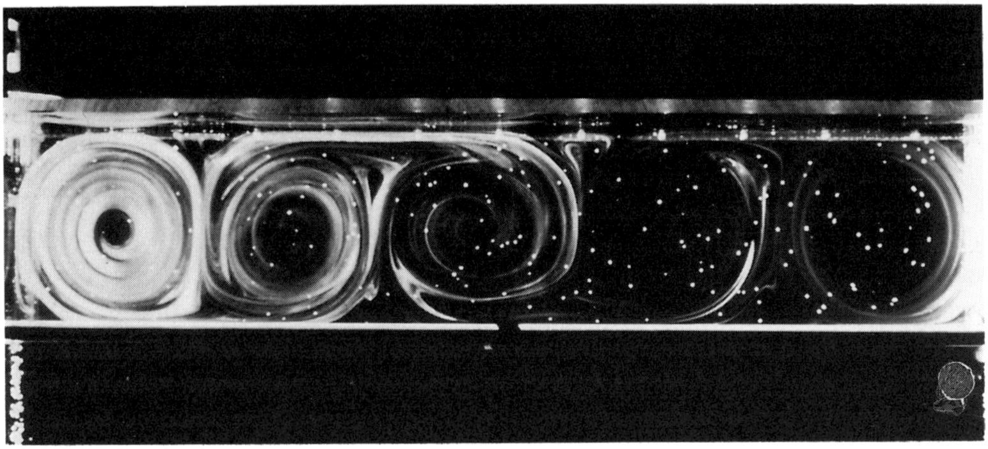

Figure 7: Plan-view dye visualization of a jet-forced flow in a rectangular tank with aspect ratio $\delta = 5$. The exchange and mixing of the dyed fluid between the different vortex regions is displayed. The flow evolution is in the late stage of decay of the vortices (Flór, 1994).

of the position of the cell centres in the tank with $\delta = 3$ have revealed oscillatory motions both in the array direction and perpendicular to it (in the direction of the long side walls). The amplitude of the oscillation is of the order of 10-20 % of the tank width. The positions of the vortex centres seem strongly correlated. The oscillation parallel to the long side walls is in phase. However, the oscillation of the centre cell, perpendicular to the long side walls, is out of phase with those of the outer cells. The experiments in the tank with aspect ratio $\delta = 5$ were performed with red and green dye in order to visualize the later stage of the decaying oscillatory motion of the five cells. At this stage particle tracking techniques do not give reliable information about the oscillatory cell motions because the flow velocities are too low. Experiments with dye as passive tracer provide this information indirectly by revealing the chaotic advection properties of the unsteady flow, see Figure 7. In this experiment red dye was put in the left cell, green dye in the right cell, and no dye in the centre cells. The separatrices between the cells are continuously deformed owing to the oscillatory motion and near the stagnation points advection of dye to neighbouring cells took place. At a later stage the dye from the left cell finally ends up in the right cell and vice versa. The dye distributions resemble qualitatively the numerical results of Solomon and Gollub (1988). It is interesting to see the process of stretching and folding of filaments of dye from cell to cell, clearly indicating a process of chaotic mixing. Another experiment in a tank with aspect ratio $\delta = 3.5$ (three cells appeared), with green dye in the central cell and red dye in the outer cells, showed strong advection of green dye to both outer cells, finally ending up in the boundary layers near the end side walls. Another feature appeared in the experiment with a tank with aspect ratio $\delta = 8$. During the course of the experiment a meandering wave-like flow with staggered vortex positions appeared.

Direct numerical simulations of 2D spin-up [3] show the same features for rectangular domains with large aspect ratio and for large Reynolds number. These meandering wave-like flows with staggered vortices near the long side walls appear as an intermediate stage before the final array of vortices arises, and have the same appearance as the channel flows studied by Jones and Montgomery (1994). Advection of passive tracers has not yet been studied for these aspect ratios, but the flow structure suggests the possibility of chaotic advection. This type of motion shows a remarkable resemblance to the analytical flow model used by Pierrehumbert (1991) in his numerical study of chaotic advection of tracers in large-scale atmospheric flows.

These relatively simple experiments on self-organization and chaotic advection in stratified flows give a good qualitative impression of the general features, but at the same time they lead to a lot of questions. For example: What are the mechanisms of the oscillations of the vortex positions? How quick and efficient will mixing take place? How can we quantify anomalous diffusion? A possible explanation for the oscillatory cell motions could be the distortion of the cells by wall-produced vorticity. The meandering flow in the container with aspect ratio $\delta = 8$ might be related to background flow induced by internal waves, as suggested by Flór (1994). Purely 2D numerical simulations, however, show the same features. Since in that case internal waves are obviously absent, the simulation results suggest a more complex process. Future research will be aimed at a more detailed experimental study of these flows in order to understand the physical processes which induce vortex wiggling, but also the effects of the presence of boundary layers on advection and three-dimensional effects like layered planar motion, vertical diffusion and the presence of secondary flows. This study should be performed in combination with high-resolution direct numerical simulations by which information about chaotic advection can be obtained by following large numbers of tracers. Hopefully this will result, in combination with analytical methods from dynamical systems theory, in a better understanding of mixing efficiency and time scales for mixing.

7 Conclusion

We have discussed several aspects of two-dimensional flows induced by nonlinear dynamics. Self-organization of 2D flows, due to nonlinear interactions leading to the so-called inverse energy cascade, is apparent. A large number of experiments, supplemented by both analytical and numerical evidence, confirm this process. Formally, self-organization is a process related with perfectly 2D flows. However, slightly viscous effects and even quasi two-dimensionality of the flow do not disrupt this simplified picture too much. In real-life experiments both aspects always play a role, and in numerical simulations one always has to introduce some amount of viscosity, although the simulations can be purely 2D. From these facts we may conclude that the process of self-organization is very robust. Another aspect, for the study of which

[3] In the 2D spin-up simulations secondary flows driven by the Ekman boundary layers are absent.

techniques originating from dynamical systems theory are being employed, concerns the Lagrangian approach to study the (chaotic) advection of passive tracers. This approach is helpful in order to gain a better understanding of the mechanisms of chaotic transport of tracers, to quantify efficiency of mixing processes and the rate of these processes. Several time-dependent flows may be described by relatively simple stream functions, and chaotic transport in this kind of flows can be studied with the theory of perturbed Hamiltonian systems. We mainly emphasized the appearance of cellular flow structures, in rectangular containers with variable aspect ratio. Chaotic advection of dye along the cells show remarkable similarities with analytical models of transport in cellular structures.

References

Aref, H. 1984 Stirring by chaotic advection. *J. Fluid Mech.* **143**, 1-24.

Batchelor, G.K. 1956 On steady laminar flow with closed streamlines at large Reynolds number. *J. Fluid Mech.* **1**, 177-190.

Behringer, R.P., Meyers, S.D. & Swinney, H.L. 1991 Chaos and mixing in a geostrophic flow. *Phys. Fluids A* **3**, 1243-1249.

Boubnov, B.M., Dalziel, S.B. & Linden, P.F. 1994 Source-sink turbulence in a stratified fluid. *J. Fluid Mech.* **261**, 273-303.

Carnevale, G.F. & Kloosterziel, R.C. 1994 Emergence and evolution of triangular vortices. *J. Fluid Mech.* **259**, 305-331.

Couder, Y. & Basdevant, C. 1986 Experimental and numerical study of vortex couples in two-dimensional flows. *J. Fluid Mech.* **173**, 225-251.

Drazin, P.G. 1961 On the steady flow of a fluid of variable density past an obstacle. *Tellus* **13**, 239-251.

Flór, J.B. 1994 *Coherent Vortex Structures in Stratified Fluids.* PhD thesis, Eindhoven University of Technology.

Flór, J.B., Fernando, H.J.S. & van Heijst, G.J.F. 1994 The evolution of an isolated turbulent region in a two-layer fluid. *Phys. Fluids A* **6**, 287-296.

Fonseka, S.V., Fernando, H.J.S. & van Heijst, G.J.F. 1994 The evolution of an isolated turbulent region in a stratified fluid. *J. Fluid Mech.* submitted.

van Heijst, G.J.F. & Flór, J.B. 1989 Dipole formation and collisions in a stratified fluid. *Nature* **340**, 212-215.

van Heijst, G.J.F. & Kloosterziel, R.C. 1989 Tripolar vortices in a rotating fluid. *Nature* **338**, 569-571.

van Heijst, G.J.F., Davies, P.A. & Davis, R.G. 1990 Spin-up in a rectangular container. *Phys. Fluids A* **2**, 150-159.

van Heijst, G.J.F., Kloosterziel, R.C. & Williams, C.W.M. 1991 Laboratory experiments on the tripolar vortex in a rotating fluid. *J. Fluid Mech.* **225**, 301-331.

Hopfinger, E.J. & van Heijst, G.J.F. 1993 Vortices in rotating fluids. *Ann. Rev. Fluid Mech.* **25**, 241-289.

Jones, W.B. & Montgomery, D. 1994 Finite amplitude steady states of high Reynolds number 2-D channel flow. *Physica D* **73**, 227-243.

Joyce, G. & Montgomery, D.C. 1973 Statistical mechanics of negative temperature states. *Phys. Fluids* **17**, 1139-1145.

Kloosterziel, R.C. & van Heijst, G.J.F. 1991 An experimental study of unstable barotropic vortices in a rotating fluid. *J. Fluid Mech.* **223**, 1-24.

Kloosterziel, R.C. & van Heijst, G.J.F. 1992 The evolution of stable barotropic vortices in a rotating free-surface fluid. *J. Fluid Mech.* **239**, 607-629.

van de Konijnenberg, J.A., Andersson, H.A., Billdal, J.T. & van Heijst, G.J.F., 1994 Spin-up in a rectangular tank with low angular velocity. *Phys. Fluids A* **6**, 1-9.

Legras, B., Santangelo, P. & Benzi, R. 1988 High-resolution numerical experiments for forced two-dimensional turbulence. *Europhys. Lett.* **5**, 37-42.

Leith, C.E. 1984 Minimum enstrophy vortices. *Phys. Fluids* **27**, 1388-1395.

Lilly, D.K. 1983 Stratified turbulence and the mesoscale variability of the atmosphere. *J. Atmos. Sci.* **40**, 749-761.

Lin, J.T. & Pao, Y.H. 1979 Wakes in stratified fluids. *Ann. Rev. Fluid Mech.* **11**, 317-338.

Matthaeus, W.H., Stribling, W.T., Martinez, D., Oughton, S. & Montgomery, D. 1991 Decaying, two-dimensional, Navier-Stokes turbulence at very long times. *Physica D* **51**, 531-538.

McWilliams, J.C. 1984 The emergence of isolated coherent vortices in turbulent flow. *J. Fluid Mech.* **146**, 21-43.

Miller, J., Weichman, P.B. & Cross, M.C. 1992 Statistical mechanics, Euler's equation, and Jupiter's Red Spot. *Phys. Rev. A* **45**, 2328-2359.

Montgomery, D., Matthaeus, W.H., Stribling, W.T., Martinez, D. & Oughton, S. 1992 Relaxation in two dimensions and the 'sinh-Poisson' equation. *Phys. Fluids A* **4**, 3-6.

Montgomery, D., Shan, X. & Matthaeus, W.H. 1993 Navier-Stokes relaxation to sinh-Poisson states at finite Reynolds numbers. *Phys. Fluids A* **5**, 2207-2216.

Morel, Y.G. & Carton, X.J. 1994 Multipolar vortices in two-dimensional incompressible flows. *J. Fluid Mech.* **267**, 23-51.

Nguyen Duc, J.M. & Sommeria, J. 1988 Experimental characterization of steady two-dimensional vortex couples. *J. Fluid Mech.* **192**, 175-192.

Onsager, L. 1949 Statistical hydrodynamics. *Nuovo Cim. suppl.* **6**, 279-287.

Ottino, J.M. 1990 Mixing, chaotic advection, and turbulence. *Ann. Rev. Fluid Mech.* **22**, 207-253.

Pasmanter, R.A. 1994 On long-lived vortices in 2D viscous flows, most probable states of inviscid 2D flows and a soliton equation. *Phys. Fluids A* **6**, 1236-1241.

Pierrehumbert, R.T. 1991 Large scale horizontal mixing in planetary atmospheres. *Phys. Fluids A* **5**, 1250-1260.

Pierrehumbert, R.T. & Yang, H. 1993 Global chaotic mixing on isentropic surfaces. *J. Atmos. Sci.* **50**, 2462-2480.

Pingree, R.D. & LeCann, B. 1992 Three anticyclonic Slope Water Oceanic eDDIES (SWODDIES) in the southern Bay of Biscay in 1990. *Deep-Sea Res.* **39**, 1147-1175.

Pointin, Y.B. & Lundgren, T.S. 1976 Statistical mechanics of two dimensional vortices in a bounded container. *Phys. Fluids* **10**, 1459-1470.

Riley, J.J., Metcalfe, R.W. & Weissman, M.A. 1981 Direct numerical simulations of homogeneous turbulence in density stratified fluids. Proc. AIP Conf. Nonlinear Properties of Internal Waves, Bruce J. West Ed., 79-112.

Robert, R. & Sommeria, J. 1991 Statistical equilibrium states in two-dimensional flows. *J. Fluid Mech.* **229**, 291-310.

Solomon, T.H. & Gollub, J.P. 1988 Chaotic particle transport in time-dependent Rayleigh-Bénard convection. *Phys. Rev. A* **38**, 6280-6286.

Sommeria, J., Staquet, C. & Robert, R. 1991 Final equilibrium state of a two-dimensional shear layer. *J. Fluid Mech.* **233**, 661-689.

Suh, Y.K. 1994 Numerical study on two-dimensional spin-up in a rectangle. *Phys. Fluids A* **6**, 2333-2344.

Taylor, G.I. 1923 Experiments on the motion of solid bodies in rotating fluids. *Proc. Roy. Soc. A* **104**, 213-218.

Thess, A., Sommeria, J. & Jüttner, B. 1994 Inertial organization of a two-dimensional turbulent vortex street. *Phys. Fluids A* **6**, 2417-2429.

Timmermans L.J.P., van de Vosse, F.N. & Minev, P.D. 1994 Taylor-Galerkin based spectral element methods for convection-diffusion problems. *Int. J. Numer. Meth. Fluids* **18**, 853-870.

Velasco Fuentes, O.U. 1994 *Two-Dimensional Vortices with Background Vorticity.* PhD thesis, Eindhoven University of Technology.

Surface tension driven cellular flows in small aspect ratio horizontally square boxes

H.A. DIJKSTRA

Institute for Marine & Atmospheric Research, University of Utrecht, Princetonplein 5, 3584CC Utrecht, The Netherlands.

Abstract

A very thin layer of motionless liquid with a free gas-liquid interface on top is situated in a horizontally square three-dimensional box. The liquid is heated from below and flow originates due to surface tension driven (Bénard-Marangoni) instabilities. In small aspect ratio containers, beautiful and unexpected finite amplitude flow patterns have been observed experimentally. Using path following techniques capable of handling very large dimensional $\mathcal{O}(10^5)$ dynamical systems, the solution structure of steady finite amplitude flow patterns and their linear stability is computed numerically for small aspect ratios of the box. Several of the patterns observed in experiments are found as linearly stable steady states.

1 Introduction

This study is motivated by the beautiful and amazing surface tension driven flow patterns obtained by Koschmieder and Prahl [7] in small aspect ratio rectangular and circular cylindrical containers. A typical experiment is, although very difficult to perform under controlled conditions, conceptually simple: an initially motionless liquid in a partially filled container is heated from below. One observes that above some critical vertical temperature gradient, cellular flow occurs. It is well-known that the flow can originate from two types of instabilities: buoyancy driven (Rayleigh-Bénard) instability and surface tension driven (Bénard-Marangoni) instability. In very thin layers, the latter effect dominates.

The physical mechanism of the Bénard-Marangoni instability is known since the work of Pearson [8]. Infinitesimally small variations in surface tension along the gas-liquid interface induce a shear stress on the interface. Because the liquid is viscous, bulk liquid motion results. The perturbations are amplified as some critical vertical temperature gradient is exceeded. Pearson [8] formulated conditions for the onset of the instability as the critical value of a dimensionless Marangoni number **Ma**, a ratio of the interfacial driving force and dissipative forces. The instability is known to be the predominant factor of the formation of the famous hexagonal cells observed by Bénard at the beginning of this century and the patterns as observed by Koschmieder and Prahl [7] in shallow layers.

The landmark paper of Pearson [8] dealt with the linear stability of a motionless horizontally infinite liquid layer due to surface tension gradients only. The smallest value of **Ma** for which instability occurs, the critical value $\mathbf{Ma}_c$, was found to be

79.6 for a no-slip bottom wall and vanishing interfacial heat transfer coefficient. For the infinite layer, the linear stability problem is degenerated; many patterns become unstable at this particular value. All patterns with a certain wavenumber $k = |\mathbf{k}|$, where $\mathbf{k}$ is the wavevector in the $x-y$ plane, satisfying a particular Helmholtz equation in that plane become unstable simultaneously [1]. Two-dimensional rolls, hexagons and bi-orthogonal cells satisfy this equation. Linear stability theory is ambigious to which pattern is most unstable.

The bifurcation behavior of the solutions near criticality was later on analysed by Scanlon and Segel [10]. They considered a vertically unbounded layer and infinite Prandtl number. By analysing the interaction of two disturbances that allow a competition between two-dimensional rolls and three-dimensional hexagons they found a transcritical bifurcation near onset. Subcritical hexagonal convection, with center upflow, was shown to be the only stable state. Cloot and Lebon [1] extended these results by examining the full range of horizontal planforms becoming unstable at criticality in a vertically bounded layer. They find that hexagonal convection is preferred near onset and bifurcates subcritically.

As much work has concentrated on the cell selection process through weakly nonlinear interactions, few have studied the influence of (distant) sidewalls on the solution structure of stable patterns. Rosenblat, Homsy and Davis [9] considered a liquid layer bounded by lateral walls on which the tangential stress is zero, the so-called 'slippery' sidewalls. In this case, the discrete spectrum at onset is given by a selection of the modes for the infinite layer which 'fit' on the box. However, they only consider two-dimensional finite amplitude flows. Dauby et al. [2] recently determined the bifurcation structure due to the interaction of two-dimensional and three-dimensional disturbances in the case of slippery sidewalls. They find stable subcritical hexagonal type convection for particular values of the aspect ratio.

In two-dimensions, Van de Vooren and Dijkstra [11] computed the neutral curve for real no-slip sidewalls. The most unstable modes differ substantially from those corresponding to 'slippery' sidewalls because a cell pattern changes along the neutral curve and the number and type of double points (where two modes have the same critical Marangoni number) changes. The solution structures for the 2D case were later on computed by Dijkstra [3] and no-slip sidewalls were shown to restrict the type of stable patterns through the occurrence of limit points.

In this chapter, pattern selection in three-dimensional Bénard convection in small aspect ratio containers is studied numerically. The aim of the present study is to determine whether some of the beautiful surface tension driven flow patterns obtained by Koschmieder and Prahl [7] can be found as linearly stable steady states of the governing system of equations. Although only a small part of parameter space is covered in the computations, two of these patterns are indeed found. Moreover, an unexpected flow pattern similar to one observed in experiments is found. The change of patterns along the branches can be understood from the bifurcation behavior of the motionless solution.

2 Formulation

A three-dimensional horizontally square box is partially filled with a Newtonian viscous liquid, having constant dynamic viscosity μ, thermal diffusivity κ and density ρ_0. Let L and H be the length and height of the liquid layer, respectively. Assume that a contant vertical temperature gradient $-\Delta T/H$, where ΔT is the difference between bottom and top temperature, is maintained over the layer and that the temperature of the bottom wall is kept at contant $T_0^* + \Delta \mathrm{T}$. The sidewalls are assumed to be perfectly insulated. The equilibrium surface tension σ_0 of the gas-liquid interface is assumed to be large, i.e. the interface does not deform. Surface tension depends linearly on the interface temperature, $\sigma = \sigma_0 - \alpha_2(T^* - T_0^*)$. Buoyancy is neglected.

We nondimensionalize velocity, pressure, time and length by $\kappa/H, \mu\kappa/H^2, H^2/\kappa$ and H, respectively. A dimensionless temperature T is given by $T^* = T\,\Delta T + T_0^*$. The dimensionless equations governing the flow in the box are

$$\mathbf{Pr}^{-1}\frac{\partial \mathbf{v}}{\partial t} + \mathbf{v}.\nabla\mathbf{v} = -\nabla p + \nabla^2\mathbf{v} \tag{1a}$$

$$\nabla.\mathbf{v} = 0 \tag{1b}$$

$$\frac{\partial T}{\partial t} + \mathbf{v}.\nabla T = \nabla^2 T \tag{1c}$$

where $\mathbf{v} = (u, v, w)$ is the velocity vector and p the pressure. At the bottom wall $(z = 0)$ there is no-slip and the temperature is kept constant,

$$T = 1, \qquad \mathbf{v} = 0 \tag{2a}$$

At the gas-liquid interface, tangential stresses are generated through horizontal temperature differences and heat is transferred from the liquid to the ambient gas. The boundary conditions are

$$w = \frac{\partial u}{\partial z} + \mathbf{Ma}\frac{\partial T}{\partial x} = \frac{\partial v}{\partial z} + \mathbf{Ma}\frac{\partial T}{\partial y} = 0, \frac{\partial T}{\partial z} = -1 \tag{2b}$$

At the north and south lateral walls $(y = 0, \mathbf{A})$ there is no-slip and no normal heat flux. The same conditions hold at the east and west lateral walls $(x = 0, \mathbf{A})$,

$$y = 0, \mathbf{A} : \mathbf{v} = \frac{\partial T}{\partial y} = 0 \; ; \; x = 0, \mathbf{A} : \mathbf{v} = \frac{\partial T}{\partial x} = 0 \tag{2c}$$

Besides the aspect ratio $\mathbf{A} = L/H$, two additional dimensionless parameters **Pr** (Prandtl) and **Ma** (Marangoni) appear. These are defined as

$$\mathbf{Pr} = \frac{\mu}{\rho_0\kappa} \quad ; \quad \mathbf{Ma} = \frac{\alpha_2 H \Delta T}{\kappa\mu} \tag{3}$$

The equations and boundary conditions commute only with the group $\mathcal{Z}_2 \oplus \mathcal{Z}_2$ [6], representing both reflections through the planes $x = \mathbf{A}/2$ and $y = \mathbf{A}/2$. A thorough analysis of the symmetries in an infinite layer was given in [6]. Note, that the translational symmetries as well as some rotational symmetries, which exist in the infinite layer, no longer exist when no-slip sidewalls are present.

3 Numerical Methods

Bifurcation analysis on the problem (1-3) is performed using path following techniques. The methodology is described extensively in [4] and only a short summary is given here. The steady equations (1-3) are discretized using a Marker and Cell finite difference method on an equidistant grid on $[0, \mathbf{A}] \times [0, \mathbf{A}] \times [0, 1]$. This leads to a system of nonlinear algebraic equations

$$\mathbf{F}(\mathbf{u}, \mathbf{p}) = 0 \tag{4a}$$

emerges. Here $\mathbf{u}$ is a d-dimensional vector consisting of the unknowns at the gridpoints (typically $d = \mathcal{O}(10^4 - 10^5)$), $\mathbf{p}$ is the n_p-dimensional vector of parameters and $\mathbf{F}$ is a nonlinear mapping from $R^d \times R^{n_p} \rightarrow R^d$. To determine branches of steady solutions of (4a) as one of the parameters (say μ) is varied, the pseudo-arclength method is used. The branches $(\mathbf{u}(s), \mu(s))$ are parametrized by an 'arclength' parameter s. An additional equation is obtained by 'normalizing' the tangent

$$\dot{\mathbf{u}}_0^T(\mathbf{u} - \mathbf{u}_0) + \dot{\mu}_0(\mu - \mu_0) - \Delta s = 0 \tag{4b}$$

where $(\mathbf{u}_0, \mu_0)$ is an analytically known starting solution or a previously computed point on a particular branch and Δs is the steplength. To solve the system of equations (4) Euler-Newton continuation is used. Limit points are detected by following $\dot{\mu}$ along a branch. Other singularities, like Hopf bifurcation points, are detected by solving the linear stability problem. For simple bifurcation points (i.e. transcritical - or pitchfork bifurcations) there are cheaper alternatives, for example the sign of the determinant of the Jacobian matrix of (4a).

The discretization of the equations governing the evolution of infinitesimally small disturbances on a particular steady state leads to a generalized eigenvalue problem

$$\mathcal{A}\mathrm{x} = \sigma\mathcal{B}\mathrm{x} \tag{5}$$

where $\mathcal{A}$ is a nonsingular, non-symmetric $d \times d$ matrix. Through Dirichlet boundary conditions and/or the incompressibility condition, $\mathcal{B}$ may become singular. Here, the instability of a certain steady flow pattern occurs only through a few number of modes. Therefore, one is only interested to compute a few eigenmodes, i.e. those with eigenvalues closest to the imaginary axis (the 'most dangerous' modes). The eigenvalue problem (5) is transformed, through a complex mapping, in such a way that the most dangerous modes become the most dominant modes (i.e. those with eigenvalues with largest norm). The simultaneous iteration algorithm, a generalized power method, is thereafter used on the transformed problem. All the linear systems of equations which appear during the computations are solved with an efficient version of pre-preconditioned conjugate gradients (details in [4]).

For the particular problem here, the code was checked with available analytical results on bifurcation points and with a solution structure as computed analytically in [2] for slippery sidewalls [5]. Based on these results, a resolution of $16 * 16 * 8$ is necessary to give sufficiently accurate solutions for aspect ratios up to 6.

4 Steady flow patterns and their linear stability.

When there is no flow in the liquid, the heat transfer is only through conduction and the temperature profile is linear, i.e. the basic state is given by

$$\bar{\psi} = 0; \; \bar{T}(z) = 1 - z \; ; \; \bar{p} = p_0 \tag{6}$$

First, bifurcations from this state are considered through the analysis of its linear stability.

4.1 Neutral curve and most unstable patterns

Values of **Ma** at neutral stability were computed in steps of 0.5 for **A** from 1 up to 8 (Fig. 1). There are several 'kinks' in the neutral curve, indicating that several branches of eigensolutions intersect. This is similar to the neutral curves in simpler situations, for example in the $2D$ situation [11]. For large **A**, a constant value is approached which is slightly larger (due to the finite resolution) than the value for the infinite layer (**Ma** $= 79.6$). For all most unstable patterns, the vertical velocity has no zeroes in the vertical, similar to the results for 'slippery' sidewalls [9]. The patterns are therefore presented (the panels $a - d$ in Fig. 1) as contour plots of the vertical velocity at $z = 0.5$. In each contour plot, the amplitude is scaled such that the maximum is 1 and contour levels are with respect to this maximum.

The most unstable patterns at the small aspect ratios consist of a 1-cell pattern, an example for **A** $= 3$ shown in panel a. The fluid rises in the centre and descends along the container wall. Note that the nearby lateral walls severely stabilize these patterns for small **A**. Another type of pattern (panel b) becomes most unstable at **A** $= 4$. As there is also a kink in the neutral curve, this flow pattern presumably comes from another eigenvalue branch. Both the patterns in the panels $a - b$ have maximum symmetry, they are both invariant with respect to reflection through $x = \mathbf{A}/2$ and to reflection through $y = \mathbf{A}/2$.

At **A** $= 5$, a pattern of lower symmetry becomes most unstable (panel c). It is only invariant to a reflection through $x = \mathbf{A}/2$ followed by a reflection through $y = \mathbf{A}/2$. The eigenvalue has algebraic multiplicity 2, because the pattern obtained through a reflection e.g. through $x = \mathbf{A}/2$ is unstable simultaneously. The main cell consists of liquid, rising to the southeast of the centre of the container and descending to the northwest of it. In addition, part of the liquid descends and rises along the surroundings of this main structure. A fully symmetric pattern is obtained again for **A** $= 6$ (panel d). The fluid descends over a large internal area within the box and rises in the centre and near the walls.

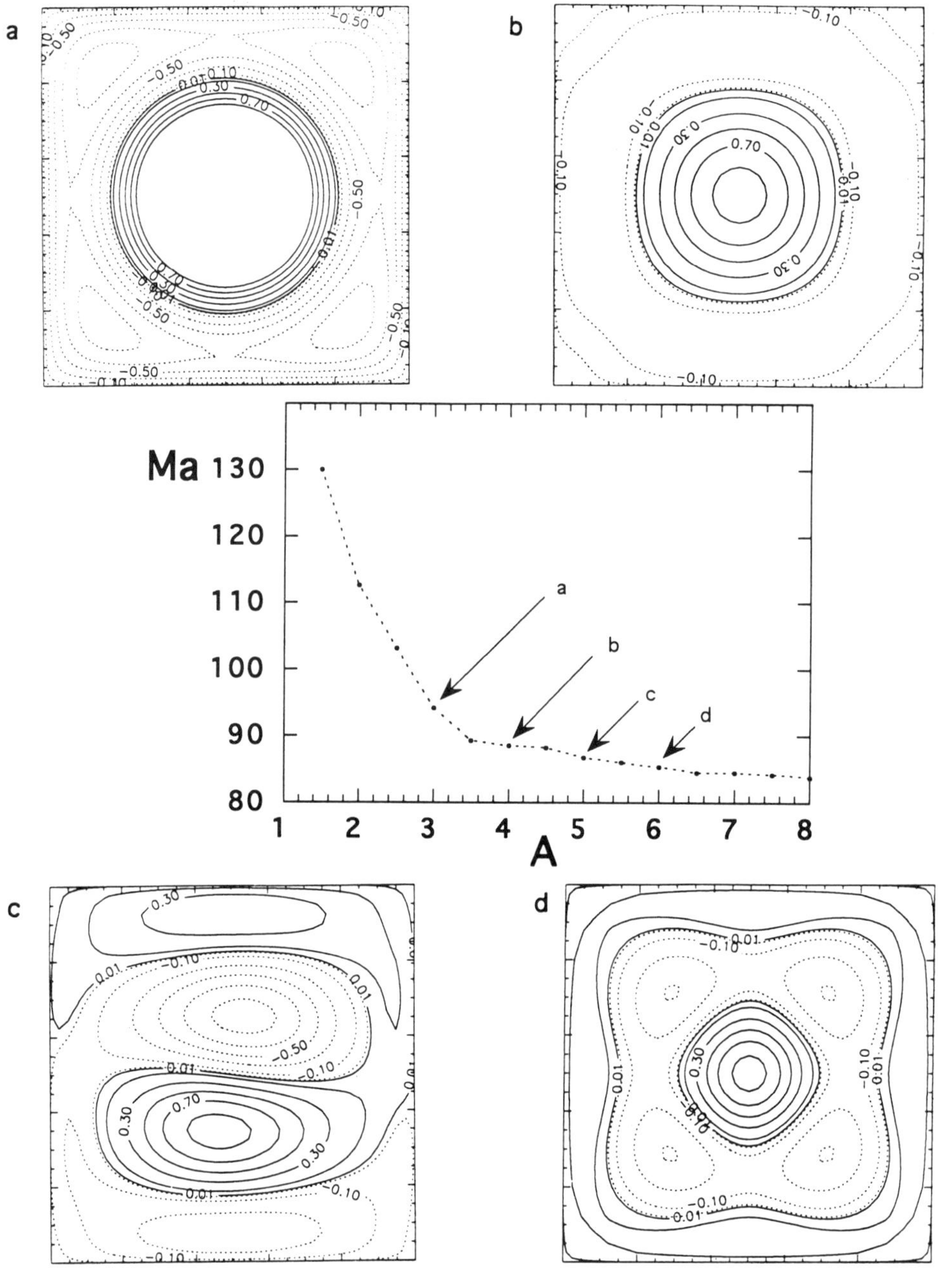

Figure 1: Values of **Ma** at neutral stability as a function of aspect ratio **A**. Contour plots of the vertical velocity at $z = 0.5$ of the most unstable patterns at marked points are shown in the panels $a - d$.

4.2 Bifurcation study

We restrict our computations to **Pr** = 1 and **A** between 4.0 and 6.0. For small **A**, the 1-cell pattern dominates, whereas for large **A**, a finer numerical resolution is required. Furthermore, we did not attempt to determine all branches within a bifurcation diagram, since it is too expensive. The number of stable (physically relevant) patterns is therefore underestimated.

The solution diagram for **A** = 4 is plotted in Fig. 2, where (dotted) drawn branches indicate (un)stable solutions. Flow patterns at selected marked points are plotted in the panels $a - d$. The most unstable pattern according to linear theory was presented in Fig. 1, panel b. This pattern bifurcates transcritically from the motionless solution as is expected according to its symmetry. The supercritical branch in Fig. 2 is stable up to a secondary bifurcation point near **Ma** = 96. The finite amplitude flow has downwelling of liquid at the center and corner structures appear for larger values of **Ma** (panels $a - b$). Along the subcritical branch, a limit point occurs and the solution stabilizes for larger **Ma**. The pattern just at the limit point (panel c) has center upflow. Along the boundaries the flow descends with largest amplitude in four symmetrically oriented centers. When **Ma** is increased along this branch, the center upflow region becomes more square-like (panel d). The bifurcation behavior at **A** = 4 shows the asymmetry between the flow patterns with center upflow and those with center downflow. This asymmetry is also present in the two-dimensional case [3] and its mathematical origin is the trivial action of the symmetry at neutral stability, giving a transcritical bifurcation. The second primary bifurcation is of pitchfork type. Only one of the supercritical branches is shown in Fig. 2. The solutions along this secondary branch are unstable over the computational domain.

The solution diagram for **A** = 5 is presented in Fig. 3, with patterns at selected points in the panels $a - d$. The first primary bifurcation is a pitchfork and one of the pair of unstable modes was shown in Fig. 1, panel c. This bifurcation is symmetry breaking and along one of the branches the pattern change is shown in the panels $a - b$. It should be stressed that the pattern in panel b is a stable finite amplitude flow pattern. Although it is not precisely the same as that found in Koschmieder and Prahl [7] (their Fig. 2c), it shows that indeed the liquid can reorganize in an unexpected way in this geometry. The second primary bifurcation point is transcritical and symmetric patterns branch off; all these patterns are unstable. At the supercritical branch the symmetric pattern in panel c is found near **Ma** = 99. The liquid moves downward in the center surrounded by 4 centers where the liquid moves upward. There is downward motion along the lateral boundaries. Along the subcritical branch, there is upward motion in the center and just at the limit point there is downflow in a hexagonal shape surrounding the circular upward flow (panel d).

The solution diagram for **A** = 6 is presented in Fig. 4. The most unstable mode found here is a symmetric pattern (panel d in Fig. 1) and the second mode to become unstable is shown in Fig. 4, panel a. The pattern reflected through $x = \mathbf{A}/2$ becomes unstable simultaneously. Hence, the first bifurcation point is transcritical and the

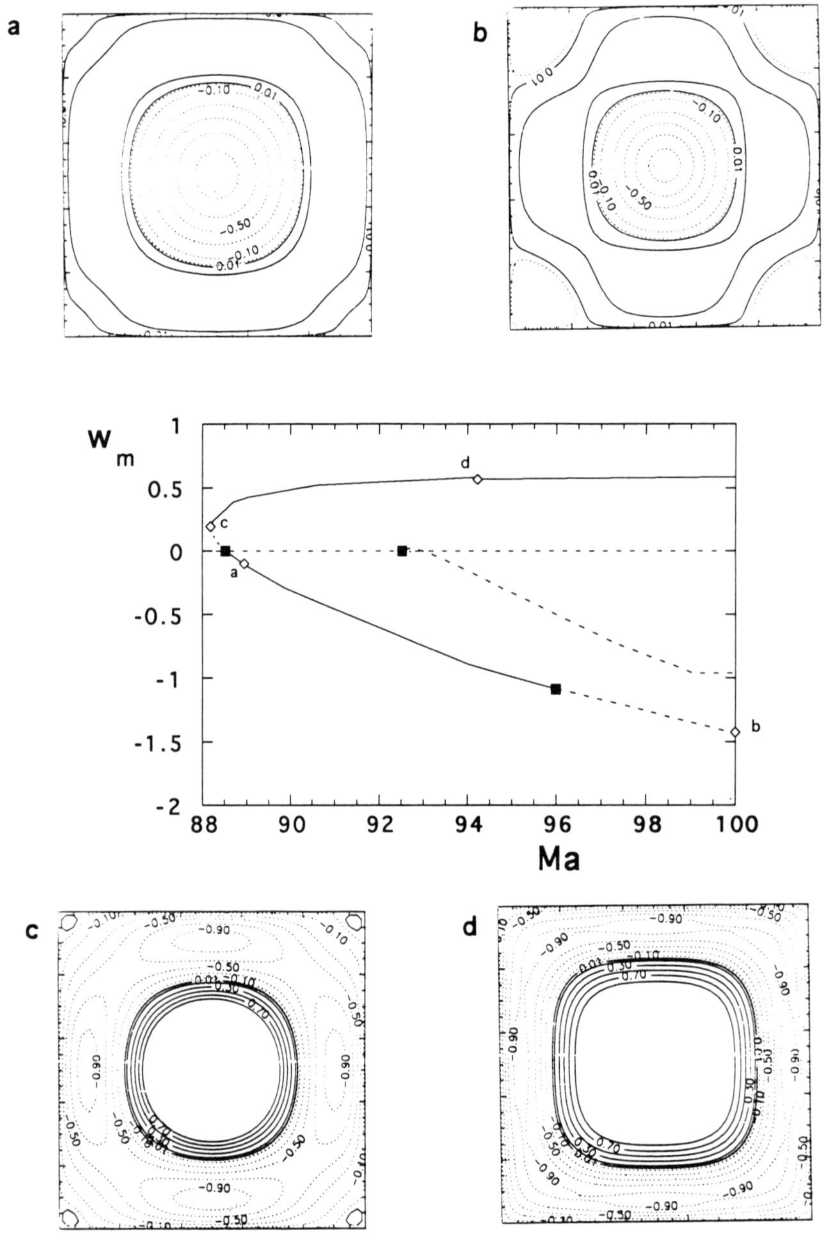

Figure 2: Solution diagram for no-slip sidewalls and $\mathbf{A} = 4$. At the vertical axis now the vertical velocity in the center (w_c) of the flow domain is plotted. Flow patterns as in Fig. 1 are shown in the panels $a - d$.

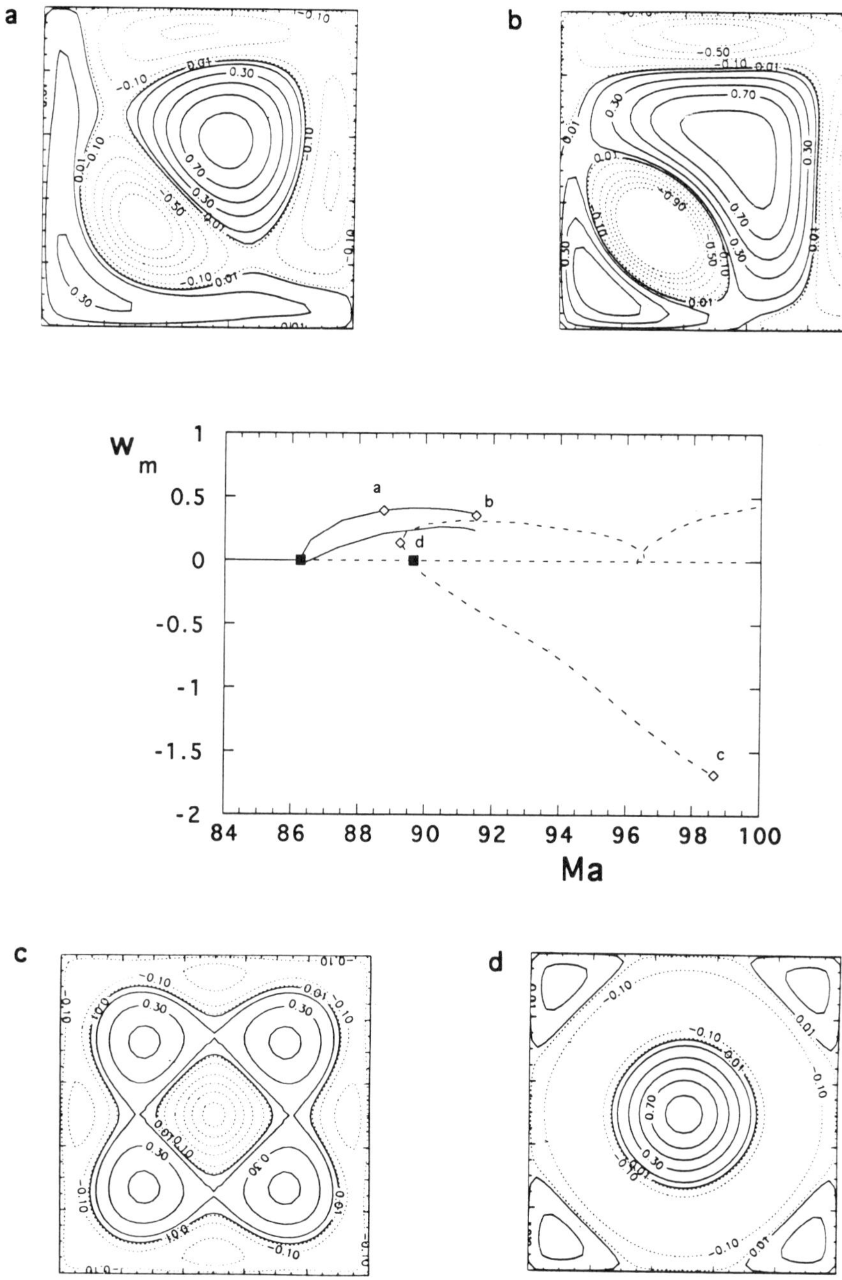

Figure 3: Solution diagram as in Fig. 2, but for $\mathbf{A} = 5$.

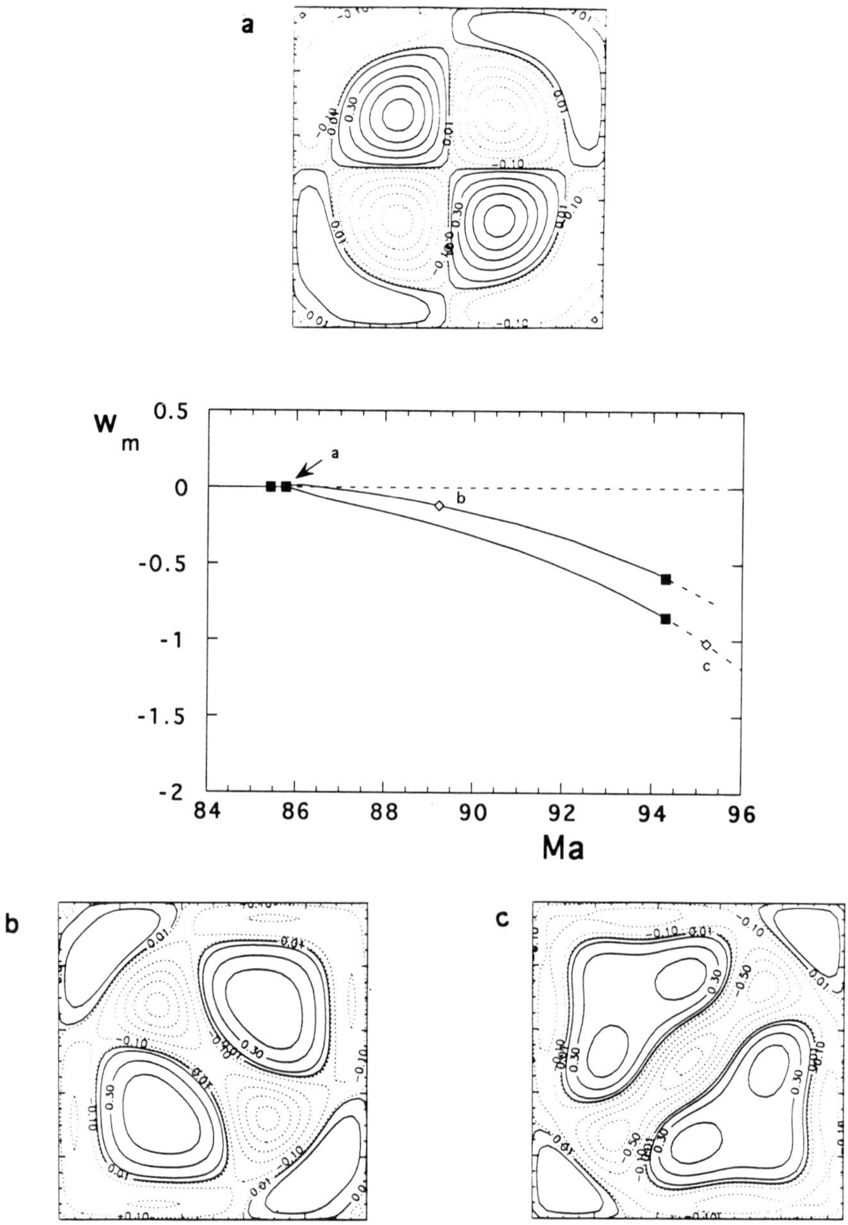

Figure 4: Solution diagram as in Fig. 2, but for $\mathbf{A} = 6$.

second one is a pitchfork. Both bifurcation points are very close together and it turned out to be hard to compute the branches from the first bifurcation point. This branch likely exist over a very short **Ma** interval. Along the branches from the second primary bifurcation point, two diagonally opposite centers merge (panel *b*) to give a 2-cell triangular structure (panel *c*) with downflow along the diagonal. The patterns on both branches are related through reflection symmetry and are stable up to a secondary bifurcation point near **Ma** = 94.

5 Discussion

Three-dimensional steady surface tension driven patterns were computed in a horizontally square container of small aspect ratio with no-slip sidewalls. The linear stability results of the basic state (4) are significantly different than those for the 'slippery' sidewall case [9]. First, the motionless state is much more stable for small **A**, a result in correspondence with experiments [7]. Second, the most unstable flow patterns are mostly three-dimensional and different from the pure roll-patterns found in [9].

The work on the bifurcation diagrams is far from complete, but two different types of bifurcation behavior from the primary branch (the motionless solution) occur. If the most unstable mode is a single mode (it then has full symmetry), the bifurcation is transcritical. Multiple stable steady states occur over an interval in **Ma** due to a limit point on the subcritical branch. However, the supercritical branch becomes unstable through a secondary bifurcation at larger **Ma**, resulting in a unique stable solution (relatively) far from onset. The flow pattern on this stable branch has a cellular structure with upwelling in the center. The bifurcation at onset thereby introduces an asymmetry between patterns with center upflow and center downflow, just as in the 2D-case [3]. If the most unstable mode has algebraic multiplicity two, two (symmetry related) solutions branch off. The bifurcation is of pitchfork type and both branches remain stable up to a secondary bifurcation point. The pitchfork reduces the symmetry of the finite amplitude patterns and several unexpected patterns appear, even as stable flows (Fig.3, panel *b*).

For **Ma** some distance from onset the velocities are of the order κ/H. The maximum vertical velocities of the stable finite amplitude flows increase with **Ma**. Since the layers depths used in experiments [7] are of the order 10^{-3} $[m]$, the velocities are about 10^{-4} $[ms^{-1}]$. For **A** = 4, the 1-cell flow pattern as observed in [7] is found as a linearly stable state. For **A** = 6, the 2-cell pattern with diagonal downflow is also found as a stable flow pattern. For **A** = 5, the strange pattern where three cells fill the container (Fig. 2c in [7]) is not found, but the pattern in Fig. 3, panel *b* shows that indeed such patterns are not unlikely. A larger **Pr** and inclusion of a realistic buoyancy strength did not have much effect on the spatial structure of the pattern. It might be that a nonzero interfacial heat transfer coefficient or the effect of surface deformation are important in experimental situations to realize these type of patterns.

Acknowledgements The author thanks Auke van der Ploeg, Eugen Botta and Jeroen

Molemaker for ongoing discussions on numerical aspects of the problem. All computations were done on the Cray $C98$ of the Academic Computing Centre in Amsterdam (SARA) during a test period of this machine and were sponsored by the NCF.

References

[1] Cloot, A., Lebon, G., A nonlinear stability analysis of the Bénard-Marangoni problem, *J. Fluid Mech.*, **145**, 447-469, (1984).

[2] Dauby, P.C., Lebon, G., Colinet, P., Legros, J.C., Hexagonal Marangoni convection in a rectangular box with slippery walls, *Q. Jl Mech. Appl. Math.*, **46**, 683-707, (1993)

[3] Dijkstra, H.A., On the structure of cellular patterns in Rayleigh-Bénard-Marangoni flows in two-dimensional containers with rigid sidewalls, *J. Fluid Mech.* **243**, 73 - 102, (1992).

[4] Dijkstra, H.A., Molemaker, M.J., Van der Ploeg, A., Botta, E.F.F., An efficient code to compute nonparallel flows and their linear stability, *Comp. & Fluids*, in press, (1995).

[5] Dijkstra, H.A, Surface tension driven patterns in three-dimensional boxes: I. Linear Stability, *Microgravity Sc. & Techn.*, VII, 4, 307-312, (1995).

[6] Golubitsky, M., Schaeffer, D.G., *Singularities and Groups in Bifurcation Theory, part I*, Springer Verlag, New York, (1985).

[7] Koschmieder, E.L., Prahl, S.A., Surface tension driven Bénard convection in small containers, *J. Fluid Mech.*, **215**, 571-583, (1990).

[8] Pearson, J.R.A., On convection cells induced by surface tension, *J. Fluid Mech.*, **4**, 489-500, (1958).

[9] Rosenblat, S., Homsy, G.M. and Davis, S.H., Nonlinear Marangoni convection in bounded layers, part 2. Rectangular cylindrical containers, *J. Fluid Mech.*, **120**, 123-138, (1982).

[10] Scanlon, J.W., Segel, L.A., Finite amplitude cellular convection induced by surface tension, *J. Fluid Mech.*, **30**, 149-162, (1967).

[11] Van de Vooren, A.I., Dijkstra, H.A., A finite-element stability analysis of the Marangoni problem in a two-dimensional container with rigid sidewalls, *Comp. Fluids*, **17**, 467-485, (1989).

Pattern formation in systems on spatially periodic domains

A. DOELMAN[1], W. ECKHAUS[1], R. KUSKE[2], R. SCHIELEN[1]
[1] *Mathematisch Instituut, Universiteit Utrecht, P.O. Box 80.010, 3508 TA Utrecht, the Netherlands.*
[2] *Department of Mathematics, Stanford University Stanford, CA 94305-2125, U.S.A.*

Abstract

In this paper we study the nonlinear evolution of patterns at near-critical conditions on domains with spatially periodic boundaries; the amplitude of the periodic effect is assumed to be small. By considering a simple one-dimensional model problem we are able to focus on the essence of the periodic effects. We find that the wave number p of the imposed periodicity has a significant influence on the behaviour of the solutions. We have to distinguish between non-resonant p, resonant p and small p. We derive a number of modulation equations which differ structurally from the classical Ginzburg-Landau equation and study the behaviour of the solutions to these equations.

1 Introduction

In the classical analysis of nonlinear stability problems on unbounded domains it is usually assumed that the geometry of the domain is simple in the unbounded direction. The problem is formulated on a cylindrical domain $\Omega \times \mathbf{R}^n$, where Ω is bounded (and $n = 1$ in most cases) – see [6] for a review. However, this assumption is not always satisfied in a physical system. An example of such a system is given in [16], where the nonlinear evolution of a river bed is studied. The river is modeled in [16] as a straight channel. The natural next step in studying this river bed problem is to allow for a small meandering effect, that is to consider the model on a domain which is periodic in the unbounded direction. By applying a transformation which rectifies the 'meandering' domain, one introduces periodic terms in the governing equations which depend explicitly on the unbounded variable (this procedure is outlined in [E & K], see [17] for a detailed description).
Other examples of physical systems which are defined on not purely cylindrical domains are: Bénard convection [2], [18]; convection in a porous layer [12], [13], [14], [15]; Taylor vortex flows [3].

In order to understand the influence of a (slightly) varying geometry on the nonlinear stability problem we study in this paper a one-dimensional model problem. The simplicity of this model enables us to focus fully on the sometimes subtle effects of the periodic terms. We believe that the one-dimensional model exhibits in essence all aspects of pattern formation in systems on periodically modulated domains. In other words: in more realistic problems with transversal dimensions the complications will be merely technical, but the modulation equations will be similar to the equations derived in this paper for the model problem.

The model problem is described by the following equation:

$$\frac{\partial \phi}{\partial t} = \mathcal{L}_R \phi + \nu f(px)\phi + \mathcal{N}(\phi) \text{ with } \mathcal{N}(\phi) = (\mathcal{P}\phi)(\mathcal{Q}\phi), \tag{1.1}$$

where $\phi = \phi(x,t)$ with $x \in \mathbf{R}$; $\mathcal{L}_R$, $\mathcal{P}$ and $\mathcal{Q}$ are linear operators in $\frac{\partial}{\partial x}$; R is the so-called control parameter, it is related to the Rayleigh number in convection models and to the width/depth ratio of the river in the river bed problem. The slightly varying geometry is modeled by a 'periodic forcing term' $\nu f(px)$, with $0 < \nu \ll 1$. It will be convenient to write f as a Fourier series:

$$f(px) = \sum_{n=-\infty}^{\infty} F_n e^{inpx} \text{ with } F_n \in \mathbf{C} \text{ and } F_{-n} = \bar{F}_n, \tag{1.2}$$

since $f(px) \in \mathbf{R}$. We assume that $F_0 = 0$, this means that the perturbation term f has 0 average, because a non-zero average term F_0 has the same influence on (1.1) as an $O(\nu)$ shift in the control parameter R. By setting $F_0 = 0$ we know that influence of f on the system is purely due to its periodicity. Note that we made some additional simplifying assumptions: the nonlinear operator $\mathcal{N}(\phi)$ consists of only one quadratic term and the periodic term νf only appears in the linear part of (1.1). Again, these assumptions only reduce the amount of the computations, more general terms will not create new phenomena.

Furthermore, we impose some extra structure on the model problem by requiring a reflection symmetry $x \rightarrow -x$. This has the consequence that when studying the linearized stability problem in the unperturbed case ($\nu = 0$) one finds as solutions $\phi = e^{ikx+\omega t}$+ c.c. (complex conjungated) with $\omega = \omega(k, R)$ real. We assume that the (unperturbed) neutral curve $\omega = 0$ has a minimum (k_c^0, R_c^0) such that the zero solution is stable for $R < R_c^0$. Now, it is possible to derive a modulation equation of Ginzburg-Landau type which describes the evolution of small solutions to (1.1) for R near R_c^0 (see section 3). The asymptotic validity of this equation has been shown in [10].

We will find that the periodic perturbation has an important influence both on the linear stability problem – and especially on the neutral curve $\omega = 0$ – and on the modulation equation(s). We will pay special attention to the influence of the wave number p of $f(px)$. There are two significantly different cases: $p = O(1)$ and $p = o(1)$. The difference is encountered immediately on the linear level. In section 2 we will see that one has to distinguish between these cases in the analysis of the linearized stability of the zero solution of the perturbed equation. Therefore, sections 2, 3, 4 and most of the remainder of this section will be divided into two parts, one on $p = O(1)$ and one on $p = o(1)$. The case $p = O(1)$ has been encountered in studies of convection in porous layers ([13], [14], [15]); the 'slowly varying case' $p = o(1)$ appeared in Bénard convection [2], [18], in convection in a porous layer [12] and in Taylor vortex flows [3].

For $p = O(1)$ there are again two sub-cases to consider: p is resonant or non-resonant. Wave number p is resonant if there exists a $N \in \mathbf{Z}$ such that $\frac{N}{2}p$ is 'close' to the critical wave number k_c^0 of the linearized stability theory (section 2). In the non-resonant case the influence of the periodic term is 'regular', that is, the outcome of the stability analysis of the perturbed problem is just an $O(\nu^2)$ modification of the unperturbed problem. However, if p is resonant, the effect of the perturbation is more dramatic. Depending on the relative magnitudes of ε and ν - where $\varepsilon = O(\sqrt{R - R_c})$ - we derive various modulation equations which differ significantly from the unperturbed (Ginzburg-Landau) equation. For instance, we find real modulations equations or (complex) equations with an extra non-symmetric term '$\bar{A}$'.

A way to understand the extra complexities which one encounters if $p = o(1)$ is by noticing that these p are always in resonance for various values of N. Instead of fully analyzing the intricate linearized stability problem, we now base the nonlinear analysis on the outcome of the unperturbed linear stability analysis. As a consequence we have to choose $\varepsilon = O(\sqrt{\nu})$ (or larger). We then derive a modulation equation with a spatially periodic coefficient.

In section 4 we study the behaviour of (some of) the solutions to the equations derived in section 3. One striking behaviour is exhibited by the space intermittent solutions which exist for $p \ll \varepsilon (= O(\sqrt{\nu}))$. These solutions consist of (spatial) periodic parts alternated with parts where the solutions are exponentionally small. By employing a detailed analysis in the boundary layer we obtain smooth solutions which connect these parts (see Figure 2 in Section 4). Another surprising phenomenon is encountered for $p = O(1)$ and $\varepsilon = O(\sqrt{\nu})$: the stationary problem for the 'non-symmetric' modulation equation turns out to be an integrable two degrees of freedom Hamiltonian system.

Remark. This paper can be regarded as a combination of two papers, one on $p = O(1)$ by Schielen and Doelman [S & D], the other on $p = o(1)$ by Eckhaus and Kuske [E & K]. We attempt here to clarify the similarities and differences between these two cases/papers. We refer to [S & D] and [E & K] for the most of the computational details.

2 The linear theory

In this section we study the linearized stability of the trivial solution, $\phi \equiv 0$, of (1.1). Instead of substituting

$$\phi_{\text{lin}}(x, t) = e^{ikx + \omega t} \quad (+ \text{ c.c.}), \tag{2.1}$$

which one does if there is no periodic term $\nu f(px)$ in (1.1), one now has to take into account the influence of f by considering quasi-periodic solutions of the structure

$$\phi_{\text{lin}}(x,t) = e^{ikx+\omega t} \sum_{n=-\infty}^{\infty} a_n e^{inpx} \quad (+ \text{ c.c.}), \tag{2.2}$$

where the Fourier coefficients a_n are yet unknown, except for the fact that we can assume that $a_0 = 1$. It is natural to assume that $a_n = o(1)$, since then one can take the limit $\nu \to 0$ in (2.2) and recover (2.1). However, we will find that this is not always the case – when we study resonances – so, we do not impose any structure on a_n. Note that so far we did not make any assumptions on the magnitude of p.

One proceeds by substituting (2.2) into (1.1) – where one neglects the nonlinear term $\mathcal{N}(\phi)$. This yields an infinite-dimensional system of algebraic equations with unknowns ω and a_n, $n \neq 0$. There are 4 parameters in this problem: the bifurcation/control parameter R, the 'dominant' wave number k of the perturbation and ν and p, respectively the amplitude and the wave number of the periodic term $\nu f(px)$. Therefore $\omega = \omega(k, R, \nu, p)$ and $a_n = a_n(k, R, \nu, p)$.

As usual, the linearized stability of $\phi \equiv 0$ is determined by the sign of ω, hence the most important outcome of the linear theory will be insight in the structure of the set $\omega = 0$. In the unperturbed case ($\nu = 0$) $\omega = 0$ is a curve in the $(k, R)-$ plane and is called the neutral stability curve. Note that $\omega(k, R) = \mu(k, R)$, the symbol of the operator $\mathcal{L}_R$, for $\nu = 0$. Due to the structure of ϕ_{lin} in (2.2), one has to be a little bit more careful if one wants to consider $\omega = 0$ as a subset of the $(k, R)-$ plane, if $\nu \neq 0$. Due to the extra periodic term in (2.2) the outcome of the linear theory will be periodic in k: for any $l \in \mathbf{N}$ we can introduce $k = \tilde{k} + lp$ in (2.2) and observe

$$\begin{aligned}
\phi_{\text{lin}} &= e^{i(\tilde{k}+lp)x+\omega t} \textstyle\sum_{n=-\infty}^{\infty} a_n(\tilde{k} + lp, R) e^{inpx} \quad (+ \text{ c.c.}) \\
&= e^{i\tilde{k}x+\omega t} \textstyle\sum_{n=-\infty}^{\infty} a_n(\tilde{k} + lp, R) e^{i(n+l)px} \quad (+ \text{ c.c.}) \\
&= e^{i\tilde{k}x+\omega t} \textstyle\sum_{m=-\infty}^{\infty} \tilde{a}_m(\tilde{k}, R) e^{impx} \quad (+ \text{ c.c.}) \text{ with } \tilde{a}_m(\tilde{k}, R) = a_{m-l}(\tilde{k} + lp, R).
\end{aligned}$$

Therefore, we must not consider $k \in \mathbf{R}$ – as in the unperturbed case – but take into account the p-periodicity in k, that is

$$\omega(k, R) = \omega(k + lp, R) \text{ for all } l \in \mathbf{N}. \tag{2.3}$$

This means geometrically that the neutral set should be considered as a (collection of) curve(s) on a cylinder. The new neutral set should be close to the unperturbed neutral curve on the p-periodic cylinder. This periodic structure can be obtained from the curve in the $(k, R)-$plane by reduction of k modulo p. Thus, we can study the linear perturbations given in (2.2) for k limited to an interval with length p. From now on, we will take $k \in [k_c^0 - p/2, k_c^0 + p/2]$, where k_c^0 is the critical wave number belonging to the unperturbed ($\nu = 0$) problem, that is (k_c^0, R_c^0) is the minimum of

$\mu(k, R) = 0.$

The equation for ω follows immediately from the substitution of (2.2) into (1.1) – with $N(\phi) \equiv 0$:

$$\omega(k, R) = \mu(k, R) + \nu \sum_{m+n=0} F_m a_n \tag{2.4}$$

(where we have used that $a_0 = 1$). The Fourier modes for $n \neq 0$ determine equations for the coefficients a_n:

$$[\omega(k, R) - \mu(k + np, R)]a_n = \nu \sum_{l+m=n} F_l a_m \tag{2.5}$$

Thus, we observe that $a_n = O(\nu), (n \neq 0)$ if $|\omega(k, R) - \mu(k + np, R)| = O(1)$. If this is the case for all $n \neq 0$, then one can iteratively solve the system (2.4), (2.5) up to any order in ν, see [S & D] for the exact expressions. Note however, that this condition can in general not be satisfied for all k: the curves $\mu(k, R)$ and $\mu(k + np, R)$ will intersect. Since we are mainly interested in the influence of the perturbation near $k = k_c^0$ and $R = R_c^0$ – the critical values of k and R in the unperturbed problem – we refer to [S & D] for the analysis of such intersections for general k. We conclude that the linearized stability problem can be solved for (k, R) <u>near</u> (k_c^0, R_c^0) if

$$\mu(k_c^0 + np, R_c^0) - \mu(k_c^0, R_c^0) = O(1) \text{ for all } n \neq 0 \tag{2.6}$$

(one finds that $|\omega - \mu| = O(\nu^2)$, [S & D]). At this point one has to distinguish between $p = O(1)$ and $p = o(1)$: if $p = o(1)$ then $k_c^0 + np$ is close to k_c^0 and therefore $\mu(k_c^0 + np, R_c^0) - \mu(k_c^0, R_c^0) = o(1)$ for $n = O(1)$. Moreover, (2.6) will also be violated for certain 'large' n (of $O(\frac{1}{p})$. This is due to the fact that μ is symmetric in k, thus $\mu(k_c^0 + np, R_c^0) = \mu(-(k_c^0 + np), R_c^0) = \mu(k_c^0, R_c^0) + o(1)$ if $-(k_c^0 + np) = k_c^0 + o(1)$, or, if there exists a $N \in \mathbf{Z}$ such that

$$k_c^0 = \frac{N}{2} p + o(1). \tag{2.7}$$

If $p = O(1)$ (2.7) can only be satisfied for certain special values of p; it is called the <u>resonance condition</u> in [S & D].

The linearized stability problem for $p = O(1)$ has been studied in detail in [S & D] – both for resonant and non-resonant values of p. We will present the outcome of this analysis in the subsection below. The analysis of the case $p = o(1)$ in [E & K] takes a somewhat different approach. The complete analysis of the linearized stability problem is circumvented by using the outcome of the unperturbed linearized stability problem as the foundation of the nonlinear analysis. The linear perturbation terms – which are of $O(\nu)$ – are then considered as 'extra' non-homogeneous terms in the derivation of the modulation equation. As a result the periodic term $\nu f(px)$ of (1.1) appears directly in the modulation equation. The penalty that one pays for this simplification is that one has to choose the distance between the control parameter R and R_c^0 in the nonlinear analysis of $O(\nu)$ or larger, since $\nu f(px)$ can not be allowed to be larger than the other non-homogeneous terms; see section 3.1.

2.1 The linearized stability theory for $p = O(1)$

This section should be divided into a part on the non-resonant and a part on the resonant values of p (see (2.7)). However, the problem for non-resonant p is a rather straightforward exercise in manipulating (2.4) and (2.5). One finds that the coordinates of the 'new' minimum of the neutral 'curve' $\omega = 0$, $(k_c, R_c) = (k_c(\nu), R_c(\nu))$, is $O(\nu^2)$ close to (k_c^0, R_c^0) (see [S & D] for the explicit expressions). The correction to R_c^0 is strictly negative, thus, the perturbation always destabilizes the system.

The resonant problem is more complicated. We will encounter a new phenomenon which will be of crucial importance in the nonlinear analysis. Due to the resonance the neutral curve will split into two curves. Passing the minimum of one of these curves corresponds to the bifurcation of a 'linearly most unstable wave'. We will find that these bifurcating waves are in essence in a 1-dimensional real family – unlike the unperturbed case where the marginally unstable waves form a 2-dimensional real (= 1-dimensional complex) family.

However, before we can describe this, we need to specify the meaning of the '$o(1)$' in the resonance condition (2.7). Relation (2.7) is based on the linearized stability theory for $\nu = 0$. The $O(\nu)$ perturbation will have an effect on k_c. For instance, k_c^0 can be in perfect resonance, and the perturbation can 'detune' the system such that k_c is not identical to $\frac{N}{2}p$. Especially in the nonlinear analysis we will find that it is more convenient to base the resonance condition on k_c instead of on k_c^0. In [S & D] it is shown that if $k_c^0 = \frac{N}{2}p + O(\nu^s)$, for some $s > \frac{1}{2}$, that $k_c = \frac{N}{2}p + O(\nu^{2s}, \nu^2)$. Thus, by considering the effect of the perturbation on k_c^0 we see that the system is closer to perfect resonance if $\frac{1}{2} < s < 2$; however, the system can never be closer to perfect resonance than $O(\nu^2)$, even when $k_c^0 = \frac{N}{2}p$. It's for this reason that we will use in the sequel that

$$k_c = k_c(\nu) = \frac{N}{2}p + O(\nu^2). \tag{2.8}$$

We refer to [S & D] for the details and for a brief analysis of the case $s \leq \frac{1}{2}$ (which describes transition between the resonant and the non-resonant case).

The origin of the above mentioned new phenomenon can be understood by analysing $\omega(k, R)$ near an intersection of $\mu(k, R)$ and $\mu(k - Np, R)$, for a certain $N \in \mathbf{Z}$. In [S & D] it is shown that if $\mu(k - Np, R) - \mu(k, R) = A\nu^s$, for certain A and $s > 0$, that there are two corrections, $\omega+$ and ω^- to ω:

$$\omega(k, R) = \omega^{\pm}(k, R) = \mu(k, R) \pm \nu|F_N| + \text{ h.o.t. (for } s > 1). \tag{2.9}$$

This 'splitting' phenomenon is sketched in Figure 1. The set $\omega(k, R)$ is a collection of curves 'near' $\mu(k, R)$ and – by the periodicity – 'near' $\mu(k - Np, R)$. We see that different branches of ω cannot intersect – unlike the branches of μ on the cylinder. We refer to [S & D] for a detailed analysis. The splitting is especially important when the intersection of $\mu(k,R)$ and $\mu(k-Np,R)$ appears near (k_c^0, R_c^0), that is, when p is resonant. Now there are two branches of the neutral set, $\omega^+(k,R)=0$ and $\omega^-(k,R)=0$, $O(\nu)$ near the unperturbed neutral curve $\mu(k,R)=0$. The $\omega^+=0$ branch is below $\mu=0$, so passing this branch causes the first bifurcation at which the trivial solution of (1.1) becomes (linearly) unstable; $\omega^+=0$ corresponds with the $\omega=0$ curve in the non-resonant case. Both branches correspond to marginally unstable perturbations (of type (2.2)). From the point of view of the linearized stability theory it is not important to study the second instability which appears when crossing the $\omega^-=0$ branch, since this branch is always $O(\nu)$ above $\omega^+=0$. However, when dealing with the nonlinear problem we will find that the splitting is of crucial importance as long as $R-R_c=O(\nu)$ and smaller. This is mainly due to the character of the marginally unstable 'waves' associated with these branches.

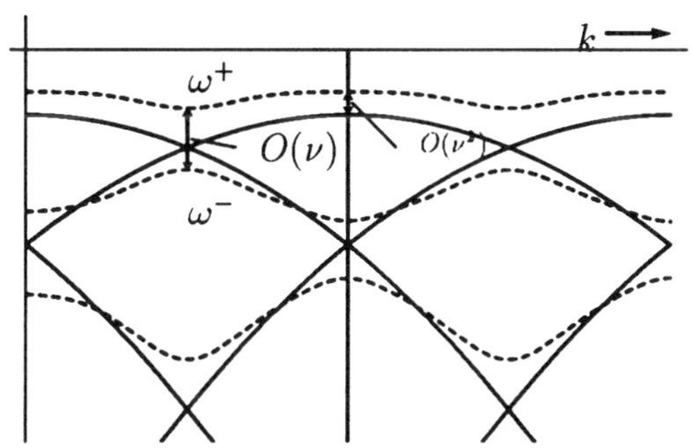

Figure 1: The unperturbed 'curve' $\mu(k,R)$ – in solid lines – and the perturbed set of curves $\omega(k,R)$ – in dotted lines – on the periodic cylinder for a fixed value of R.

This character can be obtained from solving (2.5), using (2.4) and (2.8). In [S & D] it is shown that $a_n^\pm = O(\nu)$ for $n \neq 0, -N$ and

$$a_{-N}^\pm = \pm\frac{F_{-N}}{|F_{-N}|} + \text{ h.o.t.} \Rightarrow |a_{-N}^\pm| = 1 + \text{ h.o.t.}. \tag{2.10}$$

Therefore, the highest order terms of the critical marginally unstable waves $\phi_{\text{lin}}^\pm(x,t)$ are given by (see (2.2) and (2.8)):

$$\begin{aligned}\phi_{\text{lin}}^\pm(x,t) &= Ce^{ik_cx}(1\pm\tfrac{F_{-N}}{|F_{-N}|}e^{-iNpx}) + \bar{C}e^{-ik_cx}(1\pm\tfrac{\bar{F}_{-N}}{|F_{-N}|}e^{iNpx}) \\ &= (C\pm\bar{C}\tfrac{\bar{F}_{-N}}{|F_{-N}|})e^{ik_cx} + (\bar{C}\pm C\tfrac{F_{-N}}{|F_{-N}|})e^{-ik_cx},\end{aligned} \tag{2.11}$$

on a $O(1)$ spatial scale. Although the free constant C can be chosen anywhere in $\mathbf{C}$, we observe that that's not the case for $C \pm \bar{C}e^{i\alpha}$ (with $\alpha = \arg(\bar{F}_{-N})$):

$$\begin{cases} C+\bar{C}e^{i\alpha} &= e^{\frac{1}{2}i\alpha}[2\,\mathrm{Re}(Ce^{-\frac{1}{2}i\alpha})] &= c^+e^{\frac{1}{2}i\alpha} \quad \text{for some } c^+ \in \mathbf{R} \\ C-\bar{C}e^{i\alpha} &= e^{\frac{1}{2}i\alpha}[2i\,\mathrm{Im}(Ce^{-\frac{1}{2}i\alpha})] &= c^-ie^{\frac{1}{2}i\alpha} \quad \text{for some } c^- \in \mathbf{R}. \end{cases} \tag{2.12}$$

This yields that – in resonance – the linearly most unstable waves are given by the 1-dimensional real families

$$\phi^{\pm}_{\text{lin}}(x,t) = c^{\pm}e^{\frac{1}{2}i\alpha}e^{ik_cx}\sum_{n=-\infty}^{\infty} a_n^{\pm}e^{inpx} + \text{c.c. for } c^{\pm} \in \mathbf{R}, \tag{2.13}$$

where $a_1^{\pm} = 1$ and $a_n^{\pm} = O(\nu)$ for $n \neq 0$ [S & D].

Thus far we did not pay attention to positions of the minima $(k_c^{\pm}, R_c^{\pm})$ of the two neutral branches. The exact expressions can be found in [S & D], here we only give an indication of the magnitudes of the corrections on (k_c^0, R_c^0) and of the relative positions of R_c^+ and R_c^-, and k_c^+ and k_c^-:

$$R_c^{\pm} - R_c^0 = O(\nu),\ |R_c^+ - R_c^0| = |R_c^- - R_c^0| + O(\nu^2);\ k_c^+ = k_c^- + O(\nu^2), \tag{2.14}$$

such that (2.8) holds. In the sequel we will not differentiate between k_c, k_c^+ and k_c^-.

3 Nonlinear theory: the derivation of the modulation equations

In the nonlinear stability theory we will consider the nonlinear evolution of small solutions of (1.1) for R close to R_c. Therefore, we introduce another small parameter ε and the rescaled control parameter r by $R - R_c = r\varepsilon^2$: ε measures the magnitude of $R - R_c$ and r 'controls' $R - R_c$ within the region $R - R_c = O(\varepsilon^2)$. Thus, if $p = O(1)$ there are now 2 small parameters, ε and ν; the relative magnitudes of ε and ν will cause different types of modulation equations. If $p = o(1)$ there are 3 small parameters, however, since we will base the nonlinear analysis on the outcome of the unperturbed linearized stability problem, we can not choose ε independent of ν.

The starting point of the analysis is the same in all cases. Based on the linear theory one models a small solution of (1.1) as being a 'modulation' of the 'linearly most critical wave'. In other words, one assumes that small solutions of (1.1) will 'look like' the unstable solutions found in the linear theory. Classically, this is an 'Ansatz' – see for instance [6]. Recently, Eckhaus [7] gave a rigorous foundation for this Ansatz by showing that small solutions without an imposed structure do evolve towards the structure described by the Ansatz, on a faster time-scale then the time-scale at which the nonlinear interactions take place. Thus, we expand the solution ϕ of (1.1) both in a Fourier and a power series:

$$
\begin{aligned}
\phi(x,t) = \quad & \qquad\qquad + \varepsilon^2\psi^{02} + \varepsilon^3\psi^{03} + \ldots \\
& + E[\varepsilon\psi^{11} + \varepsilon^2\psi^{12} + \varepsilon^3\psi^{13} + \ldots] + \text{ c.c.} \\
& \quad + E^2[\varepsilon^2\psi^{22} + \varepsilon^3\psi^{23} + \ldots] + \text{ c.c.} \\
& \qquad + E^3[\varepsilon^3\psi^{33} + \ldots] + \text{ c.c.} \\
& \qquad\qquad + \ldots
\end{aligned}
\tag{3.1}
$$

Note that we chose the classical scaling $\phi = O(\varepsilon) = O(\sqrt{R - R_c})$. This is not always the proper scaling; we will adapt the scaling when necessary. In this expansion E is the 'linear most critical wave': the solution to the linearized stability problem at (k_c, R_c). Thus, if $p = O(1)$, E is given by (2.2), with the a_n as computed in the above section. If $p = o(1)$ we choose $E = \exp(ik_c^0 x)$, the critical wave of the unperturbed problem. Note that there is a small difference between the definitions of ε in both cases: if $p = O(1)$ ε is related to $R_c = R_c(\nu)$, while it's related to $R_c^0 = R_c(0)$ if $p = o(1)$. The functions ψ^{ij} depend on the slow variables $\xi = \varepsilon x$ and $\tau = \varepsilon^2 t$; this scaling is again classical and will be adapted if necessary.

Before we can substitute expansion (3.1) into (1.1) we have to expand the linear operators $\mathcal{L}_R$, $\mathcal{P}$ and $\mathcal{Q}$ with respect to ε. We set

$$
\mathcal{L}_R = \mathcal{L}_{R_c + r\varepsilon^2}(\frac{\partial}{\partial x}) = \mathcal{L}_c(\frac{\partial}{\partial x}) + r\varepsilon^2 \mathcal{L}_{2,c}(\frac{\partial}{\partial x}) + O(\varepsilon^4) \tag{3.2}
$$

and decompose further by substituting $\frac{\partial}{\partial x} + \varepsilon\frac{\partial}{\partial \xi}$ for $\frac{\partial}{\partial x}$

$$
\mathcal{L}_c = \sum_{m=0}^{M} \varepsilon^m L^m(\frac{\partial}{\partial x})\frac{\partial^m}{\partial \xi^m} \tag{3.3}
$$

for some $M > 0$. Note that $L^0(e^{ikx}) = \mu(k, R_c)$. In general, we define

$$
L^m(e^{ikx}) = \mu^{(m)}(k), \quad (\text{thus } \mu(k, R_c) = \mu^{(0)}(k)). \tag{3.4}
$$

Similar expansions can be made for $\mathcal{L}_{2,c}$ – introducing the operators L_2^m, and the expressions $\mu_2^{(m)}(k)$ – $\mathcal{P}$ and $\mathcal{Q}$.

The difference between the cases $p = O(1)$ and $p = o(1)$ becomes immediately clear if one considers the equation for ψ^{11}, which follows from the $\varepsilon^1 E$-'level' of the system obtained from (1.1) by applying the above expansions:

$$
\begin{aligned}
(L^0 + \nu \textstyle\sum F_n e^{inpx}) E\psi^{11} = 0 \quad & \text{for } p = O(1) \quad \text{and } E = e^{ik_c x} \textstyle\sum a_n e^{inpx} \\
L^0 E\psi^{11} = 0 \quad & \text{for } p = o(1) \quad \text{and } E = e^{ik_c^0 x}.
\end{aligned}
\tag{3.5}
$$

If $p = o(1)$, the term $\nu f E\psi^{11}$ will appear as an inhomogeneous term at the level $\varepsilon^3 E$. In other words: if $p = o(1)$ we follow [E & K] and choose $\varepsilon^2 = O(\nu)$. Due to the above mentioned differences between the cases $p = O(1)$ and $p = o(1)$ we consider them separately and divide this section into two subsections.

3.1 Modulation equations for $p = o(1)$

In this subsection we follow [E & K]. We substitute (3.1) into (1.1) with E as in (3.5). At the εE-level we recover (3.5), therefore we conclude – as usual – that $\psi^{11} = A(\xi, \tau)$, an unknown 'amplitude' function. Due to the scaling $\varepsilon^2 = O(\nu)$, the perturbation $\nu f(px)$ has no influence on the equations, and their solutions $\psi^{02,12,22}$, on the level ε^2. Therefore, we refer to [E & K] and [S & D] for the details and proceed to the level $\varepsilon^3 E$. By introducing $\hat{\nu}$ by the scaling $\nu = \hat{\nu}\varepsilon^2$ we find that the equation for ψ^{13} reads:

$$-L^0(\psi^{13}E) = [-\frac{\partial A}{\partial \tau} + r\mu_2^{(0)}(k_c^0)A + \mu^{(2)}(k_c^0)\frac{\partial^2 A}{\partial \xi^2} + \hat{\nu}\sum_{n=-\infty}^{\infty} F_n e^{inpx} A + \beta A|A|^2]E, \quad (3.6)$$

where β is a constant which can be expressed in terms of $\mathcal{P}$, $\mathcal{Q}$ and k_c^0 (see [S & D]). We assume throughout this paper that $\beta < 0$. Note that

$$\begin{aligned} \mu_2^{(0)}(k_c^0) &= \tfrac{\partial}{\partial R}\mu(k, R)|_{k=k_c^0, R=R_c^0} &\stackrel{\text{def}}{=} \mu_R > 0 \\ \mu^{(2)}(k_c^0) &= -\tfrac{1}{2}\tfrac{\partial^2}{\partial k^2}\mu(k, R)|_{k=k_c^0, R=R_c^0} &\stackrel{\text{def}}{=} -\tfrac{1}{2}\mu_{kk} > 0, \end{aligned} \quad (3.7)$$

since (k_c^0, R_c^0) is a minimum of $\mu(k, R) = 0$. The function ψ^{13} only depends on the slow variables τ and ξ, and $L^0 E = 0$, thus (3.6) yields that the amplitude A has to satisfy the following equation:

$$\frac{\partial A}{\partial \tau} = r\mu_R A - \frac{1}{2}\mu_{kk}\frac{\partial^2 A}{\partial \xi^2} + \hat{\nu}\sum_{n=-\infty}^{\infty} F_n e^{inpx} A + \beta A|A|^2. \quad (3.8)$$

This last step seems simple, however, there is a small complication: strictly speaking the Fourier series of $f(px)$ cannot appear completely in (3.8) because the terms $F_n e^{inpx}$ may look resonant. A justification of (3.8) can be obtained by assuming that there are only a finite number of Fourier modes $F_n \neq 0$ in the expansion of $f(px)$ or that F_n decays suffiently fast with n. In this paper, we follow [E & K] and take (3.8) as outcome of the nonlinear stability theory.

3.2 Modulation equations for $p = O(1)$

As in the subsection on the linearized problem for $p = O(1)$, we won't pay much attention to the non-resonant case. The derivation of the modulation equation is somewhat less straightforward than a priori expected. However, eventually one ends up with a modulation equation for $A(\xi, \tau)(= \psi^{11})$ which has the expected form:

$$\frac{\partial A}{\partial \tau} = r\omega_R A - \frac{1}{2}\omega_{kk}\frac{\partial^2 A}{\partial \xi^2} + \beta(1 + \nu^2\beta_2)A|A|^2, \quad (3.9)$$

where ω_R and ω_{kk} are defined as in (3.7) – based on ω instead of μ. Thus, (3.9) is $O(\nu^2)$ close to the unperturbed case. The correction β_2 to the coefficient of the nonlinear term is, in general, a complex number. Therefore, stable periodic solutions to (3.9) will be (slowly) traveling (see [1]), unlike in the unperturbed case. The details can be found in [S & D]. Note that the $O(\nu^2)$ corrections only appear for $\varepsilon \ll \nu^2$, if $\varepsilon = O(\nu^2)$ (or larger) one has to neglect these effects since one already neglects the higher order terms – which are of $O(\varepsilon)$ – in the modulation equation.

In the resonant case we consider ε relative to ν by first taking it smaller than any power of ν, and then letting it increase. Note that ε measures the distance between R and $R_c = R_c^+$ in this case. This way we will pass a number of 'significant degenerations' (see [5]). The first one appears at $\varepsilon = O(\nu^2)$, that is $R - R_c = O(\nu^4)$. For $\varepsilon = o(\nu^2)$ the modulation equation is very much like the non-resonant equation (3.9). In that case one cannot apply the argument of (2.11) in order to write the linear unstable solutions as real families (2.13), since the $\xi = \varepsilon x$-scale is longer than the spatial scale $\nu^2 x$. This $\nu^2 x$-scale is caused by the fact that the system can never in perfect resonance (see section 2.1 and (2.8)). We refer to [S & D] for a discussion of this and the transition case $\varepsilon = O(\nu^2)$.

We now consider $\nu^2 \ll \varepsilon \ll \sqrt{\nu}$, thus, ε is so large that the 'linear most unstable waves' $\phi_{\text{lin}}^{\pm}$ are as in (2.13), but, ε is still so small that R is way below R_c^-: only ϕ_{lin}^+ is unstable. So, by section 3.1, we introduce the real amplitude $B(\xi, \tau)$ by setting $E\psi^{11} = B(\xi, \tau)\phi_{\text{lin}}^+$ in (3.1) – where we put $c^+ = 1$ in (2.13). We then proceed with the derivation process along the lines of the above subsection.

However, at the ε^2-level something new might happen. The term $\mathcal{N}(E\psi^{11}, E\psi^{11})$ will – of course – be expanded as $B^2 e^{2ik_c x} \times$ 'a power series in e^{inpx}'. By (2.8) we know that $Np = 2k_c$ (we can neglect the $O(\nu^2)$ error now). Thus, if N is even, we see that $k_c = 2k_c - \frac{1}{2}Np$ appears in the expansion of $\mathcal{N}(E\psi^{11}, E\psi^{11})$: there are nonlinear, non-homogeneous terms of order ν at the level $\varepsilon^2 E$ in the derivation process. This means that one derives an equation for B with a quadratic nonlinearity at the ε^3-level:

$$\frac{\partial B}{\partial \tau} = r\omega_R^+ B - \frac{1}{2}\omega_{kk}^+ \frac{\partial^2 B}{\partial \xi^2} + \frac{\nu}{\varepsilon}\gamma B^2 + \beta B^3 + \text{ h.o.t.}, \tag{3.10}$$

where β is the same as in (3.8) and (3.9); the coefficients of the linear terms are defined as in (3.7). An expression for $\gamma \in \mathbf{R}$ can be found in [S & D]. The coefficient $\gamma \equiv 0$ if N is odd. There is another significant degeneration in (3.10) at $\varepsilon = O(\nu)$ in the case N is even: if $\varepsilon \ll \nu$ one can rescale B, ξ and τ such that the quadratic term is $O(1)$ (like the linear terms) and the cubic term $o(1)$; if $\varepsilon \gg \nu$ its the other way around: the quadratic term is small compared to the other terms in (3.10). Note that the quadratic term can be neglected (that is: it is part of the 'h.o.t.' in (3.10)) if ε becomes $O(\sqrt{\nu})$. Thus, at the next significant degeneration there is no difference anymore between the cases N odd or even.

As $\varepsilon = O(\sqrt{\nu})$ we have to take into account the effect of the other unstable mode, ϕ_{lin}^-. This is because $R_c^- - R_c = R_c^- - R_c^+ = \nu d$ for a certain $0 < d \in \mathbf{R}$ (see section 3.1 and

[S & D]) thus $R - R_c = O(\varepsilon^2) = R - R_c^-$. Therefore, we have to adapt the expansion (3.1), since it now follows from the linear theory that

$$E\psi^{11} = B^+(\xi,\tau)\phi^+_{\text{lin}} + B^-(\xi,\tau)\phi^-_{\text{lin}} \tag{3.11}$$

where we put $c^+ = c^- = 1$ in (2.13). Note that B^+ corresponds to the above B. Thus, one has to remodel Ansatz (3.1) by introducing ϕ^+_{lin} and ϕ^-_{lin} through (3.11) and adapting the higher order terms. After making this extension, the derivation is process runs again along the classical lines. This means that we can derive an equation for B^+ on the $\varepsilon^3\phi^+_{\text{lin}}$-level and an equation for B^- on the $\varepsilon^3\phi^-_{\text{lin}}$-level. Together, these equations form a (nonlinearly) coupled system of equations, which reads, in highest order:

$$\left\{\begin{array}{rcrcrcl} \frac{\partial B^+}{\partial \tau} &=& r\omega_R B^+ &-& \frac{1}{2}\omega_{kk}\frac{\partial^2 B^+}{\partial \xi^2} &+& \beta B^+\left((B^+)^2 + (B^-)^2\right) \\ \frac{\partial B^-}{\partial \tau} &=& (r-d)\omega_R B^- &-& \frac{1}{2}\omega_{kk}\frac{\partial^2 B^-}{\partial \xi^2} &+& \beta B^-\left((B^+)^2 + (B^-)^2\right) \end{array}\right. \tag{3.12}$$

Note that we should have distinguished between ω^+_R and ω^-_R and ω^+_{kk} and ω^-_{kk} but the difference between these expressions disappeared into the higher order terms. Therefore, we write ω_R and ω_{kk} and identify these (in highest order) with (respectively) μ_R, $\omega^\pm_R$ and μ_{kk}, $\omega^\pm_{kk}$. The d in the second equation is exactly the same d as defined above if we choose $\varepsilon = \sqrt{\nu}$. The system can be brought into complex form by introducing the complex variable $A = A(\xi,\tau) = B^+ + iB^-$:

$$\frac{\partial A}{\partial \tau} = (r - \frac{1}{2}d)\omega_R A + \frac{1}{2}d\omega_R\bar{A} - \frac{1}{2}\omega_{kk}\frac{\partial^2 A}{\partial \xi^2} + \beta A|A|^2, \tag{3.13}$$

a Ginzburg-Landau equation with a 'non-symmetric' term (non-symmetric, since the term $\bar{A}$ is not phase-invariant). Therefore, the solutions to (3.13) are not phase-invariant, unlike the solutions to the classical Ginzburg-Landau equation. This can be understood by noticing that the phase-invariance in the classical Ginzburg-Landau equation is due to the translation symmetry $x \to x + x_0$ in the underlying system. This symmetry has been broken by $\nu f(px)$ in the perturbed model problem (1.1).

If $\varepsilon \gg \sqrt{\nu}$ not much happens: it can be seen quite easily that this corresponds to scaling d to $o(1)$ in (3.13) or (3.12); d is larger than the higher order terms as long as $\varepsilon \ll \sqrt[3]{\nu}$. As soon as $\varepsilon = O(\sqrt[3]{\nu})$ or larger one has to set $d = 0$ in (3.13) for consistency. Thus, (3.13) has become equal to the classical Ginzburg-Landau equation (in highest order): the influence of the perturbation $\nu f(px)$ can be neglected.

4 The behaviour of solutions

This section will again be divided in two subsections: one based on [E & K], $p = o(1)$, and one on [S & D], $p = O(1)$. There are significant differences between these two subsections: the analysis in the former case is the main topic of [E & K], while for the latter the behaviour of the solutions to the equations (3.10) and (3.13) has only been studied briefly in [S & D] and is the subject of current research.

4.1 $p = O(1)$: pattern selection

Again we should divide this section into a part on non-resonant and resonant values of p and again we only consider the non-resonant case very briefly, since the modulation equation (3.9) is 'just' a classical Ginzburg-Landau equation. However, there is one interesting aspect: due to the perturbations, (3.9) has a nonlinear coefficient with a small ($= O(\nu^2)$) imaginary part. The fact that the modulation equation is no longer a real Ginzburg-Landau equation corresponds to the loss of the reversibility symmetry $x \to -x$ in the model problem. This phenomenon has been studied, in a more general context, in [11] and [1]. In these papers it is shown that there exist large families of (slowly) travelling quasi-periodic solutions and, so-called, 'localized structures' to the 'weakly complex' Ginzburg-Landau equation.

There are no such detailed results in the literature on the character of the solutions to the modulation equations when p has a resonant value. As in subsection 3.2 we consider the various cases which we encounter as we let ε increase from 'very, very small', that is $\varepsilon = O(\nu^4)$ (or smaller), to $\varepsilon \gg \sqrt[3]{\nu}$. We do not pay attention to the initial stage $\varepsilon = O(\nu^4)$ and the final stage $\varepsilon \gg \sqrt[3]{\nu}$. (In the latter case the modulation equation is 'just' the real Ginzburg-Landau equation (section 3.2).)

The nonlinear behaviour of small ($= O(\varepsilon)$) solutions is governed by (3.10) as long as $\nu^2 \ll \varepsilon \ll \sqrt{\nu}$. Note that $\gamma = 0$ when N is odd, see (2.8). The fact that this equation is a real equation has a significant interpretation. Classically one searches for stationary solutions of the type $B(\xi) = \rho e^{i\theta\xi}$. There exists a family of these solutions in the real Ginzburg-Landau equation; they are stable if $|\theta| <$ a certain constant (see [4]) and correspond to periodic 'patterns'. Thus, there is no real pattern selection in the Ginzburg-Landau equation. However, due to the 'splitting' of the unperturbed 'linear most unstable wave' into ϕ^+_{lin} and ϕ^-_{lin} at the linear level (section 2.2), we observe that if $\nu^2 \ll \varepsilon \ll \sqrt{\nu}$ there can only be 2 simple non-zero, non-periodic 'patterns', since θ has to be 0 now. So, the periodic perturbation term $\nu f(px)$ introduces a selection mechanism.

The stability of the constant solutions $B = \rho \neq 0$ has been studied in [S & D]. If N is odd both solutions $B = \pm\sqrt{-\frac{r\omega_R}{\beta}}$ are stable (recall that we chose $\beta < 0$). If N is even there also 2 solutions, but, there is a difference between the case $\varepsilon \ll \nu$ and $\varepsilon \gg \nu$. In the latter case both solutions are $O(1)$ and stable, quite similar to N is odd. However, if $\varepsilon \gg \nu$ the solutions have different magnitudes, $O(\frac{\nu}{\varepsilon})$ and $O(\frac{\varepsilon}{\nu})$. Both solutions can be stable. This yields that the magnitude of the observed stable patterns depends crucially on the initial conditions. This behaviour is reminiscent of the behaviour of a certain (hyper-)degenerate modulation equation studied in [8]. It's also possible that only one of the solutions is stable, however, there is always at least one stable constant solution. We refer to [S & D] for the exact conditions on the coefficients of (3.10) which determine the stability of the constant solutions $B = \rho$.

Of course there are other types of solutions to (3.10). There is much known about the existence and stability of travelling waves to (real) equations of the type $u_t = u_{xx}+F(u)$ where F is a cubic polynomial. We do not pursue this subject further here; we refer to [9] for a general treatment of this type of equations.

If $\varepsilon = O(\sqrt{\nu})$ one has to take into account the influence of the ϕ_{lin}^--mode: equation (3.10) has to be replaced by (3.12) or (3.13), with $d = O(1)$. Parameter d becomes $o(1)$ when $\varepsilon \gg \sqrt{\nu}$, and can be put equal to zero when $\varepsilon = O(\sqrt[3]{\nu})$ or larger. The equation which one obtains for $d = 0$, the real Ginzburg-Landau equation, has a 2-parameter family of (stable) periodic solutions [4], so it is natural to pose the following question: What happens to this family if d becomes positive? From the above we know that (3.13) reduces to a real equation with at most 2 stable constant solutions in the limit $d \to \infty$ (note that (3.10) can be recovered from (3.13) by taking the limit $d \to \infty$, since $B^- \to 0$ in (3.12)). This question is the subject of current research. There are strong indications that the family exists as long as $d < r$ – that is, as long as R is above the second neutral curve $\omega_- = 0$ (section 3.2). As soon as d passes through r this family degenerates and there is pattern selection. We refer to [17] for the precise results.

Of course there are more solutions to 3.13 than the solutions described above. For simplicity we introduce some rescalings, as a result we can set: $\omega_R = 1$, $-\frac{1}{2}\omega_{kk} = 1$ and $\beta = -1$. It's natural to consider the stationary problem associated to (3.13). This is a 4-dimensional system of ordinary differential equations. Unlike the real Ginzburg-Landau equation (that is: $d = 0$ in (3.13)) one cannot reduce this system to a 3-dimensional ODE using polar coordinates (see [1]), since there is no phase-invariance in (3.13). However, this problem can still be considered as a 2-degrees-of-freedom Hamiltonian system, with Hamiltonian $\mathcal{H} = \mathcal{H}(x_1, x_2, x_3, x_4)$, where $x_1 = B^+$, $x_2 = \frac{\partial}{\partial \xi}B^+$ and $x_3 = B^-$, $x_4 = \frac{\partial}{\partial \xi}B^-$:

$$\mathcal{H}(x_1, x_2, x_3, x_4) = \frac{1}{2}[rx_1^2 + x_2^2 + (r-d)x_3^2 + x_4^2] - \frac{1}{4}(x_1^2 + x_3^2)^2. \tag{4.1}$$

A priori one should not expect that this Hamiltonian system is integrable, like the (unperturbed) real Ginzburg-Landau system. However, numerical simulations did not show the typical non-integrable behaviour, so, we started to hunt for a second integral. Using [19] we succeeded in finding the second integral:

$$\mathcal{I}(x_1, x_2, x_3, x_4) = (x_1x_4 - x_2x_3)^2 - \frac{1}{2}d[(x_1^2 - r)^2 - 2x_2^2 - (x_3^2 - r)^2 + 2x_4^2] + d^2x_3^2. \tag{4.2}$$

We refer to [17] for more information on this; there, the attention will mainly be focused on the structure of the solutions to the Hamiltonian system in the two most interesting cases: d near 0 and d near r.

4.2 Slow variations $p = o(1)$, $p^2 \gg \nu$: quasiperiodic solutions

First we consider (3.8) for $p = \sigma_0 \varepsilon$ and $\hat{\nu} \ll 1$; scaling out the $O(1)$ coefficients $r\mu_R$, $\mu_{kk}/2$, and β yields

$$\frac{\partial A}{\partial \tau} = (1 - |A|^2)A + \frac{\partial^2 A}{\partial \xi^2} + \hat{\nu} \sum_{n=-\infty}^{\infty} F_n e^{in\sigma_0 \xi} A. \tag{4.3}$$

In this subsection we consider $\sigma_0 = O(1)$; for values of $\sigma_0 \ll 1$ it is appropriate to define a new spatial scale $p\sigma_0 x$ as in the next subsection considered. The form of (4.3) suggests that it is appropriate to look for quasiperiodic stationary solutions to (4.3) of the form

$$A_S = e^{i\sigma\xi}(a_0 + \hat{\nu} \sum_{n=-\infty}^{\infty} a_n e^{in\sigma_0 \xi}). \tag{4.4}$$

The existence of such solutions A_S is determined from the following equations

$$n = 0 \ : \ a_0^2 = 1 - \sigma^2 + O(\hat{\nu}) \tag{4.5}$$

$$n \neq 0 \ : \ a_{\pm n}(1 - (\sigma \pm n\sigma_0)^2 - 2a_0^2) = a_0^2 a_{\mp n} - a_0 a_{\pm n} + O(\hat{\nu}) \tag{4.6}$$

obtained by substituting A_S into (4.3) and setting the coefficient of $e^{i(\sigma + n\sigma_0)\xi}$ equal to zero. The leading order existence criteria for solutions of the form (4.4) are:

$$\sigma^2 < 1, \tag{4.7}$$

$$\sigma^2 \neq \frac{1}{3} + \frac{n^2\sigma_0^2}{6} \text{ for } n \geq 1. \tag{4.8}$$

If $\sigma_0 > 4$ then the condition (4.8) is satisfied in the whole interval (4.7). If $\sigma_0 < 4$ then isolated values of σ in interval (4.7) appear at which (4.8) is violated. The number of these exeptional values increases if σ_0 decreases. For example, there is one exeptional value for $1 < \sigma_0 < 4$, there are two for $\frac{4}{9} < \sigma_0 < 1$. For σ violating (4.8) the construction (4.4) breaks down, indicating that for some values of n the coefficients $\hat{\nu} a_n \neq O(\hat{\nu})$. Preliminary calculations indicate that for (4.8) violated it is appropriate to consider a quasiperiodic solution in which some coefficients are of the order of a fractional power of $\hat{\nu}$. When σ_0 gets very small the interval (4.7) gets crowded with exceptional values. This is the passage to the 'very slow' case considered in the next subsection.

The stability of the solution (4.4) can be studied by writing $A = A_S + e^{i\sigma\xi}[p_0 + \sum_{n=-\infty}^{\infty} p_n e^{in\sigma_0\xi}]$ and linearizing for small p_n. Then (4.4) are stable to perturbations $p_n = p_n(\xi, t)$ if

$$\sigma^2 < \frac{1}{3}. \tag{4.9}$$

This condition is identical to the Eckhaus instability criterion [4] observed for the unperturbed case $\nu = 0$.

4.3 Very slow variations $p = o(1)$, $p^2 \ll \nu$: space intermittent solutions

We now consider the case $p^2 \ll \nu$ in (3.8) and choose ε such that $\nu = \hat{\nu}\varepsilon^2$ and $\hat{\nu} = O(1)$. Thus $p \ll \varepsilon$ and we introduce the new small parameter $0 < \delta = p/\varepsilon \ll 1$. The presence of the small parameter δ can then be included in (3.8) explicitly as

$$\frac{\partial A}{\partial \tau} = \frac{\partial^2 A}{\partial \xi^2} + (G(\hat{\xi}) - |A|^2)A + O(\delta, \nu) \tag{4.10}$$

where $\hat{\xi} = px = \delta\xi$, an 'extra-long' spatial scale, and the function $G(\hat{\xi}) = r\mu_R + \hat{\nu}f(\hat{\xi})$ is introduced for convenience. Note that this equation can also be derived from (3.8) when $\hat{\nu} = o(1)$: one then sets $r = \hat{r}\hat{\nu}$ and rescales ξ, τ and A such that $\hat{\nu}$ can be eliminated from (3.8), this rescaling is equivalent to substituting $\hat{\varepsilon} = \sqrt{\hat{\nu}}\varepsilon$ for ε in (3.1).

Equation (4.10) implies that $A(\tau, \xi, \hat{\xi})$ depends on two length scales ξ and $\hat{\xi}$. In this section we consider smooth stationary solutions of the form $A_S = \mathcal{A}(\hat{\xi})e^{i\sigma\xi}$ which are periodic on the ξ length scale. The leading order behaviour of such solutions for $A_S = O(1)$ can be determined by treating $\hat{\xi}$ as a parameter in (4.10) in the spirit of multi-scale analysis, yielding

$$A_S \sim \left(-\sigma^2 + G(\hat{\xi})\right)^{1/2} e^{i\sigma\xi}. \tag{4.11}$$

This solution exists for $G(\hat{\xi}) > \sigma^2$, so that only for $G(\hat{\xi}) > 0$ do solutions (4.11) exist. For $G(\hat{\xi}) > \sigma^2$ we expect from (4.11) that $\mathcal{A}(\hat{\xi}) \neq 0$. For $G(\hat{\xi}) < 0$ the existence condition $G(\hat{\xi}) > \sigma^2$ is violated for all ξ and σ. Then no periodic solutions (4.11) exist for the modulation equation (4.10). This range of r corresponds to values $R < R_c$ for which the basic solution is stable. We consider values of r such that

$$-\max[\hat{\nu}f(\hat{\xi})] < \mu_R r < -\min[\hat{\nu}f(\hat{\xi})]. \tag{4.12}$$

Then the criterion for the existence of solutions (4.11) given by $G(\hat{\xi}) > \sigma^2$ is satisfied only for certain values of ξ, regardless of the value of σ. For certain ranges of ξ the function $G(\hat{\xi}) < 0$, so that there are no solutions (4.11), that is $\mathcal{A}(\hat{\xi}) = 0$. In [E & K] it is concluded that only for $\sigma = 0$ there are smooth spatially periodic solutions $\phi = \varepsilon\mathcal{A}(\hat{\xi})e^{ik_c x}$ with $\mathcal{A}(\hat{\xi})$ real for all values of $\hat{\xi}$. These space-intermittent soluitions are sketched in Figure 2. We now outline the main steps of our analysis.

The regions corresponding to $G(\hat{\xi}) = O(\delta)$ are transition regions between the regions of existence for (4.11) and those for which these solutions do not exist and $A = 0$. As is shown below, the solution obtained in these layers matches with (4.11) in the appropriate limit and is the global solution for all ξ.

To analyze A in these transition layers it is convenient to define a local variable η near values of $\xi = \xi^*$ where G vanishes. For example, for $G(\hat{\xi}^*) = 0$ and $G'(\hat{\xi}^*) > 0$ we define

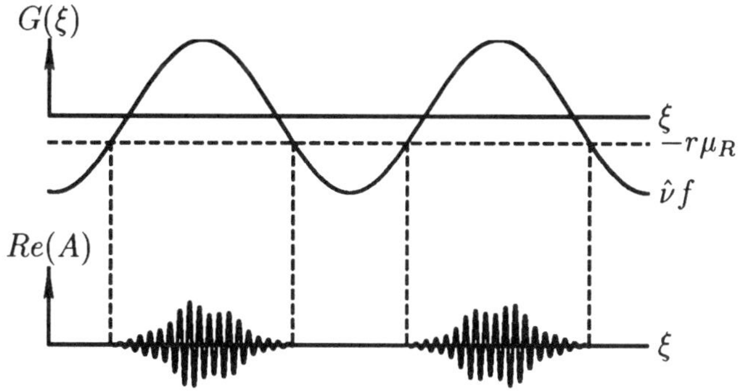

Figure 2: The space-intermittent solutions. Stationary periodic solutions only exist for $G(\hat{\xi}) = r\mu_R + \hat{\nu} f(\hat{\xi}) > 0$.

$\hat{\xi} - \hat{\xi}^* = \delta\eta$. Then $P(\delta\eta) := G(\hat{\xi}) \sim \pm G'(\hat{\xi}^*)\delta\eta$ for $\delta\eta \ll 1$ corresponding to values of ξ in the layer. The solution $A(\xi) = \psi_L(\eta)$ in the transition layer is determined in the region defined by $\eta_- \leq \eta \leq \eta_+/\delta$ for η_+ a constant appropriately chosen such that $P(\eta_+) = O(1)$, the sign of $P(\delta\eta)$ changes only as η passes through 0, and $P(\delta\eta) < 0$ for $\eta_- < \eta < 0$. That is, when we consider the behaviour of ψ_L for large η, we take the limit

$$\lim_{\substack{\delta \to 0 \\ \eta_+ \text{fixed}}} \psi_L(\eta = \eta_+/\delta). \tag{4.13}$$

We study the two cases $P(\delta\eta) < 0$ and $P(\delta\eta) > 0$ separately. The leading order equation in the transition layer is given by

$$0 = (-\psi_L^2 + P(\delta\eta))\psi_L + \frac{\partial^2 \psi_L}{\partial \eta^2} + O(\delta). \tag{4.14}$$

The equation (4.14) is solved by treating $\eta_0 := \delta\eta$ as a constant with respect to η, yielding

$$\psi_L \sim 2\sqrt{2}\frac{\sqrt{|P(\eta_0)|}K(\eta_0)e^{\sqrt{|P(\eta_0)|}\eta}}{K^2(\eta_0)e^{2\sqrt{|P(\eta_0)|}\eta} - 1} \text{ for } \eta \leq 0. \tag{4.15}$$

The coefficient $K(\eta_0)$ is constant with respect to η and is determined below by matching (4.15) to the solution for $\eta > 0$ at $\eta = 0$. One solution (to leading order) of (4.14) for $\eta > 0$ corresponds to

$$\psi_0 \; = \; \sqrt{P(\delta\eta)}\tanh(\omega(\eta)/\sqrt{2}), \quad \omega(\eta) = \int_0^\eta \sqrt{P(\delta\eta')}d\eta' \tag{4.16}$$

The solution (4.16) has the limiting behaviour

$$\lim_{\eta\to\infty} \psi_0(\eta;\eta_0) = \sqrt{P(\eta_0)}, \quad \lim_{\eta\to 0} \psi_0(\eta;\eta_0) \; = \; 0. \tag{4.17}$$

and is of particular interest because the limit $\eta \to \infty$, η_0 held constant in (4.17), yields A_S with $\sigma = 0$. The form of the leading order solution for ψ_0 has been motivated by multi-scale analysis, but only the local variable η is used, as written in (4.16). This approach is employed to avoid violating the basic assumption $\delta\eta \ll \eta$ of the multi-scale approach when examining the behaviour of ψ_L as $\eta \to \infty$. To show that (4.16) is a valid leading order approximation to ψ_L we write $\psi_L = \psi_0 + \delta\psi_1$ and demonstrate that ψ_1 is bounded for all η in the layer. To determine the leading order approximation ψ_1^0 to ψ_1, an equation for ψ_1 is obtained by substituting $\psi = \psi_0(\eta) + \delta\psi_1(\eta)$ into (4.14), employing the notation for the first and second derivative of ψ_j,

$$\begin{aligned} \psi_j' &= \psi_{(j,0)}' + \delta\psi_{(j,1)}' + \delta^2\psi_{(j,2)}' \\ \psi_j'' &= \psi_{(j,0)}'' + \delta\psi_{(j,1)}'' + \delta^2\psi_{(j,2)}'' \end{aligned} \tag{4.18}$$

so that

$$\mathcal{M}\psi_1 := \psi_{(1,0)}'' + (-3\psi_0^2 + P(\delta\eta))\psi_1 = R_0 + \delta R_1 \tag{4.19}$$

where

$$R_0(\psi_0) := -\psi_{(0,1)}'' - \delta\psi_{(0,2)},'' \quad R_1 := -\psi_{(1,1)}'' - \delta\psi_{(1,2)}'' + 3\psi_1^2\psi_0 + \delta\psi_1^3. \tag{4.20}$$

Then ψ_1^0 is determined as follows. Using the solutions to the homogeneous equation, $\mathcal{M}\psi = 0$,

$$\psi_a = \cosh^{-2}(\omega(\eta)/\sqrt{2}), \quad \psi_b := \psi_a \int_0^\eta (\psi_a(\eta'))^{-2} d\eta'. \tag{4.21}$$

the expression for ψ_1^0 is

$$\psi_1^0 := - \lim_{\substack{\delta\to 0 \\ \eta_+ \text{fixed}}} \left[\int_0^\eta \psi_a(\eta)\psi_b(\eta')R_0(\eta')d\eta' + \int_\eta^{\eta_+/\delta} \psi_b(\eta)\psi_a(\eta')R_0(\eta')d\eta' \right]. \tag{4.22}$$

The behaviour of ψ_1^0 as $\eta \to 0$ and $\eta \to \infty$ can be obtained straightforwardly. We summarize the behaviour of the $\psi_L \sim \psi_{as} := \psi_0 + \delta\psi_1^0$, which is demonstrated in detail in [E & K]:

$\psi_L \sim -\delta\eta D + O(\delta\eta^2)$ for $\eta \to 0$ $\qquad$ $\lim_{\substack{\delta\to 0 \\ \eta_+ \text{fixed}}} \psi_L(\eta = \eta_+/\delta) = \sqrt{P(\eta_+)}$

where $D = \lim_{\substack{\delta\to 0 \\ \eta_+ \text{fixed}}} \int_0^{\eta_+/\delta} \psi_a(\eta')R_0(\eta')d\eta'$

The solution (4.15) for $\eta < 0$ is matched with (4.22). Then ψ_L for $\eta < 0$ can be rewritten as in (4.16) in the more general form

$$\psi_L \sim \frac{-\sqrt{|\delta\eta|}\sqrt{|P(\delta\eta)|}De^{\int_0^\eta \sqrt{|P(\delta\eta')|}d\eta'}}{8D^2e^{2\int_0^\eta \sqrt{|P(\delta\eta')|}d\eta'} - 1} \quad \text{for } \eta_- < \eta < 0. \tag{4.23}$$

Above we have determined the first two terms in the asymptotic approximation to the solution of (4.14) for $P(\delta\eta) > 0$. In [E & K] the asymptotic validity of $\psi_{as} = \psi_0 + \psi_1^0$ as approximation of the solution ψ_L of (4.14) is shown using a fixed point argument is an appropriate Banach space and taking $P(\delta\eta) = \tanh(\delta\eta)$. Employing this substitution avoids complications due to the limit (4.13) but does not sacrifice the qualitative features of (4.14), since $\tanh(\delta\eta)$ changes sign only for η passing through zero, and remains bounded as $\eta \to \infty$. The function $P(\delta\eta)$ has this same behaviour for values of η in the layer for which ψ_L was determined above.

To demonstrate that ψ_{as} is attracting it is convenient to write

$$\psi = \psi_{as}[1 + \rho(\eta, t)] \tag{4.24}$$

where $\rho(\eta, t) \ll 1$. Then we have the following linear PDE for ρ [E & K]:

$$\frac{\partial \rho}{\partial t} = \frac{2}{\eta}\frac{\partial \rho}{\partial \eta} + \frac{\partial^2 \rho}{\partial \eta^2}. \tag{4.25}$$

For ρ to be bounded in the limit (4.13), $\rho(\eta) = e^{-\lambda t}\sin(\sqrt{\lambda}\eta)/\eta$ with $\lambda > 0$. Thus $\lim_{t\to\infty} \rho = 0$, so that ψ_L is stable for $\delta\eta \ll 1$.

5 Discussion

We have shown that in problems on domains with slightly varying geometry the pattern formation through instability and bifurcations exhibits new phenomena. We found that the wave number p of periodic term f has a crucial influence on the behaviour of the system. There are three structurally different cases to consider: p is non-resonant, p is resonant (see (2.7)) and $p = o(1)$. In the first case the influence of νf is regular. In the second case we derived various modulation equations (at near-critical conditions) which differ from the classical Ginzburg-Landau equation. Based on a careful treatment of the linearized stability problem we succeeded in constructing the appropriate modulation equation for any choice of $\varepsilon = \sqrt{R - R_c}$ compared to ν, the amplitude of the spatial period effect. The last case, $p = o(1)$, is the most difficult case. Therefore, we focused on the situation in which $\varepsilon = O(\sqrt{\nu})$ (or larger). In this case it is possible to derive the modulation equation based on the unperturbed ($\nu = 0$) linearized theory. As a consequence, νf appears in the modulation equation (3.8).

Our treatment is certainly not exhaustive: various questions are only touched upon. For instance, one would like to understand the behaviour of the system if $p = o(1)$ and $\varepsilon \ll \sqrt{\nu}$. Therefore, it is necessary to perform a detailed analysis of the linearized stability problem, just as is done in this paper for the more simple case $p = O(1)$ and resonant. Related to this is the question of what happens in the transition between $p = O(1)$ and $p = o(1)$: this is not at all clear from the results in this paper. Another aspect which we did not consider in this paper is the influence of the assumption that the basic system (1.1) is reflection symmetric in $x \to -x$: if this is not the case the unperturbed Ginzburg-Landau equation will have complex coefficients and the solutions of this equation correspond to travelling solutions of (1.1). It is fairly possible that this will have an important effect on the outcome of the analysis of the system with varying geometry.

So far, we only mentioned questions which are related to the derivation of the appropriate modulation equations. The very important next step is of course the analysis of the behaviour of the solutions to these equations. In this paper we took some first steps in this direction, but the equations are not fully understood, yet.

These questions are the subject of current and future research.

References

[E & K] W. Eckhaus and R. Kuske (1994), Pattern formation in systems with slowly varying geometry, *preprint 857*, University of Utrecht.

[S & D] R. Schielen and A. Doelman (1994), Nonlinear stability analysis in unbounded, weakly oscillating domains, *preprint 866*, University of Utrecht.

[1] A. Doelman (1993) Traveling waves in the complex Ginzburg-Landau equation, *J. Nonlinear Sc.* **3**, 225-266.

[2] P.M. Eagles (1980) A Bénard problem with a perturbed lower wall, *Proc. Roy. Soc. Ser. A* **371**, 359-379.

[3] P.M. Eagles and K. Eames (1983) Taylor vortices between almost cylindrical boundaries, *J. Eng. Math.* **17**, 263-280.

[4] W. Eckhaus (1965) *Studies in Nonlinear Stability Theory*, Springer-Verlag.

[5] W. Eckhaus (1979) *Asymptotic Analysis of Singular Perturbations*, North Holland.

[6] W. Eckhaus (1992) On modulation equations of the Ginzburg-Landau type, in *ICIAM 91: Proc. 2nd Int. Conf. Ind. Appl.* (R.E. O'Malley ed.), 83-98.

[7] W. Eckhaus (1993) The Ginzburg-Landau manifold is an attractor, *J. Nonlinear Sci.* **3**, 329-348.

[8] W. Eckhaus and G. Iooss (1989) Strong selection or rejection of spatially periodic patterns in degenerate bifurcations, *Physica* **39D**, 124-146.

[9] P.C. Fife (1979) *Mathematical Aspects of Reacting and Diffusing Systems*, Lecture Notes in Biomathematics **28**, Springer-Verlag.

[10] A. van Harten (1991) On the validity of Ginzburg-Landau's equation, *J. Nonlinear Sci.* **1**, 397-422.

[11] P. Holmes (1986) Spatial structure of time-periodic solutions of the Ginzburg-Landau equation, *Physica* **23D**, 84-90.

[12] D.A.S. Rees (1990) The effect of long-wavelength thermal modulations on the onset of convection in an infinite porous layer heated from below, *Phys. of Fluids A* **2**, 56-62.

[13] D.A.S. Rees and D.S. Riley (1986) Free convection in an undulating saturated porous layer: resonant wavelength excitation, *J. Fluid Mech.* **166**, 503-530.

[14] D.A.S. Rees and D.S. Riley (1989) The effects of boundary imperfections on convection in a saturated porous layer: near resonant wavelength excitation, *J. Fluid Mech.* **199**, 133-154.

[15] D.N. Riahi (1993) Preferred pattern of convection in a porous layer with a spatially nonuniform boundary temperature, *J. Fluid Mech.* **246**, 529-543.

[16] R. Schielen, A. Doelman and H.E. de Swart (1993) On the nonlinear dynamics of free bars in straight channels, *J. Fluid Mech.* **252**, 325-356.

[17] R. Schielen, thesis, to appear.

[18] I.C. Walton (1982) Effects of slow spatial variations on Bénard convection, *Q.J. Mech. Appl. Math.* **35**, 33-48.

[19] E.T. Whittaker (1917) *A Treatise on the Analytical Dynamics of Particles and Rigid Bodies*, Cambridge University Press.

Solitary wave interactions with external forces

R. GRIMSHAW

Department of Mathematics, Monash University, Calyton, Vic 3168, Australia.

Abstract

Using the forced Korteweg-de Vries equation as a canonical model, we investigate the interaction of a solitary wave with a moving external force of small amplitude. The analysis is a multi-scale asymptotic procedure, which leads to a pair of autonomous ordinary differential equations for the solitary wave amplitude and position. These are analysed in detail for the case when the lengthscale of the external force is much greater than that of the solitary wave. For the case of a single isolated force this theory predicts that the main regimes are **passage**, when the solitary wave passes through the external force with some amplitude modulation, **repulsion**, when the solitary wave is reflected by the external force with a significant amplitude change, and **trapping**, when the solitary wave remains in the vicinity of the external force location for an extended period of time. This theory is then extended to the case when the external force has two forcing centres, and we show that there is then a similar set of regimes but with rather more complicated dynamical behaviour.

1 Introduction

The solitary wave solutions of the Korteweg-de Vries equation represent perhaps the simplest canonical model of coherent phenomena. In a recent paper, Grimshaw et. al (1994) have studied the interaction of a solitary wave with an isolated external force of small amplitude using the forced Korteweg-de Vries equation as a model. They showed that there are three typical scenarios. These are **passage**, when the solitary wave passes through the external force with an amplitude modulation, **repulsion** when the solitary wave is reflected by the external force with a significant amplitude change, and **trapping** when the solitary wave remains in the vicinity of the external force for an extended period of time. Each of these regimes represents an interesting example of how a coherent structure, in this case a Korteweg-de Vries solitary wave, can interact with an external feature and produce significant modifications. Motivation for this study is provided by the expectation that applications of this concept may prove useful in studying coherent phenomena in the atmosphere and ocean. Some steps in this direction have already been taken by Patoine and Warn (1982), Warn and Brasnett (1983), and Mitsudera and Grimshaw (1991, 1993 and 1994).

Our purpose here is to review succinctly the analysis of Grimshaw et. al (1994), and then to extend it to the case when the solitary wave interacts with an external force which has several forcing centres. In fact it is sufficient to consider the case of an external force with just two forcing centres to give an indication of the complex dynamics that can result. As in Grimshaw et. al (1994) the governing equation is the forced Korteweg-de Vries equation

$$u_t + \Delta u_x + 6uu_x + u_{xxx} = f_x(x) \tag{1.1}$$

Here x and t are the space- and time-like variables respectively, $f(x)$ is the external force and Δ is a parameter which measures the difference between the linear long wave speed of the unforced system and the speed of the force. Note that (1.1) has been written in the frame of reference of the force. The forced Korteweg-de Vries equation (1.1) is now widely recognised as a canonical model for the resonant generation of solitary waves (see, e.g., Grimshaw et. al (1994) for a summary of the recent literature). Here we are concerned with the situation when a solitary wave (i.e. the solitary wave solution of (1.1) when $f(x)$ is zero) encounters an isolated force.

As in Grimshaw et. al (1994) we shall assume that $f(x)$ has a small amplitude, and then use a perturbation analysis to describe the interaction. This analysis is presented in section 2, and leads to a pair of coupled ordinary differential equations for the solitary wave amplitude and position. In section 3 these perturbation equations are solved for the case of broad forcing when the external force has a length scale much greater than the solitary wave. In section 3 we summarize the results of Grimshaw et. al (1994) when the external force has a single forcing centre, and then in section 4 we consider the case when there are two forcing centres.

2 Perturbation equations

When the external force $f(x)$ in (1.1) has a small amplitude relative to the solitary wave, then equation (1.1) can be regarded as a perturbed KdV equation. For this situation the equations which govern the evolution of a solitary wave are well known. Here, we shall give a brief outline of their derivation following the multi-scale analysis of Grimshaw and Mitsudera (1993) (see also Grimshaw et. al, 1994). In this section it is useful to replace f with εf where ε is a small, positive parameter. We can then expect a solitary wave to evolve on a time scale ε^{-1}. Thus, we seek asymptotic expansions of the form

$$u = u_0 + \varepsilon u_1 + \varepsilon^2 u_2 + \ldots \tag{2.1a}$$

$$q = q_0 + \varepsilon q_1 + \varepsilon^2 q_2 + \ldots \tag{2.1b}$$

where

$$u_0 = a(T)\operatorname{sech}^2\{y(T)\xi\}, \tag{2.2a}$$

$$\xi = x - \phi(T) \tag{2.2b}$$

$$\phi(T) = x_0 + \frac{1}{\varepsilon}\int_0^T q(T')\,dT' \tag{2.2c}$$

and

$$T = \varepsilon t. \tag{2.2d}$$

Substitution of these expressions into (1.1) confirms that u_0 is given by (2.2a-d) provided that

$$a = 2\gamma^2, \tag{2.3a}$$

and

$$q_0 = \Delta + 2a. \tag{2.3b}$$

Note that u_0 is just the well-known solitary wave solution of the KdV equation.

At the next order we obtain the following linear inhomogeneous equation for u_1,

$$Lu_1 \equiv (\Delta - q_0)u_{1\xi} + 6(u_0 u_1)_\xi + u_{1\xi\xi\xi} = F_1, \tag{2.4a}$$

where

$$F_1 = f_\xi - u_{0T} + q_1 u_{0\xi}. \tag{2.4b}$$

Here we have used (2.2b) so that $f = f(\xi + \phi(T))$. The linear operator L is the linearized KdV-operator, and it may be shown that in order for (2.4a) to have a bounded solution for u_1, the following compatibility condition must be satisfied.

$$\int_{-\infty}^{\infty} F_1 u_0 \, d\xi = 0 \tag{2.5}$$

Substituting (2.4b) into (2.5) we obtain the well-known result expressing the leading order energy balance.

$$\frac{d}{dT} \int_{-\infty}^{\infty} \frac{1}{2} u_0^2 \, d\xi = \int_{-\infty}^{\infty} u_0 f_\xi \, d\xi \tag{2.6}$$

Then, on substituting (2.2a) into (2.6) we get

$$\frac{da}{dT} = \gamma \int_{-\infty}^{\infty} \operatorname{sech}^2(\gamma\xi) f_\xi(\xi + \phi) \, d\xi. \tag{2.7}$$

Equation (2.7) is one of the two perturbation equations for the two unknowns $a(T)$ and $\phi(T)$, these being the solitary wave amplitude and position respectively. The second equation is obtained from (2.2c) and is

$$\varepsilon \frac{d\phi}{dT} = \Delta + 2a + \varepsilon q_1, \tag{2.8}$$

where we have used (2.3b) for q_0.

In utilising perturbation equations such as (2.7) and (2.8) it is often assumed that the term εq_1 in (2.8) can be neglected. However, Grimshaw et. al (1994) have shown that in the present context there is considerable improvement if this term is retained. To obtain an expression for q_1 it is necessary to solve (2.4a) for u_1 and then apply a compatibility condition similar to (2.5) to the linear inhomogeneous equation determining u_2. We shall not give details here (but see Grimshaw and Mitsudera, 1993, or Grimshaw et. al, 1994), and will instead note some essential points and quote the main result. First, when solving (2.4a) for u_1 we find that $u_1 \to u_1^{\pm}$ as $\xi \to \pm\infty$, where only the difference $u_1^+ - u_1^-$ can be determined from (2.4a, b). Here we follow the usual practice of putting $u_1^+ = 0$ on the grounds that the solitary wave will propagate faster than any small amplitude wave, and hence no pedestal represented by u_1^+ can form ahead of the wave. Then it can be shown that

$$u_1^- = -\frac{1}{4\gamma^2}\int_{-\infty}^{\infty} \operatorname{sech}^2\gamma\xi f_\xi(\xi+\phi)\,d\xi, \tag{2.9}$$

where since the right-hand side is proportional to da/dT (see (2.7)) it is not zero in general, and hence the pedestal represented by u_1^- is also not zero in general. This is now a well known result and is a consequence of the fact that while the fKdV equation (1.1) conserves mass, the evolving solitary wave while conserving energy through (2.6) does not conserve mass by itself. The result is the formation of a trailing long wave behind the solitary wave (Kaup and Newell, 1978, or Grimshaw, 1979).

Next, it can be shown that the equation determining q_1 can be interpreted as a second-order expression in the energy balance. The final expression for q_1 is quite complicated, and here we follow the usual practice of assuming that it is sufficient to evaluate q_1 for small times T (strictly $\varepsilon \ll T \ll 1$). The result is

$$q_1 = -\frac{1}{2\gamma}\int_{-\infty}^{\infty} \operatorname{sech}^2(\gamma\xi)\{1+(1-\gamma\xi)\tanh\gamma\xi\}f(\xi+\phi)\,d\xi. \tag{2.10}$$

3 Analysis for a single force

The perturbation equations which govern the interaction of a solitary wave with an external force are thus (2.7) and (2.8) where q_1 is given by (2.9). Henceforth we remove the small parameter ε and revert to the original time variable t. The equations then form a system of two autonomous equations for the amplitude a and the position ϕ. In general this system is not explicitly integrable. However, it was shown by Grimshaw et. al (1994) that in the limit of broad forcing the system becomes Hamiltonian and is integrable, while remaining broadly representative of the general case. Hence we shall examine this case here.

For broad forcing, the solitary wave acts like a delta-function, and so (2.7) and (2.8) become

$$\frac{da}{dt} = 2f_\phi(\phi), \tag{3.1a}$$

$$\frac{d\phi}{dt} = \Delta + 2a - \frac{f(\phi)}{a}. \tag{3.1b}$$

This system is an integrable Hamiltonian system with Hamiltonian

$$H = \frac{8}{3}\Delta\gamma^3 + \frac{32}{5}\gamma^5 - 4\gamma f(\phi), \tag{3.2}$$

where we recall from (2.3a) that $a = 2\gamma^2$. It is pertinent to note that in this limit of broad forcing, u_1^- is effectively zero and it is this circumstance which causes the system to become Hamiltonian. Further H is an invariant and hence the integrals of (3.1a, b) are defined by (3.2) with H equal to a constant.

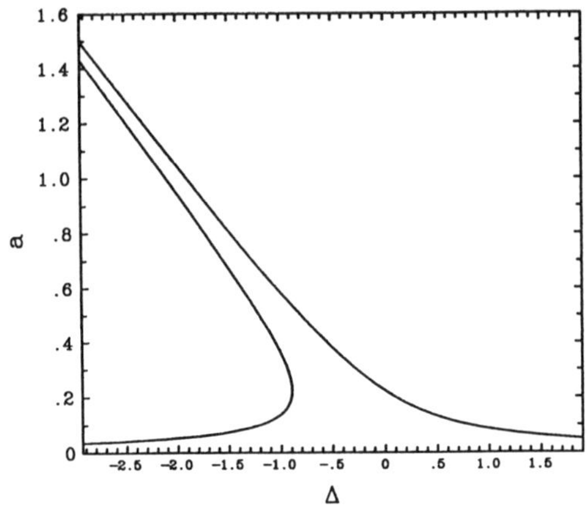

Figure 1: Plot of the equilibrium state (3.6a) for the solitary wave amplitude a versus Δ. The upper (lower) branch is for $b > 0$ (< 0)

In this section we shall consider the case of an symmetric isolated external force $f(x)$ for which the function

$$f(x) = b\text{sech}^2(x/l) \tag{3.3}$$

is representative. We then put

$$\Theta = \frac{\phi}{l}, \quad A = \frac{2a}{|\Delta|}, \quad Q = \frac{8b}{\Delta^2} \tag{3.4}$$

with these substitutions, (3.2) becomes

$$\frac{3}{4}Q\text{sech}^2\Theta = \pm A + \frac{3}{5}A^2 + \frac{K}{A^{\frac{1}{2}}}, \tag{3.5}$$

where $\pm$ is sign Δ, and the constant K is proportional to H in (3.2). Equation (3.5) enables us to plot the $A - \Theta$ phase plane where the independent parameters are Q and sign Δ.

First we consider the equilibrium states of (3.1a, b) which are given by

$$\phi = 0, \quad a(\Delta + 2a) - b = 0, \tag{3.6a}$$

or

$$\phi = 0, \quad \pm A + A^2 - \frac{1}{4}Q = 0. \tag{3.6b}$$

A plot of the solitary wave amplitude a as a function of Δ for fixed forcing amplitude b is shown in Figure 1, and we see that it has the classical structure for the forced periodic states of a nonlinear oscillator. For positive forcing ($b > 0$ or $Q > 0$) there is just one solution for a for any value of Δ, while for negative forcing ($b < 0$ or $Q < 0$) solutions exist only if $\Delta < -(-8b)^{\frac{1}{2}}$ (i.e. $-1 < Q < 0$) and then there are two solutions. The stability of these equilibrium states are determined by calculating

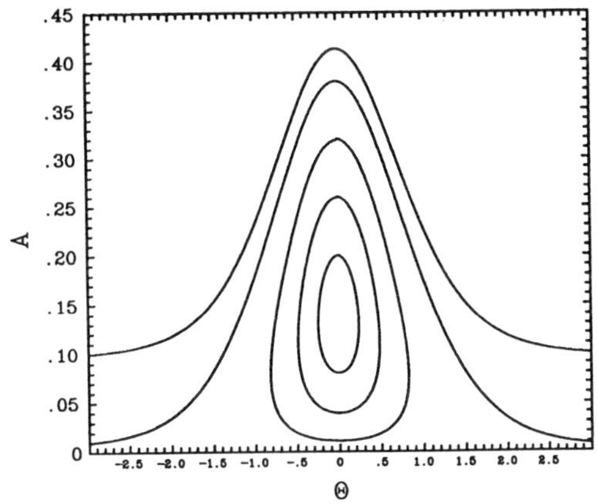

Figure 2: Phase portrait of the system (3.1a, b) for the single force (3.3) for $\Delta = 0.36$ and $Q = 0.62$

the eigenvalues λ of the linearized operator corresponding to perturbations about each equilibrium state. We find that

$$\lambda^2 = -\frac{4b}{l^2}(4 + \frac{\Delta}{a}). \tag{3.7}$$

Hence for $b > 0$ (the upper branch in Figure 1) the equilibrium state is a centre, while for $b < 0$ it is a saddle point for the middle branch in Figure 1 and a centre for the lower branch. It is interesting to note that these equilibrium states correspond to an exact stationary solution of the fKdV equation (1.1) given by $u = a\text{sech}^2(x/l)$ where $a = 2/l^2$ and $f(x)$ is given (3.3). This exact solution was noted by Wu (1987) and its stability discussed by Camassa and Wu (1991). The present stability results obtained from the perturbation theory are consistent with the more general theory of Camassa and Wu (1991).

Next we return to the integral (3.5) and plot some representative phase portraits. A more complete description of the phase plane is given by Grimshaw et. al (1994). Here we emphasise those cases where solitary wave trapping may occur. In Figure 2 we show a case when $\Delta > 0$ and $Q > 0$. These are two possible régimes, **passage** or **trapping** which are separated by the "homoclinic" orbit in which $K = 0$ in (3.5) (since $A \to 0$ as $\Theta \to \pm\infty$). All solitary waves which reach infinity belong to the passage régime, in which there is a local increase in amplitude as the wave passes over the external force. On the other hand, those solitary waves with amplitudes close to the equilibrium value remain trapped and their amplitudes oscillate about this value. When $\Delta > 0$ but $Q < 0$ there are no equilibrium states, and only **passage** can occur with the solitary wave amplitude decreasing to a minimum over the external force.

The case $\Delta < 0$ and $Q > 0$ is shown in Figure 3. These are again two régimes, **passage** and **trapping**, separated by the "homoclinic" orbit on which $K = 2/5$ in

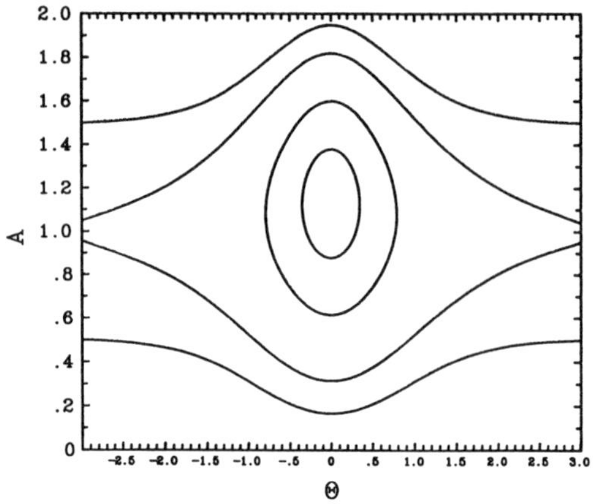

Figure 3: As for Figure 2, but for $\Delta = -0.36$ and $Q = 0.62$

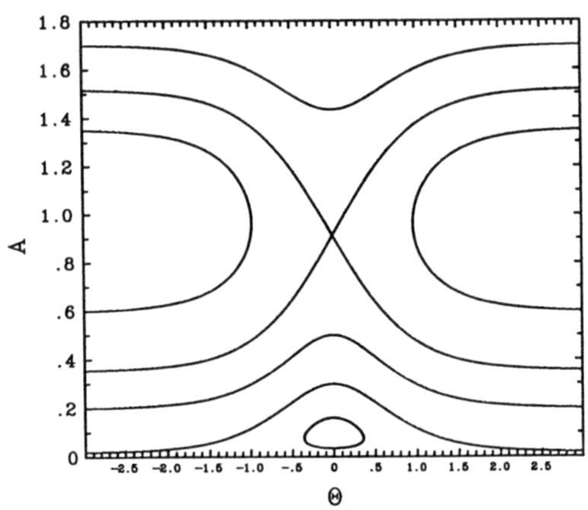

Figure 4: As for Figure 2, but for $\Delta = -0.5$ and $Q = -0.32$

(3.5), and on which $A \to 1$ as $\Theta \to \pm\infty$. In the passage régime, solitary waves with amplitudes $A > 1$ (< 1) at infinity increase (decrease) to a local maximum (minimum) as they pass over the external force.

If $\Delta < 0$ and $Q < 0$ these are three distinct cases. The case $-\frac{5}{9} < Q < 0$ is shown in Figure 4. There are two equilibrium states, a large-amplitude saddle point and a small-amplitude centre. This leads to three régimes, **passage, repulsion** and **trapping**. The passage régime has two parts, a large-amplitude part where the solitary wave amplitude locally decreases and a small-amplitude part where the solitary wave régime locally increases. The repulsion régime is separated from the passage régime by the separatrices through the saddle point, and has two parts. Small amplitude solitary waves in front of the force are repelled and propagate back to infinity with increased amplitude, while large amplitude solitary waves behind the force are repelled

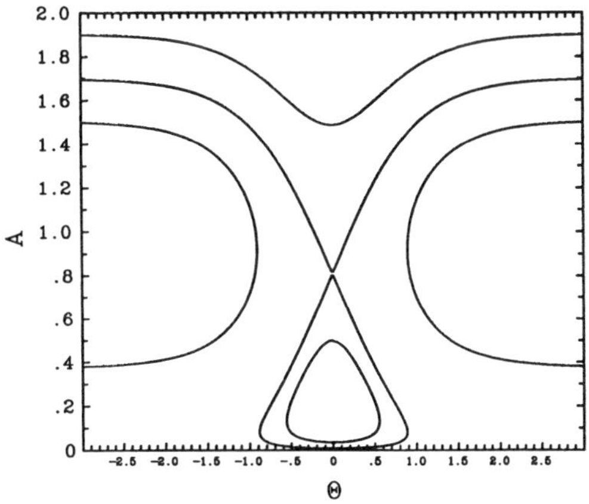

Figure 5: As for Figure 2, but for $\Delta = -0.36$ and $Q = -0.62$

and propagate back to infinity with a decreased amplitude. The trapping régime is separated from the small-amplitude passage régime by the "homoclinic" orbit in which $K = 0$ in (3.5) (since $A \to 0$ as $\Theta \to \pm\infty$). The case $-1 < Q < -\frac{5}{9}$ is shown in Figure 5. It is similar to Figure 4 except that the lower separatrix is now closed, and encloses the centre. For $Q < -1$ these are no equilibrium states, and just two régimes, passage and repulsion.

4 Analysis for multiple forcing

We again consider the case of broad forcing for which the perturbation equations become (3.1a, b), with Hamiltonian (3.2). Now, however, we suppose that the external force consists of two separated but otherwise identical isolated terms, and so in place of (3.3) we let

$$f(x) = b\{\text{sech}^2((x - L)/l) + \text{sech}^2(x + L)/l\}. \tag{4.1}$$

Using the non-dimensionalization (3.4), the integral (3.2) now becomes

$$\frac{3}{4}Q\{\text{sech}^2(\Theta - \beta) + \text{sech}^2(\Theta + \beta)\} = \pm A + \frac{3}{5}A^2 + \frac{K}{A^{\frac{1}{2}}}, \tag{4.2a}$$

where
$$\beta = \frac{L}{l}. \tag{4.2b}$$

Our main interest in this case is in the possibility of an interaction between the two external forces leading to a new mechanism for solitary wave trapping. Indeed, we shall show that a scenario can arise when each force, considered separately, acts as a repelling force, and hence a solitary wave lying between the two forces will be trapped in this region. In general, when the two external forces are widely separated ($\beta \gg 1$), they act independently of each other and the phase portraits near each force are similar to those described in section 3. On the other hand, when the two external forces are

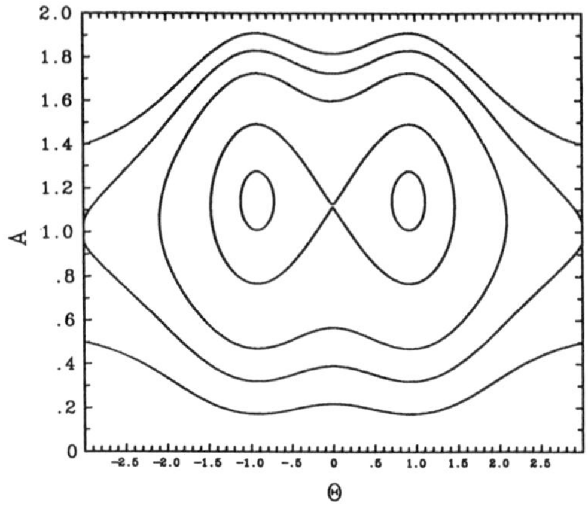

Figure 6: Phase portrait of the system (3.1a, b) for the multiple force (4.1) for $\Delta = -0.36$, $Q = 0.62$ and $\beta = 1$

close together ($\beta \ll 1$), they will act essentially as a single force, and again the phase portraits are similar to those of section 3. Hence, here we shall consider only the intermediate case when β scales with unity.

First we consider the equilibrium states of (3.1a, b) which are given by

$$f_\phi(\phi) = 0, \quad \text{and} \quad a(\Delta + 2a) - f(\phi) = 0. \tag{4.3}$$

The stability of any equilibrium state is determined by the eigenvalue λ where

$$\lambda^2 = 2f_{\phi\phi}(\phi)(4 + \frac{\Delta}{a}). \tag{4.4}$$

For $3\text{sech}2\beta > 2$ the only solution of (4.3) is $\Theta = 0$, and in this case the equilibrium states are given by

$$\Theta = 0, \qquad a(\Delta + 2a) - 2b\text{sech}^2\beta = 0, \tag{4.5}$$

while $f_{\phi\phi}(0) = -4bl^{-2}\text{sech}^2\beta(3\text{sech}^2\beta - 2)$. Comparison with (3.6a) and (3.7) shows that the equilibrium states, and hence the phase portraits, are essentially the same as for a single isolated external force. In particular Figure 1 again holds, where the equilibrium state is a centre for the upper branch ($b > 0$), a saddle point for the middle branch ($b < 0$, $\Delta < 0$), and a centre for the lower branch ($b < 0$, $\Delta < 0$).

For $3\,\text{sech}^2\beta < 2$, there are three solutions of (4.3) for Θ, $\Theta = 0$ and $\Theta = \pm\Theta_0$. For each of these the corresponding solutions for a are given by the second equation in (4.3), and the plot of the amplitude a as a function of Δ is similar to that shown in Figure 1. However, now the equilibrium state $\Theta = 0$, defined by (3.5) is such that it is a saddle point for the upper branch ($b > 0$), a centre for the middle branch ($b < 0$, $\Delta < 0$) and a saddle point for the lower branch ($b < 0$, $\Delta < 0$). In contrast the equilibrium states $\Theta = \pm\Theta_0$ are centres for the upper branch ($b > 0$), saddle points

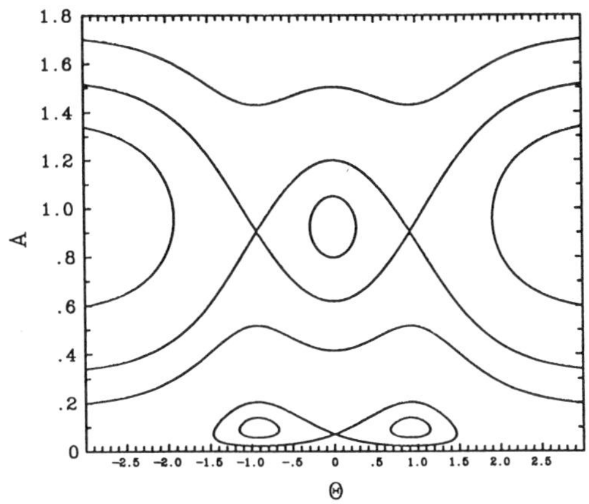

Figure 7: As for Figure 6, but for $\Delta = -0.5$, $Q = -0.32$ and $\beta = 1$

for the middle branch ($b < 0$, $\Delta < 0$), and centres for the lower branch ($b < 0$, $\Delta < 0$). Here Θ_0 (i.e. ϕ_0/l) is defined as the positive solution of

$$\text{sech}^2(\Theta_0 - \beta)\tanh(\Theta_0 - \beta) + \text{sech}^2(\Theta_0 + \beta)\tanh(\Theta_0 + \beta) = 0, \qquad (4.6)$$

and it may be shown that $0 < \Theta_0 < \beta$, where $\Theta_0 \to 0$ as $3\text{sech}^2\beta \to 2$ and $\Theta_0 \to \beta$ as $3\text{sech}^2\beta \to 0$.

Next we return to the integral (4.2) and plot some representative phase portraits. For brevity, we consider only the case when $\Delta < 0$, and in Figure 6 show a case when $Q > 0$. Here there are just three equilibrium states, a saddle point at $\Theta = 0$ and two centres at $\Theta = \pm\Theta_0$. There are now three **trapping** régimes, two associated with the centres at $\Theta = \pm\Theta_0$ and one in which the solitary wave passes around both these centres. They are separated by the homoclinic orbit through the saddle point at $\Theta = 0$. Outside these trapping regimes there are two **passage** régimes, separated by the "homoclinic" orbit on which $K = \frac{2}{5}$ in (4.2) and $A \to 1$ as $\Theta \to \pm\infty$ so that the passage régime has two parts; a large-amplitude part with $A > 1$ as $\Theta \to \pm\infty$ and a small-amplitude part with $A < 1$ as $\Theta \to \pm\infty$. Note that the parameters in Figure 6 correspond to those of Figure 3, and the phase portraits of the centres $\Theta = \pm\Theta_0$ correspond to those of the single centre at $\Theta = 0$ in Figure 3.

In Figure 7 we show a case when $0 > 2Q\text{sech}^2\beta > -\frac{5}{9}$ for which the phase portraits near $\Theta = \pm\Theta_0$ correspond to those shown in Figure 4. Thus at $\Theta = \pm\Theta_0$ there are two equilibrium states, a large-amplitude saddle point and a small-amplitude centre, while at $\Theta = 0$ there are also two equilibrium states, a large-amplitude centre and a smallamplitude saddle point. There are now three **trapping** régimes, a large-amplitude régime associated with the centre at $\Theta = 0$ which is formed by the repelling equilibrium states at $\Theta = \pm\Theta_0$ and bounded by the heteroclinic orbits connecting these saddle points, and two small-amplitude régimes associated with the centres at $\Theta = \pm\Theta_0$ and each bounded by the homoclinic orbit through the saddle point at $\Theta = 0$. There is a

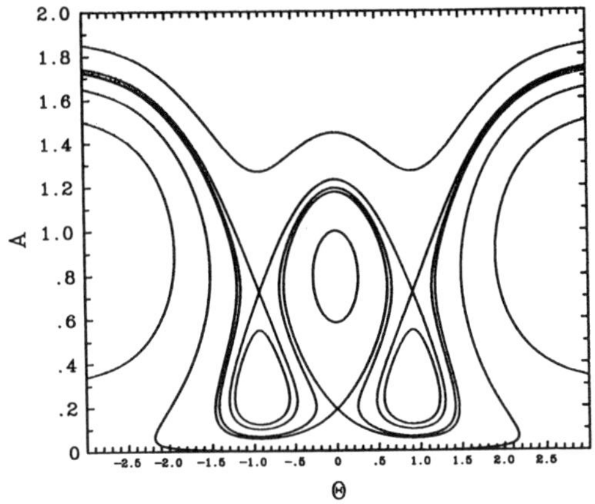

Figure 8: As for Figure 6, but for $\Delta = -0.36$, $Q = -0.74$ and $\beta = 1$

further trapping régime, a small-amplitude régime in which the solitary wave passes around both the centres at $\Theta = \pm\Theta_0$ This régime lies outside the homoclinic orbits through the saddle point at $\Theta = 0$, and below the "homoclinic to zero" orbit defined by $K = 0$ in (4.2a). Between the large- and small-amplitude trapping régimes, there is a small-amplitude **passage** régime, bounded below by the "homoclinic to zero" orbit and above by the heteroclinic orbits through the saddle points at $\Theta = \pm\Theta_0$. There is also a large-amplitude passage régime above these heteroclinic orbits, and two **repulsion** régimes on the outer sides of the saddle points at $\Theta = \pm\Theta_0$.

In Figure 8 we show a case when $-\frac{5}{9} > 2Q\text{sech}^2\beta > -1$ for which the phase portraits near $\Theta = \pm\Theta_0$ correspond to those shown in Figure 5. This case can be compared to Figure 7, and the essential difference is that now the homoclinic orbit through the small-amplitude saddle point at $\Theta = 0$ encloses the large-amplitude centre at $\Theta = 0$, and likewise the homoclinic orbits through the large-amplitude saddle points at $\Theta = \pm\Theta_0$ enclose the small-amplitude centres at $\Theta = \pm\Theta_0$. There are again three **trapping** régimes, a large-amplitude régime associated with the centre at $\Theta = 0$, and two small-amplitude régimes associated with the centres at $\Theta = \pm\Theta_0$. In addition there are various passage régimes, a small-amplitude régime bounded above by the separatrix orbits through the saddle point at $\Theta = 0$, a large-amplitude régime bounded below by the separatrix orbits through the saddle points at $\Theta = \pm\Theta_0$, and an intermediate passage régime in which the solitary wave reaches a small amplitude at $\Theta = \pm\Theta_0$ and a largeamplitude at $\Theta = 0$. Finally, there are two **repulsion** régimes on the outer sides of the saddle points at $\Theta = \pm\Theta_0$.

5 Summary

In this paper we have discussed the interaction of a Korteweg-de Vries solitary wave with an external force in the context of the forced Korteweg-de Vries equation (1.1).

Using a perturbation analysis for the case when the external force $f(x)$ has a small amplitude we have reduced the system to a pair of coupled ordinary differential equations (2.7) and (2.8) (where q_1 is given by (2.9)) for the solitary wave amplitude and position. These are solved for the case of broad forcing when the external force has a length scale much greater than the solitary wave. The case of an external force with a single forcing centre (i.e. $f(x)$ is given by (3.3)) has been discussed in detail by Grimshaw et. al (1994), and the results presented here are a representative summary of those. Here we have also extended that discussion to consider the case when the external force has two forcing centres (i.e. $f(x)$ is given by (4.1)).

Even though the model presented here is quite simple, it is evident that the interaction of a solitary wave with an external force can produce complicated dynamics. This is the case even for an external force which has a single forcing centre, and the complexity increases when the external force has two, or more, forcing centres. Not surprisingly, Grimshaw and Tian (1994a, b) have shown that the interaction of a solitary wave with a periodic external force can produce very complicated dynamics, and possibly spatial chaos. We also note there that although the present analysis has been confined to the perturbation equations, Grimshaw et. al (1994) have shown that direct numerical simulations of (1.1) have confirmed the validity of the perturbation analysis.

References

Camassa, R and Wu, T Y (1991) *Stability of forced steady solitary waves*, Phil. Trans. Roy. Soc. Lond A, **337**, 429-466

Grimshaw, R (1979) *Slowly varying solitary waves 1. Korteweg-de Vries equation, Proc. Roy. Soc. Lond. A,* **368,** 359-375

Grimshaw, R and Mitsudera, H (1993) *Slowly-varying solitary wave solutions of the perturbed Korteweg-de Vries equation revisited*, Stud. Appl. Math., **90**, 75-86

Grimshaw, R; Pelinovsky, E. and Tian, X. (1994) *Interaction of a solitary wave with an external force*, Physica D, **77**, 405-433

Grimshaw, R and Tian, X (1994a) *Periodic and chaotic behaviour in a reduction of the perturbed Korteweg-de Vries equation*, Proc. Roy. Soc., **445A**, 1-21

Grimshaw, R and Tian, X (1994b) *k-soliton structures, period-doubling and chaos in the periodically forced Korteweg-de Vries equation*, Differential Equations and Dynamical Systems (submitted)

Kaup, D J and Newell, A C (1978) *Solitons as particles, oscillators and in slowly changing media: a singular perturbation theory*, Proc. Roy. Soc. Lond. A, **361**, 412-446

Mitsudera, H and Grimshaw, R (1991) *Generation of mesoscale variability by resonant interaction between a baroclinic current and localized topography*, J. Phys. Ocean, **21**, 737-765

Mitsudera, H and Grimshaw, R (1993) *Effects of friction on a localized structure in a baroclinic current*, J. Phys. Ocean., **23**, 2265-2292

Mitsudera, H and Grimshaw, R (1994) *Capture and resonant forcing of solitary waves by the interaction of a baroclinic current with topography*, J. Phys. Ocean., **24**, 2217-2244

Patoine, A and Warn, T (1982) *The interaction of long, quasi-stationary baroclinic waves with topography*, J. Atmos. Sci. **39**, 1018-1025

Warn, T and Brasnett, B (1983) *The amplification and capture of atmospheric solitons by topography: a theory of the onset of regional blocking*, J. Atmos. Sci. **40**, 28-38

Wu, T Y (1987) *Generation of upstream advancing solitons by moving disturbances*, J. Fluid Mech., **184**, 75-99

Breaking the Dimension of a Steady Wave: Some Examples

M. HĂRĂGUŞ and K. KIRCHGÄSSNER
Mathematisches Institut A, Universität Stuttgart, Pfaffenwaldring 57, 70569 Stuttgart, Germany

Abstract

The appearance of crosswise modulation of steady one- or two-dimensional waves is studied analytically. The corresponding mathematical question, when a solution of a nonlinear elliptic equation living in $\mathbf{R}^1$ or $\mathbf{R}^2$ breaks its homogeneity in a new unbounded direction is answered for three cases, all related to water waves. First the Kadomtsev-Petviashvili (KP) equation is considered. The problem is treated from a local point of view near the quiescent state by looking for patterns which are periodic in one direction and noncompact in the transversal one. The second example is of Long-Yih-type, and it shows that the appearance of turning- or fold-points in a family of solutions may generate breaking of dimension in extended systems. In the third example we prove that for a one-parameter family of periodic solutions of the Ginzburg-Landau equation a breaking of dimension occurs between the neutral curve of stability and the Eckhaus boundary.

1 Introduction

In this paper we describe various analytical scenarios which can cause a two-dimensional steady wave to break its homogeneity in a direction transversal to its propagation. We do not consider external causes like the geometric shape of the flow region, oblique interactions of 2d waves, rotations and others. A recent article by Akylas [1] surveys some of these phenomena and may be consulted for these aspects of the theory. Our interest lies exclusively in the spontaneous "breaking of the dimension" of a steady wave.

Three examples are treated, all related to equations describing water wave dynamics, and in each we find another reason for the generation of the necessary resonance. The first one treats the case of travelling capillary-gravity waves. As a basis for the mathematical description we use the so called KP-equation, which model the case of long 2 dimensional waves, when the three dimensional effects are small. We choose surface tension to be large, i.e. Bond number b is greater than one third. We construct waves which are periodic in the main direction of propagation and possibly noncompact but bounded crosswise. Both effects are of equal order in their amplitudes, and therefore can be detected already at the level of linearization about the quiescent state. Previous work in this spirit can be found e.g. in [4] for the Taylor-Couette problem. Using $\mu = c/\omega^2$ as a parameter, where c is the wave speed and ω the frequency of the periodic part, one obtains an infinite sequence of dimension-breaking bifurcations with increasing complexity. The first one is discussed in some detail.

The second example shows how fold points, or turning points, of solution branches can generate breaking of dimension in an extended system. The equations used are of the Long-Yih type for waves in density stratified media. We treat the most elementary case, when just one eigenvalue of the linearization along the lower-dimensional solution branch changes sign. As the amplitude of the carrier wave is in general large compared to its crosswise modulation, this approach should be suitable to construct wave solutions of large amplitude as bifurcations from fold points.

The third example treats the Ginzburg-Landau equation in $\mathbf{R}^2$. First, a family of periodic solutions in one variable is constructed, parametrized by their frequencies. Then, bifurcations are sought from this family which break the dimension. It turns out that, between the "neutral" curve of stability and the Eckhaus boundary, there is a cascade of curves where modulation in crosswise direction occurs. The corresponding bifurcations show increasing complexity when approaching the Eckhaus stability boundary.

2 Local Breaking and the Role of Fold Points

The appearance of 3d waves can be studied from a purely local point of view near the quiescent state. One can adopt the method by Iooss and Mielke [4] in the study of 2π-time periodic solutions of the Taylor-Couette problem. Here we use this method to investigate dimension breaking of water waves using, instead of the full Euler equation, the so called KP modulation equation (Kadomtsev-Petviashvili). It describes the dynamics of long 3d waves when the weak three dimensional and the dispersive effects are of equal order, cf. [1]. Here we consider the case of capillary-gravity waves. The equation for the free surface $\eta(t,x,z)$ reads

$$\partial_x(\partial_t\eta + \frac{3}{2}\eta\partial_x\eta + \frac{1}{2}(\frac{1}{3} - b)\partial^3_{xxx}\eta) + \frac{1}{2}\partial^2_{zz}\eta = 0. \tag{2.1}$$

The propagation is in x-direction; z denotes the crosswise coordinate; b the Bond number measuring the effect of surface tension. The basic domain is $\mathbf{R}^2$. Therefore, we see no chance to determine the totality of bounded solutions via unfolding of the singularity at the bifurcation point. Instead we use a "dynamic" approach for the unbounded variable z and consider a priori periodic solutions in x. We look for travelling waves $\eta(t,x,z) = \eta(x+ct,z)$, and replace $x+ct$ by x again, to obtain

$$(c\eta + \frac{3}{4}\eta^2 - \beta\eta_{xx})_{xx} + \frac{1}{2}\eta_{zz} = 0, \tag{2.2}$$

where $\beta = \frac{1}{2}(b - \frac{1}{3})$. This equation changes its type from elliptic ($\beta > 0$) to hyperbolic ($\beta < 0$), a fact, which seems to be not realistic, since the corresponding Euler equation for the potential flow has only a singularity of codimension 2 at $b = \frac{1}{3}$, when the Froude number is 1, cf. [5], but stays elliptic otherwise. For this reason we restrict our considerations to the case $\beta > 0$ henceforth.

We look for periodic waves of frequency ω in x and determine their modulation in z by interpreting z as a timelike variable. We determine solutions of small amplitude, in order to be able to use center manifold reduction, cf [7]. Scale x to ωx and thus normalize the periodicity in x to 2π. Then we obtain the following system

$$\begin{aligned} \eta_z &= \eta_1 \\ \eta_{1z} &= 2\omega^4\beta\eta_{xxxx} - 2\omega^2 c\eta_{xx} - \tfrac{3}{2}\omega^2 c(\eta^2)_{xx} \end{aligned} \tag{2.3}$$

or shortly for $\underline{\eta} = (\eta, \eta_1)$

$$\partial_z\underline{\eta} = A(\omega^2, c, \partial_x)\underline{\eta} + \omega^2\underline{f}(\partial_x, \eta). \tag{2.3'}$$

This system possesses translational and the reflectional symmetries $z \to z+z_0$, $z \to -z$ and also $x \to -x$. It is reversible in the sense that A and $\underline{f}$ anticommute with the matrix

$$R = \begin{pmatrix} 1 & 0 \\ 0 & -1 \end{pmatrix}.$$

We shall call a solution $\underline{\eta}$ reversible if $\underline{\eta}(\cdot, -z) = R\underline{\eta}(\cdot, z)$ holds for $z \in \mathbf{R}$. From a qualitative point of view these symmetries and the central part of the spectrum of A determine the totality of small solutions.

We discuss (2.3) for reversible solutions under the additional constraint $[\eta] = 0$, where $[\cdot]$ denotes the mean value of η. Hence η is even and its mean value vanishes. The eigenvalues of A: $\sigma = iq$, which have zero real part follow from

$$\beta\eta^{(4)} - \mu\eta^{(2)} + q^2\eta = 0, \quad [\eta] = 0,$$

where $\mu = c/\omega^2$, $q \in \mathbf{R}$. Set $\eta = \exp(ik)$, $k \in \mathbf{N}_0$, it follows $k \neq 0$. Moreover, setting $\mu_k = -\beta k^2$, $k \in \mathbf{N}$, then, if $\mu_{k+1} < \mu < \mu_k$, the spectrum ΣA has k conjugate complex pairs of purely imaginary, simple eigenvalues. The rest of the spectrum is real, consists of simple eigenvalues which appear in pairs $(\sigma, -\sigma)$ due to reversibility. The parameter values $\mu_k = -\beta k^2$ form the set of bifurcation points. There the spectrum of A has an additional eigenvalue 0 which has multiplicity 2 and is not semisimple.

A complete analysis of the possible modulation in y-direction of a wave would require to solve the $2k$-dimensional system on the center manifold. For $k = 2$ the reader can consult [5] for this task. For $k = 1$ one can easily calculate the lowest order terms. Setting

$$\eta(x, z) = \sum_{k=1}^{\infty} c_k(\omega\sqrt{2}z)\cos kx,$$

and $\gamma = 3/4\omega^2$, $\mu = -\beta + \delta$, $\zeta = z\omega\sqrt{2}$, one obtains in lowest order

$$\begin{aligned} c_1'' &= \delta c_1 - \tfrac{\gamma}{2}c_1c_2 \\ c_2'' &= (12 + 4\delta)c_2 - \tfrac{\gamma}{2}c_1^2. \end{aligned}$$

Using the fact that $c_2 = h(c_1, c_1', \delta)$, where $h = \mathcal{O}(\delta c_1 + \delta c_1' + c_1^2 + {c_1'}^2)$, and h is even in c_1' due to reversibility, one obtains

$$\begin{aligned} c_1'' &= \delta c_1 - \tfrac{\gamma}{2} c_1 \left(\tfrac{\gamma}{24} c_1^2 + \tfrac{\gamma}{144} {c_1'}^2\right) \\ c_2 &= \tfrac{\gamma}{24} c_1^2 + \tfrac{\gamma}{144} {c_1'}^2 . \end{aligned} \tag{2.5}$$

Now, conclude that, up to order δ inclusively, η reads

$$\eta(x, z) = c_1(\omega\sqrt{2}z)\cos x + c_2(\omega\sqrt{2}z)\cos 2x,$$

where c_1, c_2 are given by (2.5). For $\delta > 0$, c_1 consists of two solutions homoclinic to 0; for $\delta < 0$ all solutions are periodic (see Figures 1 and 2).

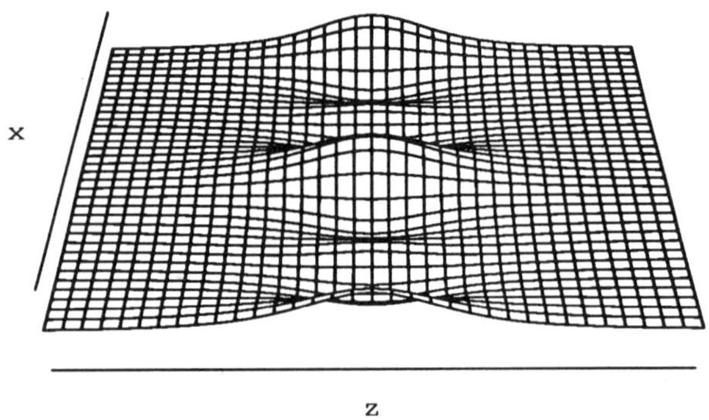

Figure 1: $\delta > 0$

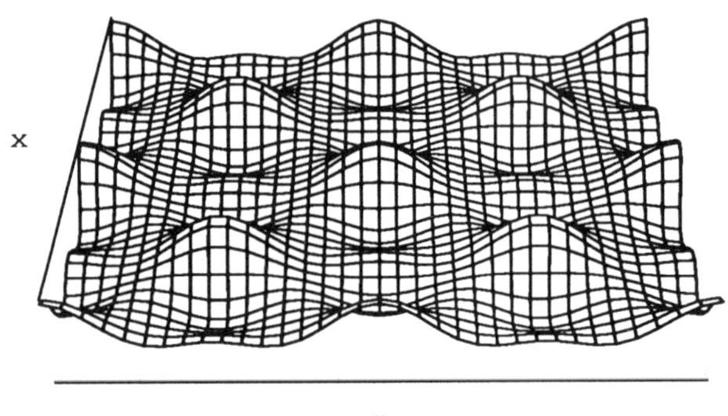

Figure 2: $\delta < 0$

While the previous method solves the local problem of breaking the homogeneity in crosswise direction, the question, whether there are global indicators of a spontaneous breaking is answered for a simple case now. We show that the appearance of turning-

or fold-points in a family of solutions, may generate modulations of the solutions in the crosswise direction. Consider e.g. the following boundary-value problem in $(\underline{x}, y) \in \Omega \times \mathbf{R}$

$$\begin{aligned} u_{yy} + \Delta u + g(u, \underline{x}, \lambda) &= 0 \\ u(\underline{x}, y) &= 0 \quad , \quad (\underline{x}, y) \in \partial\Omega \times \mathbf{R} \end{aligned} \tag{2.6}$$

where $\Omega \subset \mathbf{R}^n$ is a bounded smooth domain, Δ the Laplacean in $\underline{x}$; g is a smooth function of its arguments, λ a real parameter. Observe that g is not allowed to depend on y. We could equally well replace u_{yy} by $u_{yyyy} - u_{yy}$ as long as the problem stays elliptic. We state the problem in an abstract form.

H1: Given a real Banach space $(X, \|\cdot\|)$ and a closed linear operator $-L : D(L) \to X$, which is sectorial, $D(L)$ dense in X. Let $(u, \lambda) \in X^\gamma \times I$:

$$g : X^\gamma \times I \to X$$

be C^2 for some $\gamma \in [0, 1]$ and some nonempty open interval I. If $\gamma = 1$ (the quasilinear case) we assume that X is a Hilbert space; for an explanation cf. [7].

H2: Consider the equation

$$Lu + g(u, \lambda) = 0, \tag{2.7}$$

and assume, there exists a C^2-curve $(\lambda^*(p), u^*(p)) \in J \times X^\gamma$, $J = (-p_0, p_0) \subset I$, such that $u^*(p)$ solves (2.7) for $\lambda = \lambda^*(p)$ and every $p \in J$. Moreover, for all $p \in J$

$$\| \frac{du^*}{dp}(p) \|_{X^1} + | \frac{d\lambda^*}{dp}(p) | \neq 0, \quad \frac{d\lambda^*}{dp}(0) = 0, \quad \frac{d^2\lambda^*}{dp^2}(0) \neq 0,$$

holds in X^1. Without loss of generality we may assume

$$\lambda^*(-p) = \lambda^*(p) \text{ and } \frac{d^2\lambda^*}{dp^2}(0) > 0. \tag{2.8}$$

This means the fold is open to the right. Define the family of linear operators

$$T(p) = L + D_u g(u^*(p), \lambda^*(p)),$$

then $T(0)$ has a zero eigenvalue with eigenfunction $dv^*/dp(0)$. For the spectrum $\Sigma T(p)$ we suppose

H3: $\Sigma T(p) = \{\sigma_0(p)\} \cup \Sigma_-$, where

$$\Sigma_- \subset \{\sigma \in \mathbf{C} / \text{Re}\sigma < -\beta\},$$

for some positive β. $\sigma_0(p)$ is a simple eigenvalue with $\sigma_0(0) = 0$, $\sigma_0'(0) \neq 0$.

In the following we shall take, pars pro toto, $\sigma_0'(0) < 0$. It follows that $(\lambda^*(p), u^*(p))$ defines a fold, open to the right in the (λ, u)-space. Its lowest branch has exactly one positive eigenvalue; it is simple. On the upper branch $T(p)$ has only eigenvalues with strictly negative real parts. Consider the extended problem

$$\frac{d^2}{dy^2}u + Lu + g(u,\lambda) = 0. \tag{2.9}$$

We construct solutions $u \in Y = C_b^0(\mathbf{R}, X^1) \cap C_b^2(\mathbf{R}, X) \cap \{u \text{ even in } y\}$. In analogy to the ODE-case we call a solution (λ, u) of (2.9) homoclinic to a solution (λ^*, u^*) of (2.7), if $\lim_{|y|\to\infty} u(y) = u^*$ holds in X^1. On the other hand (λ^*, u^*) is called a center (relative to (2.9)), if there exists a continuous family $(\lambda, u(q))$, $q \in (0, q_0)$, of (2.9), such that $u(q)$ is nonconstant and periodic in y and $\lim_{q\to 0+} u(q) = u^*$ holds in Y.

Theorem *Let the assumptions H1-H3 be satisfied. Moreover, assume $\sigma_0'(0) < 0$. Then the following conclusions can be drawn for sufficiently small value of p:*
(i) to $(\lambda^*(p), u^*(p))$ there exists a unique C^1-family of solutions $(\lambda^*(p), U^*(p))$ of (2.9) homoclinic to $(\lambda^*(p), u^*(p))$. The convergence as $|x| \to \infty$ of $U^*(p)$ to $u^*(p)$ is exponential in X^1.
(ii) $(\lambda^*(p), u^*(-p))$ is a center relative to (2.9).

The proof of this theorem is contained in a paper by R. Schaaf and one of the authors [6]. We refer the reader to that manuscript. The theorem gives a method to construct solutions of (2.9) semiglobally. However, the situation is usually more complicated than stated in H1-H3. One has to generalize it to cover cases where more than one eigenvalue of $T(p)$ has a positive real part. Then, the method can be used to construct waves via the behaviour of its crossection even for large amplitudes. The difficulty is usually the verification of the spectral assumptions.

As an example consider

$$\begin{aligned} \Delta u + \lambda e^u &= 0 \quad \text{in } B_1(0) \subset \mathbf{R}^n \\ u &= 0 \quad \text{on } \partial\Omega \end{aligned} \tag{2.10a}$$

and

$$\begin{aligned} u_{yy} + \Delta u + \lambda e^u &= 0 \\ u &= 0 \qquad \text{for } (y, \underline{x}) \in \mathbf{R} \times \partial\Omega. \end{aligned} \tag{2.10b}$$

If $n = 1, 2$ resp. $3 \le n \le 9$ the solution set of (2.10a) is given by Figure 3 resp. Figure 4, cf. [3].

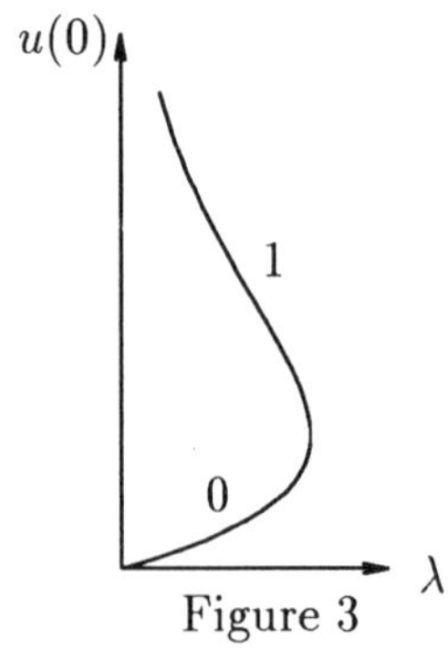

Figure 3

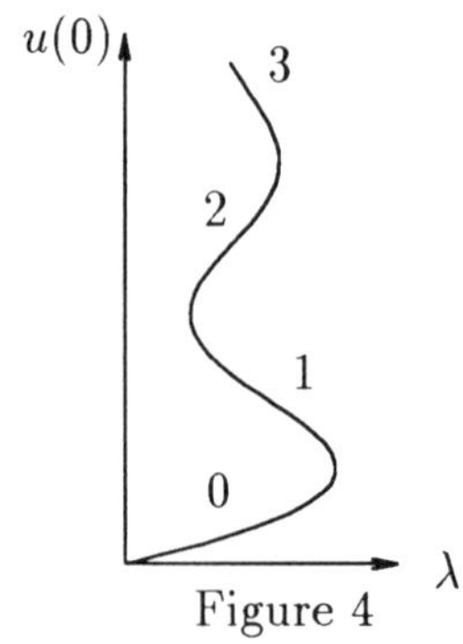

Figure 4

The numbers on the branches indicate the number of positive eigenvalues of $T(p)$. Thus, Figure 3 gives a conclusion similar to the one of the preceding theorem, Figure 4 should have conclusions similar to those for the KP equation.

3 Dimension Breaking of a Family of Periodic Solutions

In the last section it was shown that dimension breaking can be the result of the appearance of a turning- or fold-point in a family of solutions. Here we show that the existence of a one-parameter family of periodic solutions may also generate, via resonance, modulations of the solutions in the crosswise direction. We consider the two-dimensional Ginzburg-Landau equation

$$u_{xx} + u_{yy} + \lambda u - \mid u \mid^2 u = 0, \tag{3.1}$$

in which x and y are both unbounded directions, u is a complex-valued function and λ a real parameter. In the KP case we started with the quiescent state and kept c/ω^2 as the parameter. Here we construct first, for each $\lambda > 0$, a family of 1d periodic solutions, parametrized by their frequencies. Then, we look for bifurcations of 2d solutions, from these periodic solutions, by taking their frequency as new parameter.

We suppose the 1-dimensional solutions to be y-independent so they satisfy the following equation

$$u_{xx} + \lambda u - \mid u \mid^2 u = 0. \tag{3.2}$$

If $\lambda < 0$ this equation has only one fixed point, $u_0 = 0$, which is a saddle. Therefore no bounded solutions exist in that case. If $\lambda > 0$ in addition to $u_0 = 0$, we obtain a circle of fixed points $u_\lambda = \sqrt{\lambda} e^{i\omega}$. Then (3.2) has, for each $\lambda > 0$, a family of periodic solutions $u^*(x) = \alpha e^{i\omega x}$, where $\alpha \in \mathbb{C}$ and $\omega \in \mathbb{R}$ are such that $\mid \alpha \mid^2 + \omega^2 = \lambda$.

Now, look for solutions of (3.1) which are 2-dimensional perturbations of these periodic solutions, i.e. determine solutions of the form

$$u(x, y) = u^*(x) + v(x, y).$$

Since u^* and v are complex-valued functions, we can write $u^* = u_1^* + i u_2^*$ and $v = v_1 + i v_2$. Then (3.1) yields

$$\begin{aligned} v_{1yy} + v_{1xx} + \omega^2 v_1 - 2u_1^{*2} v_1 - 2u_1^* u_2^* v_2 - 3u_1^* v_1^2 - 2u_2^* v_1 v_2 & \\ -u_1^* v_2^2 - v_1(v_1^2 + v_2^2) = 0 & \\ v_{2yy} + v_{2xx} + \omega^2 v_2 - 2u_2^{*2} v_2 - 2u_1^* u_2^* v_1 - 3u_2^* v_2^2 - 2u_1^* v_1 v_2 & \\ -u_2^* v_1^2 - v_2(v_1^2 + v_2^2) = 0. & \end{aligned} \tag{3.3}$$

In order to find a "dynamical" formulation for this system we scale x to ωx which makes the coefficients to be all 2π-periodic functions in x. Then the solution u^* will be $U^*(x) = \alpha e^{ix}$, where still $\mid \alpha \mid^2 + \omega^2 = \lambda$ holds and (3.3) becomes

$$\begin{aligned} v_{1yy} + \omega^2 v_{1xx} + \omega^2 v_1 - 2U_1^{*2} v_1 - 2U_1^* U_2^* v_2 - 3U_1^* v_1^2 - 2U_2^* v_1 v_2 & \\ -U_1^* v_2^2 - v_1(v_1^2 + v_2^2) = 0 & \\ v_{2yy} + \omega^2 v_{2xx} + \omega^2 v_2 - 2U_2^{*2} v_2 - 2U_1^* U_2^* v_1 - 3U_2^* v_2^2 - 2U_1^* v_1 v_2 & \\ -U_2^* v_1^2 - v_2(v_1^2 + v_2^2) = 0. & \end{aligned} \tag{3.4}$$

We see that in this system the parameter $\lambda > 0$ does not play any role; no bifurcation in λ occurs. Moreover, we can scale with $\sqrt{\lambda}$. Then (3.4) and U^* remain unchanged; the only difference is that $|\alpha|^2 + \omega^2 = 1$ should hold. From now on, the parameter will be ω.

We write (3.4) as

$$\underline{V}_y = A(\omega)\underline{V} + F(\underline{V}), \tag{3.5}$$

where

$$\underline{V} = \begin{pmatrix} v_1 \\ v_2 \\ w_1 \\ w_2 \end{pmatrix}, \quad A(\omega)\underline{V} = \begin{pmatrix} w_1 \\ w_2 \\ -\omega^2 v_{1xx} - \omega^2 v_1 + 2U_1^{*2} v_1 + 2U_1^* U_2^* v_2 \\ -\omega^2 v_{2xx} - \omega^2 v_2 + 2U_2^{*2} v_2 + 2U_1^* U_2^* v_1 \end{pmatrix},$$

$$F(\underline{V}) = \begin{pmatrix} 0 \\ 0 \\ 3U_1^* v_1^2 + 2U_2^* v_1 v_2 + U_1^* v_2^2 + v_1(v_1^2 + v_2^2) \\ 3U_2^* v_2^2 + 2U_1^* v_1 v_2 + U_2^* v_1^2 + v_2(v_1^2 + v_2^2) \end{pmatrix}.$$

This system is reversible i.e. A and F anticommute with the symmetry

$$R = \begin{pmatrix} 1 & 0 & 0 & 0 \\ 0 & 1 & 0 & 0 \\ 0 & 0 & -1 & 0 \\ 0 & 0 & 0 & -1 \end{pmatrix}.$$

We apply the center manifold reduction, cf. [7].

In order to simplify the analysis, we restrict ourselves to the case $\alpha > 0$, i.e. $U_1^*(x) = \sqrt{1-\omega^2}\cos x$, $U_2^*(x) = \sqrt{1-\omega^2}\sin x$. We assume $\underline{V}$ to be reversible, i.e. v_1, w_1 are even in x, and v_2, w_2 are odd functions in x. Since the coefficients in (3.5) are 2π-periodic functions in x, the solutions $\underline{V}$ will be considered as $2k\pi$-periodic functions, for some $k \in \mathbf{N}_0$. We remark that k should be fixed.

Choose as the function space $X^{(k)} = H^{1,e}_{per} \times H^{1,o}_{per} \times L^{2,e}_{per} \times L^{2,o}_{per}$, where

$$\begin{aligned} H^{1,e}_{per} &= \{f \in H^1_{loc}(\mathbf{R}) \mid 2k\pi - \text{periodic},\ f(x) = f(-x)\} \\ H^{1,o}_{per} &= \{f \in H^1_{loc}(\mathbf{R}) \mid 2k\pi - \text{periodic},\ f(x) = -f(-x)\} \end{aligned}$$

and $L^{2,e}_{per}$ resp. $L^{2,o}_{per}$ are the closure of $H^{1,e}_{per}$ resp. $H^{1,o}_{per}$ in L^2_{loc}. Then $A(\omega)$ is a closed linear operator in $X^{(k)}$ with domain $Y^{(k)} = H^{2,e}_{per} \times H^{2,o}_{per} \times H^{1,e}_{per} \times H^{1,o}_{per}$, the spaces $H^{2,e}_{per}$ and $H^{2,o}_{per}$ being defined like $H^{1,e}_{per}$ and $H^{1,o}_{per}$. Moreover, $A(\omega)$ has compact resolvent; so its spectrum consists only of isolated eigenvalues, which we denote by σ.

The eigenvalue problem yields

$$\begin{aligned} -\omega^2 v_{1xx} - \omega^2 v_1 + 2U_1^{*2} v_1 + 2U_1^* U_2^* v_2 &= \sigma^2 v_1 \\ -\omega^2 v_{2xx} - \omega^2 v_2 + 2U_2^{*2} v_2 + 2U_1^* U_2^* v_1 &= \sigma^2 v_2. \end{aligned} \tag{3.6}$$

To solve (3.6) we write it as a first order system

$$\underline{U}_x = B(\omega, x)\underline{U}, \tag{3.7}$$

where $\underline{U} = (v_1, v_2, v_{1,x}, v_{2,x})$ and then use the Floquet ansatz

$$\underline{U} = Q^*\underline{W}, \quad Q^* = \begin{pmatrix} \cos x & -\sin x & 0 & 0 \\ \sin x & \cos x & 0 & 0 \\ 0 & 0 & \cos x & -\sin x \\ 0 & 0 & \sin x & \cos x \end{pmatrix}.$$

Now, $\underline{U}$ is a $2k\pi$-periodic solution of (3.7) if $\underline{W}$ is a $2k\pi$-periodic solution of

$$\underline{W}_x = C(\omega)\underline{W}, \tag{3.8}$$

where $C(\omega)$ is the constant matrix

$$C(\omega) = \begin{pmatrix} 0 & 1 & 1 & 0 \\ -1 & 0 & 0 & 1 \\ -3 + \frac{2-\sigma^2}{\omega^2} & 0 & 0 & 1 \\ 0 & -1 - \frac{\sigma^2}{\omega^2} & -1 & 0 \end{pmatrix}.$$

We conclude that σ is an eigenvalue for $A(\omega)$, in $X^{(k)}$, if (3.8) has $2k\pi$-periodic solutions $\underline{W} = (w_1, w_2, w_3, w_4)$ such that w_1, w_3 are even functions in x, and w_2, w_4 are odd functions in x. After some elementary calculations we obtain that $\sigma \neq 0$ is an eigenvalue for $A(\omega)$ if and only if

$$\frac{\sigma^4}{\omega^4} + \frac{2\sigma^2}{\omega^2}(\lambda^2 + 1 - \frac{1}{\omega^2}) + \lambda^4 + 2\lambda^2(3 - \frac{1}{\omega^2}) = 0, \tag{3.9}$$

for $\lambda^2 \in \{-\frac{n^2}{k^2}, n = 0, 1, 2, ...\}$, and that $\sigma = 0$ is an eigenvalue for $A(\omega)$ if and only if

$$\omega^2 = \omega_n^2(k) := \frac{2k^2}{6k^2 - n^2}, \quad n = 1, 2, ..., 2k.$$

Then, if $0 < \omega < \omega_1(k)$, the spectrum $\Sigma A(\omega)$ consists of nonzero simple real eigenvalues while, if $\omega_n(k) < \omega < \omega_{n+1}(k) \leq 1$, it has n conjugate complex pairs of simple eigenvalues on the imaginary axis, the rest of the spectrum is real. The values $\omega = \omega_n(k)$ form the set of bifurcation points. For these values, $\Sigma A(\omega)$ has in addition an eigenvalue 0 of multiplicity 2 which is not semisimple. For fixed k, the first bifurcation occurs for the smallest value of $\omega_n(k)$, i.e. for $\omega_1(k)$. We find a sequence $\omega_1(1) = \sqrt{2/5}$, $\omega_1(2) = \sqrt{8/23}$,..., $\omega_1(k)$,... of critical values of ω corresponding to bifurcations of 2d solutions, which are $2k\pi$-periodic in x. Since this sequence is decreasing in ω it follows that the first bifurcations occur for large k, i.e. for large periods. Moreover $\omega_1(k) \to 1/\sqrt{3}$, as $k \to \infty$, so that for $\omega < 1/\sqrt{3}$, no bifurcations are possible.

Whenever $\omega > 1/\sqrt{3}$ bifurcating 2d solutions, which are $2k\pi$-periodic, may appear, for $k > k_0$, k_0 depending on the value of ω. The bound $1/\sqrt{3}$, where modulations in the y-direction become possible, is exactly the bound where the Eckhaus instability for 1d solutions occurs, cf. [2]. We can visualize the bifurcation as diagram in Figure 5.

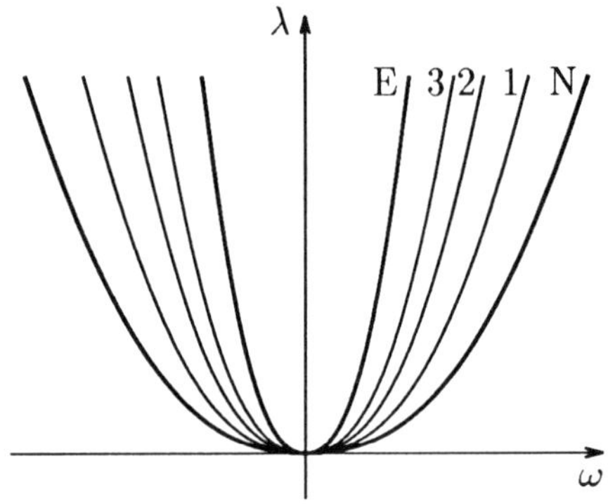

Figure 5: The cascade of bifurcating crosswise modulated waves for $k \in \mathbb{N}$, between the neutral curve of stability N and the Eckhaus boundary E

In order to get more information about the bifurcating solutions, we compute the reduced system in lowest order. We suppose k fixed and ω close to $\omega_1(k)$. Since the operator $A(\omega_1(k))$ has only one eigenvalue on the imaginary axis, namely 0 being of multiplicity 2, the reduced system is 2-dimensional. Let $\underline{e}_0$ resp. $\underline{e}_1$ be the eigenfunction resp. generalized eigenfunction to the eigenvalue 0 and set

$$\mu = \omega^2 - \omega_1(k)^2.$$

Then all bounded small solutions of (3.5) are of the form

$$\underline{V} = a_0\underline{e}_0 + a_1\underline{e}_1 + h(a_0, a_1, \mu),$$

where $h = \mathcal{O}(\mu a_0 + \mu a_1 + a_0^2 + a_1^2)$. One obtains the reduced system in lowest order

$$\begin{aligned} a_{0,y} &= a_1 \\ a_{1,y} &= -\tfrac{6k^2-1}{k^2(1+4k^2)}\mu a_0 - 3a_0^3 + \alpha a_0 a_1^2 \end{aligned} \tag{3.10}$$

where α is some constant which is not needed explicitly. Due to reversibility, one obtains the validity of the following phase portrait for the full reduced system.

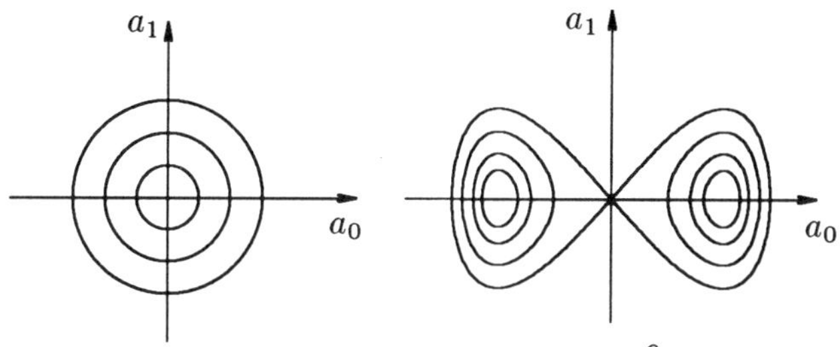

One concludes that for $\omega < \omega_1(k)$ the modulations in the y-direction are homoclinic; for $\omega > \omega_1(k)$ they are periodic. In both cases the solutions are $2k\pi$-periodic in x. The solutions $v_1 = \text{Re}v$ of (3.3) look as in Figure 1 if $\omega < \omega_1(k)$ and like in Figure 2 if $\omega > \omega_1(k)$. Similar pictures can be drawn for $v_2 = \text{Im}v$.

References

[1] Akylas, T.R. (1994) Three-dimensional long water-wave phenomena, *Annu. Rev. Fluid Mech.* **20**, 191.

[2] Eckhaus, W. (1965) Studies in Non-Linear Stability Theory, *Springer Tracts in Nat. Phil.* **6**, Springer-Verlag.

[3] Joseph, D.D. and Lundgren, T.S. (1973) Quasilinear Dirichlet Problems Driven by Positive Sources, *Arch. Rat. Mech. Anal.* **46**, 241.

[4] Iooss, G. and Mielke, A. (1991) Bifurcating time-periodic solutions of Navier-Stokes equations in infinite cylinders, *J. Nonlinear Sci.* **1**, 107.

[5] Iooss, G. and Kirchgässner, K. (1992) Water waves for small surface tension: an approach via normal form, *Proc. R. Soc. Edinburgh* **122** A, 267.

[6] Kirchgässner, K. and Schaaf, R. (1994) Homoclinic bifurcation from fold points for nonlinear elliptic problems on cylindrical domains, *manuscript.*

[7] Mielke, A. (1988) Reduction of quasilinear elliptic equations in cylindrical domains with applications, *Math. Meth. Appl. Sci.* **10**, 51.

Research supported by the Deutsche Forschungsgemeinschaft under Ki 131/12-1.

Remarks on the Use and Misuse of the Ginzburg-Landau Equation

E. KNOBLOCH

Department of Physics, University of California, Berkeley CA 94720, U.S.A.

Abstract

The use of envelope equations to describe travelling waves in both finite and infinite systems is reviewed, with emphasis on the differences between asymptotic and nonasymptotic descriptions.

Envelope equations are a well-known and powerful tool in the study of patterns. As such they have seen applications in a great variety of physical systems. In fact such equations often provide the only alternative to numerical simulations. Typically these equations are used in one of three ways, as model systems believed to capture some essence of the phenomenon under study, as amplitude equations describing the growth and equilibration of a particular pattern-forming instability, and as an asymptotically exact description of such an instability. Of these different usages, which are infrequently distinguished, the first is by far the most common. Here the equations are used to some extent as a "crutch" to help understand the problem at hand. The information that is sought is qualitative, an essential first step in describing phenomena where perhaps even the governing field equations are poorly known. An excellent example of this use of envelope equations is the work of Coullet et al (1989) on the dynamics of defects arising from a parity-breaking instability of a spatially periodic stationary pattern, and the work of Riecke (1992) explaining the slow drift of localized travelling wave pulses in binary fluid convection. In both cases the understanding gained is qualitative; the basic equations are not derived in any strict sense of the word, but are constructed to contain the essential physical picture, and be consistent with the prescribed physical symmetries. The second usage is restricted to cases in which the envelope equation can be justified by some type of formal, usually multiscale, expansion. The well-known Ginzburg-Landau equation (both real and complex) provides an example. This equation describes the stability of patterns arising from steady state or oscillatory instabilities in translation-invariant systems, and will form the focus of this article. In many applications of these equations no attempt is made to calculate the coefficients from the field equations, so that once again a quantitative understanding of the pattern-forming instability is impossible. However, one knows at least that the equations used are correct in some asymptotic (usually small amplitude) limit. The third usage, in which the coefficients of the equations are computed in terms of the physical parameters of the system is relatively rare, but only this usage can provide a quantitative description of the instability. Two-dimensional oscillatory binary fluid convection is one of the few systems for which the necessary computations have actually been done (Cross and Kim 1988; Schöpf and Zimmermann 1989; Clune and Knobloch 1992).

In this article we focus on the dynamics of travelling and standing waves in systems on a line and emphasize generalizations of the Eckhaus (1965) sideband instability condition to these systems. This is perhaps the simplest problem illustrating some of the difficulties associated with the various uses of envelope equations: used as a model the equations may appear plausible but in fact be incorrect, either in predicting different stability properties or having too much symmetry. We also discuss problems associated with applying these techniques to systems in finite but large domains. Throughout the theme will be that one has to pay careful attention to the order in which various limits are (implicitly) taken.

1 Derivation of the Ginzburg-Landau equation for travelling and standing waves

In this section we consider a translation-invariant system with a trivial translation-invariant equilibrium undergoing an oscillatory instability. Since we shall typically be interested in systems with reflection symmetry admitting standing waves, we begin by discussing the translation- and reflection-invariant problem.

We start from a system of field equations written in the symbolic form

$$L\Big(\frac{\partial}{\partial t}, \frac{\partial}{\partial x}; \epsilon^2\Big)\psi = N[\psi], \tag{1}$$

where L is the linear operator, N denotes nonlinear terms and ϵ is a small parameter denoting the distance from onset of the instability. If ψ is a scalar then L must be of at least second order in time; if ψ is a vector L can be taken to be of first order. In the following we suppose (for simplicity only) that ψ is a scalar. As usual we introduce the slow variables

$$X = \epsilon x, \quad T = \epsilon t, \quad T' = \epsilon^2 t, \tag{2}$$

and expand

$$\psi(x,t) = \epsilon\psi_1(x,t;X,T,T') + \epsilon^2\psi_2(x,t;X,T,T') + \dots \tag{3}$$

Because of the reflection symmetry the solution of the linear problem $L(\partial_t, \partial_x; 0)\psi_1 = 0$ is assumed to be of the form

$$\psi_1 = \ \mathrm{Re}\,\{A_1(X,T,T')e^{i(kx+\omega t)} + B_1(X,T,T')e^{i(kx-\omega t)}\}, \tag{4}$$

where A_1, B_1 describe the slowly varying amplitudes of left- and right-travelling waves, respectively. At leading order one obtains the wavenumber k_c of the instability, and from the dispersion relation $\omega = \omega(k)$ one finds the corresponding oscillation frequency ω_c. The dispersion relation defines the group velocity c_g of wave packets. The envelope equations which result depend crucially on whether (at onset) c_g is $\mathcal{O}(1)$ (the generic case) or $\mathcal{O}(\epsilon)$. In the former case one obtains at second order the following solvability conditions:

$$\frac{\partial A_1}{\partial T} - c_g\frac{\partial A_1}{\partial X} = 0, \quad \frac{\partial B_1}{\partial T} + c_g\frac{\partial B_1}{\partial X} = 0, \tag{5}$$

showing that $A_1 = A_1(\eta, T')$, $B_1 = B_1(\xi, T')$, where $\eta = X + c_g T$, $\xi = X - c_g T$. At next order one obtains solvability conditions of the form

$$\begin{aligned}
\frac{\partial A_2}{\partial T} - c_g \frac{\partial A_2}{\partial X} &= -\frac{\partial A_1}{\partial T'} + \lambda A_1 + \alpha |A_1|^2 A_1 + \beta |B_1|^2 A_1 + \gamma A_{1XX} && \text{(6a)} \\
\frac{\partial B_2}{\partial T} + c_g \frac{\partial B_2}{\partial X} &= -\frac{\partial B_1}{\partial T'} + \bar{\lambda} B_1 + \bar{\alpha} |B_1|^2 B_1 + \bar{\beta} |A_1|^2 B_1 + \bar{\gamma} B_{1XX}. && \text{(6b)}
\end{aligned}$$

The resulting expansion will be asymptotic for $(x, t) = \mathcal{O}(\epsilon^{-2})$ if and only if A_2, B_2 remain of order one on these scales. Consequently the right hand sides of equations (6a,b) must vanish when averaged over the corresponding comoving variables:

$$\begin{aligned}
\frac{\partial A_1}{\partial T'} &= \lambda A_1 + \alpha |A_1|^2 A_1 + \beta \langle |B_1|^2 \rangle A_1 + \gamma A_{1\eta\eta} && \text{(7a)} \\
\frac{\partial B_1}{\partial T'} &= \bar{\lambda} B_1 + \bar{\alpha} |B_1|^2 B_1 + \bar{\beta} \langle |A_1|^2 \rangle B_1 + \bar{\gamma} B_{1\xi\xi}. && \text{(7b)}
\end{aligned}$$

Here $\lambda_r(\equiv \operatorname{Re}\lambda)$ is to be identified with the bifurcation parameter, $\lambda_i(\equiv \operatorname{Im}\lambda)$ denotes the correction to the onset Hopf frequency, and α, β and γ are complex coefficients. The resulting equations will be referred to as the mean-field Ginzburg-Landau (hereafter MFGL) equations and are the required asymptotically exact envelope equations for this problem. They were derived by Knobloch and De Luca (1990) and their properties are discussed in detail by Knobloch (1992). Generalizations of the MFGL equations to systems in which the amplitudes can couple to mean displacements (as in the propagation of unstable fronts) or to mean flows (as in the theory of water waves) have also been obtained (Matkowsky and Volpert 1992; Pierce and Knobloch 1994a,b). In the following we will assume the nondegeneracy conditions

$$\alpha_r \neq 0, \qquad \alpha_r + \beta_r \neq 0, \qquad \alpha_r - \beta_r \neq 0. \tag{8}$$

In the special case $c_g = \mathcal{O}(\epsilon)$ there is no need to introduce the intermediate timescale T and a similar expansion yields the envelope equations

$$\begin{aligned}
\frac{\partial A_1}{\partial T'} - v \frac{\partial A_1}{\partial X} &= \lambda A_1 + \alpha |A_1|^2 A_1 + \beta |B_1|^2 A_1 + \gamma A_{1XX} && \text{(9a)} \\
\frac{\partial B_1}{\partial T'} + v \frac{\partial B_1}{\partial X} &= \bar{\lambda} B_1 + \bar{\alpha} |B_1|^2 B_1 + \bar{\beta} |A_1|^2 B_1 + \bar{\gamma} B_{1XX}, && \text{(9b)}
\end{aligned}$$

where $v \equiv c_g/\epsilon$ is $\mathcal{O}(1)$ by assumption. These equations are called the coupled complex Ginzburg-Landau (hereafter CCGL) equations and have been used as *model* equations in a number of problems involving wave propagation in which the requirement $c_g = \mathcal{O}(\epsilon)$ fails to hold. Note that since the value of the group velocity is a consequence of the linear problem one expects the requirement $v = \mathcal{O}(1)$ to always fail at sufficiently small ϵ unless c_g can be tuned by varying an additional parameter.

It should be noted that when one restricts attention to a wave propagating in one direction only (e.g., to the left) the amplitude of the wave satisfies the complex

Ginzburg-Landau (CGL) equation, written in terms of the comoving variable η,

$$\frac{\partial A_1}{\partial T'} = \lambda A_1 + \alpha |A_1|^2 A_1 + \gamma A_{1\eta\eta}. \tag{10}$$

In this problem, like the one above, there are in fact two slow time scales. However, for a wave travelling in one direction we can always go into an appropriate comoving frame, thereby eliminating the intermediate slow time. This device is not available for the case of counterpropagating waves, with the result that one envelope "sees" the counterpropagating one moving through it rapidly relative to the timescale on which it evolves. This is the physical reason why the MFGL equations are the correct envelope equations when the group velocity is $\mathcal{O}(1)$.

The difference between the MFGL and CCGL equations is more than cosmetic. Knobloch and De Luca (1990) point out that the two sets of equations make different predictions for the stability of standing wavetrains with respect to long wavelength perturbations (see section 2). Knobloch (1992) traces the cause of this difference to the following effect. In both equations the stability condition is obtained by examining the limit of vanishingly small modulational wavenumbers q. However, in the former this limit is formally taken *after* the limit $\epsilon \to 0+$, while in the latter v (and hence ϵ) is kept finite while the limit $q \to 0$ is taken. Thus the limit $\epsilon \to 0+$ is only taken after the limit $q \to 0$. Evidently these two limiting procedures do not commute. We shall find this to be a recurring theme in the rest of this article. From the derivation it is clear, however, that the MFGL equations are the correct ones sufficiently close to the onset of the instability, and this has now been shown rigorously in particular cases by Pierce and Wayne (1994) and Schneider (1994b). In particular these authors show that there exists an $\epsilon^\star$ such that the solutions of the MFGL equations track those of the original field equations for a time of order ϵ^{-2}, for all $\epsilon < \epsilon^\star$. This is an important result since it extends the formal validity of the MFGL equations to *finite* ϵ.

Method of Reconstitution The CCGL equations are the favorite model equations for systems with reflection symmetry, largely because they are perceived to be in some sense "valid" for finite ϵ (at which of course one wants to use them). I believe the argument behind this belief is based on eschewing rigorous asymptotics (i.e., taking the limit $\epsilon \to 0$ order by order) and instead carrying out a kind of resummation (reconstitution) procedure that goes as follows. Suppose we add to equation (5a) ϵ times equation (6a):

$$\frac{\partial A_1}{\partial T} + \epsilon\frac{\partial A_2}{\partial T} + \epsilon\frac{\partial A_1}{\partial T'} - c_g\left(\frac{\partial A_1}{\partial X} + \epsilon\frac{\partial A_2}{\partial X}\right) = \epsilon\lambda A_1 + \epsilon\alpha|A_1|^2 A_1 + \epsilon\beta|B_1|^2 A_1 + \epsilon\gamma A_{1XX} \tag{11}$$

with a similar equation for the right-travelling waves. In terms of the *reconstituted* amplitudes $A \equiv \epsilon A_1 + \epsilon^2 A_2 + ...$, $B \equiv \epsilon B_1 + \epsilon^2 B_2 + ...$, now assumed to depend on x and a combined slow time τ such that $(\partial/\partial\tau) = \epsilon(\partial/\partial T) + \epsilon^2(\partial/\partial T')$, these equations take the form

$$\frac{\partial A}{\partial \tau} - c_g\frac{\partial A}{\partial x} = (R - R_c + i\Omega)A + \alpha|A|^2 A + \beta|B|^2 A + \gamma A_{xx} + ... \tag{12a}$$

$$\frac{\partial B}{\partial \tau} + c_g \frac{\partial B}{\partial x} = (R - R_c - i\Omega)B + \bar{\alpha}|B|^2 B + \bar{\beta}|A|^2 B + \bar{\gamma} B_{xx} + \dots \quad (12b)$$

Here R denotes the actual bifurcation parameter, R_c its value at threshold and Ω the frequency shift due to $R - R_c$. Note that (i) the small parameter ϵ is absent from these equations (in particular the h.o.t. terms are now formally $\mathcal{O}(1)$), and that (ii) the resulting equations (if *arbitrarily* truncated!) are again of the CCGL form, although in terms of different variables. It is clear that as $R \to R_c$ these equations have the *same* asymptotic expansion as equation (1). In particular both are described by the MFGL equations for all $\epsilon < \epsilon^*$. There is no reason to believe that the CCGL look-alike equations will do better outside this regime than any other set of equations with the correct small ϵ asymptotics. Of course equations (12a,b) have one advantage over the MFGL equations: they are *local* while the latter are *nonlocal.*

In a recent paper Clune et al (1994) found a more interesting example of this kind of behavior in binary fluid convection. In a problem concerning the competition among long wavelength oscillatory patterns in a plane the asymptotically exact procedure (due to Pismen 1988) predicts that a two-dimensional oscillatory pattern called alternating rolls is stable, while the reconstituted partial differential equations predict that at most travelling rolls (i.e., a one-dimensional pattern) is stable. Why is this? The reason is that in this problem there are again two small parameters, ϵ measuring in effect the wavenumber of the pattern, and ε measuring the amplitude of the pattern. These two small parameters bear in general no relation to one another. In the asymptotically exact method one takes the limit $\epsilon \to 0+$ *before* computing the stability of the patterns. These stability calculations are themselves carried out using a perturbation expansion in powers of ε, with the eigenvalues of interest being of order ε^2. In contrast the reconstitution procedure requires one to take ϵ finite, and calculate the $\mathcal{O}(\varepsilon^2)$ eigenvalues in the limit of small amplitude. In effect it requires that one first takes the limit $\varepsilon \to 0+$ *before* taking the limit $\epsilon \to 0+$. The asymptotic procedure takes these limits in the opposite order, and the results of Clune et al indicate that these two limiting procedures once again do not commute. We will meet another example in the next section.

2 Stability properties of travelling and standing waves

In this section we discuss in some detail the stability properties of spatially uniform solutions to the CCGL equations (9a,b) paying particular attention to the case of small v. In these equations we may assume, without loss of generality, that λ is real since its imaginary part can always be absorbed into a time-dependent phase of A_1, B_1.

These equations have two types of spatially independent solutions

$$\text{TW}: \quad A_1 = Re^{i\Omega T'}, \quad B_1 = 0, \quad (13a)$$

$$\text{SW}: \quad A_1 = Re^{i\Omega T'}, \quad B_1 = Re^{-i\Omega T'}, \quad (13b)$$

corresponding to spatially uniform travelling waves (TW) and standing waves (SW), respectively. To study the stability of TW we set $A_1 = Re^{i\Omega T'}(1 + a(X,T'))$, $B_1 =$

$Re^{-i\Omega T'}b(X,T')$, where $|a| \ll 1$, $|b| \ll 1$, and substitute into equations (9a,b). The perturbations a, b satisfy the linearized equations

$$\frac{\partial a}{\partial T'} - va_X = \alpha R^2(a+\bar{a}) + \gamma a_{XX}, \tag{14a}$$
$$\frac{\partial b}{\partial T'} + vb_X = (\bar{\beta}-\bar{\alpha})R^2 b + \bar{\gamma} b_{XX}. \tag{14b}$$

We expect the first equation to have one small eigenvalue. This is because the TW have a continuous symmetry: if $(A_1, 0)$ is a solution so is $(A_1 e^{i\phi}, 0)$, for any real ϕ. The presence of this symmetry implies that a spatially uniform TW has a simple zero eigenvalue; this eigenvalue may lead to instability if the wave is spatially modulated. To find this eigenvalue let $a(X,T') = a^+(T')e^{iqX} + \bar{a}^-(T')e^{-iqX}$, where q is the modulation wavenumber (assumed small). Then $a^\pm(T')$ satisfy

$$\frac{\partial a^+}{\partial T'} - iqva^+ = \alpha R^2(a^+ + a^-) - \gamma q^2 a^+, \tag{15a}$$
$$\frac{\partial a^-}{\partial T'} - iqva^- = \bar{\alpha} R^2(a^+ + a^-) - \bar{\gamma} q^2 a^-. \tag{15b}$$

The characteristic equation has the solutions $s = 2\alpha_r R^2 + \mathcal{O}(q)$, and $s = iqv - \frac{q^2}{2\alpha_r}(\alpha\bar{\gamma} + \bar{\alpha}\gamma)$. In view of the fact that stable uniform TW require $\alpha_r < 0$, the condition for modulational instability is given by

$$\alpha\bar{\gamma} + \bar{\alpha}\gamma > 0. \tag{16}$$

This is the well-known Benjamin-Feir criterion. Equation (14b) has two eigenvalues $s = (\beta_r - \alpha_r)R^2 + \mathcal{O}(q)$ describing stability of LTW with respect to perturbations in the form of RTW. A uniform LTW is thus stable if and only if $\alpha_r < 0$, $\beta_r - \alpha_r < 0$, and is then modulationally unstable if (16) holds.

A similar calculation can be carried out for the MFGL equations. Here one finds that the criterion (16) not only describes the instability of travelling waves, but *also* of standing waves (Knobloch and De Luca 1990, Knobloch 1992).

It is of interest to do a similar analysis for the SW within the CCGL equations. As before, we set $A_1 = Re^{i\Omega T'}(1 + a(X,T'))$, $B_1 = Re^{-i\Omega T'}(1 + b(X,T'))$. The linearized equations for the amplitudes $a^\pm(T')$, $b^\pm(T')$ now take the form

$$\frac{\partial a^+}{\partial T'} - iqva^+ = \alpha R^2(a^+ + a^-) + \beta R^2(b^+ + b^-) - \gamma q^2 a^+, \tag{17a}$$
$$\frac{\partial a^-}{\partial T'} - iqva^- = \bar{\alpha} R^2(a^+ + a^-) + \bar{\beta} R^2(b^+ + b^-) - \bar{\gamma} q^2 a^-, \tag{17b}$$
$$\frac{\partial b^+}{\partial T'} + iqvb^+ = \bar{\alpha} R^2(b^+ + b^-) + \bar{\beta} R^2(a^+ + a^-) - \bar{\gamma} q^2 b^+, \tag{17c}$$
$$\frac{\partial b^-}{\partial T'} + iqvb^- = \alpha R^2(b^+ + b^-) + \beta R^2(a^+ + a^-) - \gamma q^2 b^-. \tag{17d}$$

In contrast to the TW the SW have two *distinct* continuous symmetries $(A_1, B_1) \to (A_1 e^{i\phi}, B_1 e^{i\phi})$ and $(A_1, B_1) \to (A_1 e^{i\varphi}, B_1 e^{-i\varphi})$ corresponding to spatial translations and phase shifts in time. Consequently a uniform SW will have two zero eigenvalues and two modes can be destabilized by spatial inhomogeneities. In contrast to the TW case the group velocity v plays a fundamental role. We consider first the case $v = 0$ for which the characteristic equation factorizes:

$$s^2 - 2s(\alpha_r + \beta_r - \gamma_r q^2) - 2(\alpha_r + \beta_r)\gamma_r q^2 - 2(\alpha_i + \beta_i)\gamma_i q^2 + |\gamma|^2 q^4 = 0, \quad (18a)$$
$$s^2 - 2s(\alpha_r - \beta_r - \gamma_r q^2) - 2(\alpha_r - \beta_r)\gamma_r q^2 - 2(\alpha_i - \beta_i)\gamma_i q^2 + |\gamma|^2 q^4 = 0. \quad (18b)$$

It follows that in addition to the two $\mathcal{O}(1)$ eigenvalues $s = 2(\alpha_r \pm \beta_r)R^2 + \mathcal{O}(q^2)$, there are two $\mathcal{O}(q^2)$ eigenvalues

$$s = -\gamma_r q^2 - \frac{\alpha_i + \beta_i}{\alpha_r + \beta_r}\gamma_i q^2, \quad s = -\gamma_r q^2 - \frac{\alpha_i - \beta_i}{\alpha_r - \beta_r}\gamma_i q^2. \quad (19)$$

Since $\alpha_r \pm \beta_r < 0$ for stability of a uniform SW, it follows that there is modulational instability if at least one of the following conditions holds:

$$(\alpha + \beta)\bar{\gamma} + (\bar{\alpha} + \bar{\beta})\gamma > 0, \quad (\alpha - \beta)\bar{\gamma} + (\bar{\alpha} - \bar{\beta})\gamma > 0. \quad (20)$$

Note that there can be either one or two unstable modes, both appearing through steady state bifurcations. In contrast when $v \neq 0$ the characteristic equation no longer factors:

$$[(s - iqv)^2 - 2(s - iqv)(\alpha_r + \beta_r - \gamma_r q^2) - 2(\alpha_r + \beta_r)\gamma_r q^2 - 2(\alpha_i + \beta_i)\gamma_i q^2 + |\gamma|^2 q^4]$$
$$\times[(s + iqv)^2 - 2(s + iqv)(\alpha_r - \beta_r - \gamma_r q^2) - 2(\alpha_r - \beta_r)\gamma_r q^2 - 2(\alpha_i - \beta_i)\gamma_i q^2 + |\gamma|^2 q^4]$$

$$+ \; 4iqv\{i\gamma_i q^2[\bar{\beta}(s + iqv + \gamma q^2) + \bar{\alpha}\beta - \alpha\bar{\beta}]$$
$$+ \; (s - iqv + \bar{\gamma}q^2)[(s + iqv)\beta_r + (\beta_r\gamma_r + \beta_i\gamma_i)q^2]\} = 0. \quad (21)$$

When $v = \mathcal{O}(1)$ but q is small it follows that there are two roots $s = \pm iqv + \mathcal{O}(q^2)$, each doubled. By computing the $\mathcal{O}(q^2)$ term Coullet et al (1985) found that for both signs modulational instability is present whenever

$$-\gamma_r - \frac{\alpha_r\alpha_i - \beta_r\beta_i}{\alpha_r^2 - \beta_r^2}\gamma_i > 0. \quad (22)$$

Note that this condition is independent of v. Note also that it bears no relation to conditions (20) derived for $v = 0$. To understand the relation between these two results it is necessary to consider the case of small v, specifically $v = \mathcal{O}(q)$. The eigenvalues of interest are then $s = \mathcal{O}(vq) = \mathcal{O}(q^2)$, and are given by

$$\left(s - iqv + \gamma_r q^2 + \frac{\alpha_i + \beta_i}{\alpha_r + \beta_r}\gamma_i q^2\right)\left(s + iqv + \gamma_r q^2 + \frac{\alpha_i - \beta_i}{\alpha_r - \beta_r}\gamma_i q^2\right) = 2iqv\left(\frac{\alpha_r\beta_i - \alpha_i\beta_r}{\alpha_r^2 - \beta_r^2}\right)\gamma_i q^2. \quad (23)$$

Observe that as $v \to 0$ this expression indeed factors, yielding the instability conditions (20). When $v = \mathcal{O}(1)$, we write $s = \pm iqv + s_\pm$, and find (to leading order)

$$s_\pm = -\gamma_r q^2 - \frac{\alpha_r\alpha_i - \beta_r\beta_i}{\alpha_r^2 - \beta_r^2}\gamma_i q^2, \tag{24}$$

as obtained by Coullet et al. In the general case $v = \mathcal{O}(q)$ we have the exact expression

$$s^2 + 2s\left(\gamma_r + \frac{\alpha_r\alpha_i - \beta_r\beta_i}{\alpha_r^2 - \beta_r^2}\gamma_i\right)q^2 + q^2v^2 + \left(\gamma_r + \frac{\alpha_i + \beta_i}{\alpha_r + \beta_r}\gamma_i\right)\left(\gamma_r + \frac{\alpha_i - \beta_i}{\alpha_r - \beta_r}\gamma_i\right)q^4 = 0. \tag{25}$$

It follows that the condition (22) *suffices* for instability, but even when it fails modulational instability can still be present provided

$$\frac{v^2}{q^2} + \left(\gamma_r + \frac{\alpha_i + \beta_i}{\alpha_r + \beta_r}\gamma_i\right)\left(\gamma_r + \frac{\alpha_i - \beta_i}{\alpha_r - \beta_r}\gamma_i\right) < 0. \tag{26}$$

When $v = 0$ this requirement reduces to one of the factorization conditions (20); when $v \neq 0$ it indicates that the effect of the group velocity is *stabilizing*. Note that when (22) and (26) both hold, there is one unstable mode; in contrast when (26) fails the condition (22) implies there are two unstable modes and these become oscillatory when

$$\frac{v^2}{q^2} > \left(\frac{\alpha_r\beta_i - \beta_r\alpha_i}{\alpha_r^2 - \beta_r^2}\right)^2 \gamma_i^2. \tag{27}$$

Thus the group velocity determines both the number of unstable modes and their character, other quantities remaining fixed. It is appropriate to note that in the limit $|q| \to 0$ the condition (26) cannot be satisfied. We see therefore that the limits $v \to 0$, $q \to 0$ once again fail to commute. Of course physically the modulational wavenumber q is at our disposal while either $v = 0$ or $v \neq 0$. In the latter case for $q \ll \mathcal{O}(v)$ the instability condition is (22), while for $q \gg \mathcal{O}(v)$ it is (20). Thus in any given situation with a small group velocity v *both* conditions apply, but they describe instability with respect to perturbations of different wavelengths. Since in either case the growth rate of the instability is $\mathcal{O}(q^2)$ it is in fact possible that the finite wavelength instability will be realized, and hence that in practice the criterion (22) is not relevant.

3 Ginzburg-Landau equation for shear instabilities

The CGL equation (10) arises in two distinct types of problems. It describes the large scale spatial modulation of an otherwise spatially uniform oscillation (e.g., Cross and Hohenberg 1993) that arises through a Hopf bifurcation in a system with translation and reflection symmetry. It was first derived, however, for a class of systems, like viscous shear flows, in which the reflection symmetry $x \to -x$ is absent (Stewartson and Stuart 1971, Pierce 1992). With imposed periodic boundary conditions the resulting envelope equation has the symmetry SO(2)×O(2) specified by the following operations:

$$\begin{aligned}
\text{rotations} \quad R_\phi \in SO(2) \quad &: \quad A_1 \to A_1 e^{i\phi} && \text{(28a)}\\
\text{translations} \quad T_\ell \in O(2) \quad &: \quad \eta \to \eta + \ell, \quad A_1 \to A_1 && \text{(28b)}\\
\text{reflection} \quad \kappa \in O(2) \quad &: \quad \eta \to -\eta, \quad A_1 \to A_1. && \text{(28c)}
\end{aligned}$$

Note that the rotation symmetry is present because A_1 is the envelope of a travelling wavetrain. The resulting high degree of symmetry has a number of important consequences (cf. Crawford 1991) which are nongeneric. By construction the equation describes the primary Hopf bifurcation from the trivial solution $A_1 = 0$. When λ is taken to be real this bifurcation occurs at $\lambda = \lambda_c \equiv \gamma_r q^2$ and has frequency $\Omega_c \equiv -\gamma_i q^2$. Here $q = 2\pi n/L$ for imposed periodicity with wavelength L and n is an integer. When $n \neq 0$ the Hopf bifurcation within the CGL equation breaks the O(2) symmetry and the reflection symmetry in O(2) then implies that $e^{\pm iq\eta}$ are both eigenfunctions of the same eigenvalue $i\Omega_c$. Consequently the multiplicity of the eigenvalue is doubled and one seeks solutions of the form

$$A_1 = \epsilon' A_{11} + \epsilon'^3 A_{13} + \ldots \qquad \lambda = \lambda_c + \epsilon'^2 \lambda_2 + \ldots, \tag{29}$$

where

$$A_{11} = v(\tau) e^{i(\Omega_c T' + q\eta)} + \bar{w}(\tau) e^{i(\Omega_c T' - q\eta)} \tag{30}$$

and $\tau = \epsilon'^2 T'$. Thus v and w are amplitudes of waves travelling to left and right relative to the original wave. Evolution equations for $v(\tau)$ and $w(\tau)$ are obtained at $\mathcal{O}(\epsilon'^3)$:

$$\begin{aligned}
v_\tau &= \lambda_2 v + \alpha |v|^2 v + 2\alpha |w|^2 v && \text{(31a)}\\
w_\tau &= \bar{\lambda}_2 w + \bar{\alpha} |w|^2 w + 2\bar{\alpha} |v|^2 w. && \text{(31b)}
\end{aligned}$$

These equations have been analyzed before (e.g., Crawford and Knobloch 1991). When $\alpha_r < 0$ one finds that of the two types of nontrivial solutions (v, w) the solutions $(v, 0)$ is stable, while (v, v) is unstable. However, because of the reflection symmetry κ this conclusion implies that *both* $(v, 0)$ and $(0, v)$ are stable *simultaneously* near $\lambda = \lambda_c$. Consequently equation (10) has two *distinct* solutions; which is stable depends on initial conditions. In terms of the original field equation this implies that there are two distinct solutions,

$$\psi(x,t) = \operatorname{Re} A_1(\eta, T') e^{i(k_c x + \omega_c t)} = \operatorname{Re} |v| e^{i(k_c \pm \epsilon q)x + i\omega' t}, \tag{32}$$

both of which have the same amplitude and stability properties at a given value of λ.

The above observation is in conflict with the general result that in the original SO(2) invariant system a symmetry-breaking bifurcation is generically a simple Hopf bifurcation leading to a single *rotating* wave (e.g., Ecke et al 1992, Knobloch 1994). To reconcile these two results we observe that the SO(2)$\times$O(2) symmetry of equation (10) is an artifact of the asymptotic derivation. If higher order terms in ϵ are retained the O(2) symmetry is broken to SO(2) with the result that the Hopf bifurcation with double multiplicity eigenvalue splits into a pair of simple Hopf bifurcations, each giving rise to a different rotating wave (hereafter RW). Since one or other of these will set in first, the first instability will be to a single RW as predicted by the generic theory.

To demonstrate this result we retain $\mathcal{O}(\epsilon)$ corrections to the usual CGL equation:

$$\frac{\partial A_1}{\partial T'} = \lambda A_1 + \alpha |A_1|^2 A_1 + \gamma A_{1\eta\eta} + \epsilon\delta_0 A_{1\eta} + \epsilon\delta_1 A_{1\eta\eta\eta} + \epsilon\delta_2 |A_1|^2 A_{1\eta} + \epsilon\delta_3 A_1^2 \bar{A}_{1\eta}. \quad (33)$$

Proceeding as before we find that when $\epsilon \sim \epsilon'^2$ the $\mathcal{O}(\epsilon'^3)$ solvability equations are modified, and take the form

$$v_\tau = (\lambda_2 + i\delta_0 q - i\delta_1 q^3)v + \alpha|v|^2 v + 2\alpha|w|^2 v \quad (34a)$$

$$w_\tau = (\bar{\lambda}_2 + i\bar{\delta}_0 q - i\bar{\delta}_1 q^3)w + \bar{\alpha}|w|^2 w + 2\bar{\alpha}|v|^2 w. \quad (34b)$$

Thus the main effect of $\delta_1 \neq 0$ is to split the Hopf bifurcations to the $(v, 0)$ and $(0, w)$ solutions: if $\delta_{1i} > 0$ the RW $(v, 0)$ sets in when $\lambda_{2r} = \delta_{0i}q - \delta_{1i}q^3$, while the RW $(0, w)$ sets in when $\lambda_{2r} = -\delta_{0i}q + \delta_{1i}q^3$, i.e., later. The first of these is stable, while the second is initially unstable but gains stability at a secondary bifurcation to a branch of quasiperiodic solutions. This branch takes the place of the single frequency standing wave branch in the symmetric problem, and like it is unstable. Owing to the translation invariance no frequency locking takes place on this secondary branch. Note that all reflection symmetry-breaking effects are captured by the linear symmetry-breaking terms (cf. van Gils and Mallet-Paret 1986).

The above theory is based on the assumption that equation (33) holds for finite but small ϵ. Whether this is the case here remains an open question.

4 Travelling waves in finite containers

Large aspect ratio systems are commonly treated as perturbations of the unbounded, translation-invariant system. This idea forms the basis, for example, of Cross' approach to travelling wave convection in binary mixtures (Cross 1986, 1988). Starting with equations (9a,b) describing the interaction between left-and right-travelling waves, Cross appends to them the phenomenological boundary conditions (see Dangelmayr 1994)

$$A_1 - \epsilon(\mu A_{1X} + \nu \bar{B}_{1X}) = 0, \quad B_1 - \epsilon(\mu B_{1X} + \nu \bar{A}_{1X}) = 0 \quad \text{at } X = L, \quad (35a)$$

$$A_1 + \epsilon(\bar{\mu} A_{1X} + \bar{\nu} \bar{B}_{1X}) = 0, \quad B_1 + \epsilon(\bar{\mu} B_{1X} + \bar{\nu} \bar{A}_{1X}) = 0 \quad \text{at } X = -L. (35b)$$

These are the generic boundary conditions that are linear and depend only on the amplitudes and their first derivatives. They are to be thought of in terms of the unscaled variable x in the model equations (12a,b); it is for this reason that the parameter ϵ appears explicitly, in addition to the (complex) "reflection" coefficients μ, ν. Similar equations have also been used to discuss the interaction between spiral vortices in the counter-rotating Taylor-Couette system (Knobloch and Pierce 1992), and that between counterpropagating waves generated by the oscillatory instability of convection rolls (Cross and Kuo 1992).

The boundary conditions (35a,b) are applied at $X = \pm L$, where L is an $\mathcal{O}(1)$ constant. Consequently the physical length of the system is $2L/\epsilon$, and hence is directly related to both the supercriticality of the system and the wave amplitude. This is an essential feature of the formulation. If the physical length is kept fixed then sufficiently near the first bifurcation the system is effectively infinite, while for sufficiently large ϵ it is finite in the sense that one clearly cannot think of the walls as a small perturbation. One can now show, using an expansion similar to that in the preceding section (Dangelmayr et al 1991), that in the limit $\epsilon \to 0+$ the problem (12,35) reduces to the equations

$$v_\tau = \lambda_1 v + a|v|^2 v + b|w|^2 v + d\bar{w}, \tag{36a}$$

$$w_\tau = \bar{\lambda}_1 w + \bar{a}|w|^2 w + \bar{b}|v|^2 w + \bar{d}\bar{v} \tag{36b}$$

studied by Dangelmayr and Knobloch (1987,1991). Here v, w are again the amplitudes of left- and right-travelling waves, while λ_{1r} is the bifurcation parameter and $\lambda_{1i} = \mathcal{O}(\lambda_{1r})$. The term involving the complex coefficient d arises from the forcing due to the (distant) sidewalls. Explicitly, one finds (Dangelmayr et al 1991) that the primary bifurcation takes place at

$$\lambda_c = \operatorname{Re}\gamma\Big[\Big(\frac{\pi}{2L}\Big)^2 + \Big(\frac{v}{2\gamma}\Big)^2\Big], \tag{37}$$

while

$$\lambda_{1r} = \frac{(\lambda_r - \lambda_c)}{\epsilon} - \Big(\frac{\pi^2}{2L^3}\Big)\mu_r, \qquad d = -\frac{\pi^2}{4L^3}\gamma(\nu e^{\gamma^{-1}vL} + \bar{\nu}e^{-\gamma^{-1}vL}), \tag{38a}$$

$$(a,b) = K(\alpha,\beta), \qquad K \equiv \frac{48g^4}{(f^2+4g^2)(f^2+16g^2)}\,\frac{\sinh fL}{fL}. \tag{38b}$$

Here $f = \frac{\gamma_r v}{|\gamma|^2}$, $g = \frac{\pi}{2L}$. The bifurcating physical field takes the form

$$\psi(x,t) = \operatorname{Re}\epsilon^{3/2}\{v(\tau)\exp(-\frac{vX}{2\gamma})e^{i\omega' t} + w(\tau)\exp(\frac{vX}{2\bar{\gamma}})e^{-i\omega' t}\}e^{ik_c x}\cos\frac{\pi X}{2L} + \mathcal{O}(\epsilon^{5/2}), \tag{39}$$

i.e., the spatial eigenfunctions are no longer simple trigonometric functions. Observe that the group velocity is responsible for the asymmetry between the eigenfunctions for left- (LTW') and right-travelling waves (RTW'). In particular if $v > 0$ the eigenfunction of the LTW' peaks in the left half of the domain, while that of the RTW' peaks in the right half. It is this asymmetry between the eigenfunctions that is responsible for the complexity of the spatial patterns which can be described by equations (36a,b) and (39).

In the absence of sidewalls ($d \equiv 0$) equations (36a,b) possess the symmetries

$$(v,w) \rightarrow e^{ik_cl}(v,w), \quad (v,w) \rightarrow (\bar{w},\bar{v}), \quad (v,w) \rightarrow (e^{i\varphi}v, e^{-i\varphi}w), \tag{40}$$

corresponding to translations $x \rightarrow x + l$, reflections $x \rightarrow -x$ and phase shifts $t \rightarrow t + \varphi/\omega_c$. When sidewalls are present ($d \neq 0$) the translation symmetry is broken, and the double multiplicity eigenvalues are split apart producing two successive bifurcations to even and odd parity *standing* waves (hereafter $\mathrm{SW}_{0,\pi}$, respectively). Because of the asymmetry between the LTW' and RTW' eigenfunctions these modes take the form of waves propagating from the center of the container to the sides: the even modes are in phase at the two ends, the odd modes are exactly out phase. Analogues of travelling waves bifurcate from the $\mathrm{SW}_{0,\pi}$ in secondary pitchfork bifurcations, much as in section 3. Studies of equations (36a,b) by Dangelmayr and Knobloch (1991) and Hirschberg and Knobloch (1994) or the numerical integration of equations (9,35) with arbitrarily chosen coefficients reveal a large variety of solutions with good qualitative correspondence with experiments (Kolodner et al 1989; Steinberg et al 1989), and hence serve as a good *model* of the experiments. But is the model correct?

Recent work by Landsberg and Knobloch (1994) suggests not. Landsberg and Knobloch pose the problem in a *finite* container and conclude that in order to describe patterns that propagate one must study the interaction between odd and even modes of the container. When the aspect ratio of the container is *large* the odd and even modes are nearly degenerate. As a consequence the equations for the mode interaction also have an approximate interchange symmetry $(z_o, z_e) \rightarrow (z_e, z_o)$, where $z_{o,e}$ are the amplitudes of the modes. Since, by assumption, reflection in the center of the container acts by $(z_o, z_e) \rightarrow (-z_o, z_e)$ while the normal form symmetry acts by $(z_o, z_e) \rightarrow e^{i\varphi}(z_o, z_e)$, the truncated amplitude equations take the form:

$$\frac{dz_o}{dt} = \delta z_o + K_1|z_o|^2 z_o + K_2|z_e|^2 z_o + K_3\bar{z}_o z_e^2 \tag{41a}$$

$$\frac{dz_e}{dt} = \delta' z_e + K_1'|z_e|^2 z_e + K_2'|z_o|^2 z_e + K_3'\bar{z}_e z_o^2. \tag{41b}$$

Here $\delta \equiv \mu + i\Omega, \delta' \equiv \mu' + i\Omega'$ represent the linear growth rates and frequencies of the two modes and K_1', K_2', K_3' are complex coefficients that are close to K_1, K_2, K_3 because of the near-degeneracy of the two modes. The relevance of equations (41a,b) to the study of oscillatory instabilities in finite containers was recognized already by Bestehorn et al (1989) and Nagata (1991). These equations describe a Hopf bifurcation with approximate D_4 symmetry owing to the near-degeneracy between the two modes, and bear a close relation to equations (36a,b). If we write them in terms of the amplitudes (v, w) of the left- and right-travelling waves, $z_o = v - \bar{w}$, $z_e = v + \bar{w}$, they take the form

$$\frac{dv}{dt} = \lambda v + a|v|^2 v + b|w|^2 v + c\bar{v}\bar{w}^2 + d\bar{w}, \tag{42a}$$

$$\frac{dw}{dt} = \bar{\lambda} w + \bar{a}|w|^2 w + \bar{b}|v|^2 w + \bar{c}\bar{v}^2\bar{w} + \bar{d}\bar{v}, \tag{42b}$$

where $\lambda = \frac{1}{2}(\delta + \delta')$, $d = \frac{1}{2}(\delta' - \delta)$, $a = K_1 + K_2 + K_3$, $b = 2K_1 - 2K_3$ and $c = K_1 - K_2 + K_3$. In writing these equations we have ignored any differences between K_j and K'_j, $j = 1, 2, 3$.

These equations differ from (36a,b) in the presence of new *cubic* term, suggesting that the problem in a finite container cannot be formulated as a perturbation of the unbounded system, *however far apart the walls are.* Explicit computations for doubly diffusive convection in a Hele-Shaw cell show that all three cubic coefficients remain $\mathcal{O}(1)$ as $L \to \infty$ (Landsberg and Knobloch 1995). Moreover, in addition to reproducing solutions of the type found in studies of equations (36a,b) the new equations also exhibit, for appropriate choices of coefficients, a new type of solution called a "collapse" state or "repeated transient". In this state a small amplitude travelling wave grows in amplitude (maintaining its spatial profile) and then undergoes a rapid collapse to a small amplitude state, a process that repeats at irregular intervals and depends sensitively on the aspect ratio of the system. Such states have been found in experiments on binary fluid convection (Sullivan and Ahlers 1988, Kolodner 1993) as well as in numerical simulations (Jacqmin and Heminger 1993), and could not be explained in terms of either the formulation (9,35) or (36a,b). Since for these experiments the coefficients have once again not been calculated, the new equations (42a,b) provide another, but better, model of the experiments.

Modulational instabilities of the SW states. For a unidirectional wave equations (42a,b) reduce to the normal form for a Hopf bifurcation. In an extended system we know from the linear stability problem that such a wave should be described by the CGL equation (10), once local (but slow) variations of its envelope are taken into account. It is likely, therefore, that the dynamics of travelling waves in a large container permitting modulational instabilities should be described by equations of the form

$$\frac{\partial A}{\partial T'} - v\frac{\partial A}{\partial X} = \mu A + \alpha|A|^2 A + \beta|B|^2 A + \gamma \bar{A}\bar{B}^2 + \delta A_{XX} \quad (43a)$$

$$\frac{\partial B}{\partial T'} + v\frac{\partial B}{\partial X} = \bar{\mu} B + \bar{\alpha}|B|^2 B + \bar{\beta}|A|^2 B + \bar{\gamma}\bar{B}\bar{A}^2 + \bar{\delta} B_{XX}, \quad (43b)$$

subject to, as before, the boundary conditions (35a,b). Here we have assumed, as in section 1, that c_g is $\mathcal{O}(\epsilon)$. As before these equations can be reduced by an asymptotic expansion to equations of the form (42a,b) and thence to equations (41a,b). Consequently equations (43a,b) can be thought of as a larger (though still small) amplitude generalization of equations (41a,b). Note, however, that now the linear terms $(d\bar{w}, \bar{d}v)$ are no longer the only terms breaking translation invariance; when $\gamma \neq 0$ this symmetry is already broken. For completeness such a reduction yields a scaling factor K for the new cubic term, given by (38b) with $a = \frac{(\gamma_r - 2i\gamma_i)v}{|\gamma|^2}$.

In this section we ignore the role played by the boundary conditions, and use equations (43a,b) to determine the generalization of the Benjamin-Feir instability criterion for the two types of standing wave admitted by these equations. These are the pure parity states SW and the mixed parity states ASW. When boundary conditions are imposed the former split into $\mathrm{SW}_{0,\pi}$, while the latter no longer bifurcate from the trivial state and thus represent an additional type of finite amplitude solutions (cf. Nagata 1991). Thus the effect of the boundary conditions on the ASW is similar to that already described for the TW. The stability properties of both sets of solutions relative to long wavelength perturbations will of necessity be much different from those derived in section 2 for the SW in an unbounded system. This is because the SW states now have only one continuous symmetry (the phase shift symmetry) instead of two. Consequently as the (small) wavenumber q of the perturbation varies at most *one* eigenvalue can (generically) be destabilized. Since the spatially uniform case of equations (43a,b) is a special case of the equations studied by Silber and Knobloch (1991) it is completely understood. In particular both SW and ASW bifurcate simultaneously with the TW. Moreover, we know that in addition to these three solutions there can be, in appropriate parameter ranges, an additional primary branch of solutions. These solutions are either periodic modes of the form (A, B), $|A| \neq |B|$, or solutions that are *quasiperiodic* in time and bifurcate directly from the trivial solution simultaneously with the periodic solutions. The former, if present, are unstable. The latter are either "rotations" or "librations" (Clune and Knobloch 1994) and can be stable. We expect that the boundary conditions will promote frequency locking of these solutions.

As in section 2 we first consider the case $v = 0$. The resulting equations have two types of spatially independent solutions

$$\mathrm{SW}: \quad A_1 = Re^{i\Omega T'}, \quad B_1 = Re^{-i\Omega T'}, \tag{44a}$$
$$\mathrm{ASW}: \quad A_1 = iRe^{i\Omega T'}, \quad B_1 = Re^{-i\Omega T'}, \tag{44b}$$

in addition to the TW. To study the stability of SW we set $A_1 = Re^{i\Omega T'}(1 + a(X, T'))$, $B_1 = Re^{-i\Omega T'}(1 + b(X, T'))$, where $|a| \ll 1$, $|b| \ll 1$, and substitute into equations (43a,b). The perturbations a, b satisfy the linearized equations

$$\frac{\partial a}{\partial T'} = \alpha R^2(a + \bar{a}) + \beta R^2(b + \bar{b}) + \gamma R^2(\bar{a} - a + 2\bar{b}) + \delta a_{XX}, \tag{45a}$$
$$\frac{\partial b}{\partial T'} = \bar{\alpha} R^2(b + \bar{b}) + \bar{\beta} R^2(a + \bar{a}) + \bar{\gamma} R^2(\bar{b} - b + 2\bar{a}) + \bar{\delta} b_{XX}. \tag{45b}$$

To find the near-marginal eigenvalue we again let $a(X, T') = a^+(T')e^{iqX} + \bar{a}^-(T')e^{-iqX}$. Then $a^\pm(T')$ satisfy

$$\frac{\partial a^+}{\partial T'} = \alpha R^2(a^+ + a^-) + \beta R^2(b^+ + b^-) + \gamma R^2(a^- - a^+ + 2b^-) - \delta q^2 a^+, \tag{46a}$$
$$\frac{\partial a^-}{\partial T'} = \bar{\alpha} R^2(a^+ + a^-) + \bar{\beta} R^2(b^+ + b^-) + \bar{\gamma} R^2(a^+ - a^- + 2b+) - \bar{\delta} q^2 a^-, \tag{46b}$$
$$\frac{\partial b^+}{\partial T'} = \bar{\alpha} R^2(b^+ + b^-) + \bar{\beta} R^2(a^+ + a^-) + \bar{\gamma} R^2(b^- - b^+ + 2a^-) - \bar{\delta} q^2 b^+, \tag{46c}$$
$$\frac{\partial b^-}{\partial T'} = \alpha R^2(b^+ + b^-) + \beta R^2(a^+ + a^-) + \gamma R^2(b^+ - b^- + 2a+) - \delta q^2 b^-. \tag{46d}$$

At this point it is useful to refer to Silber and Knobloch (1991) to identify the linear change of coordinates which diagonalizes these equations when $q = 0$. Since in this case there is one zero eigenvalue we can use these coordinates to find this eigenvalue when $q \neq 0$. We find that this eigenvalue is

$$-\frac{(\alpha+\beta+\gamma)\bar{\delta}+(\bar{\alpha}+\bar{\beta}+\bar{\gamma})\delta}{\alpha_r+\beta_r+\gamma_r}\frac{q^2}{2}+\mathcal{O}(q^4). \tag{47}$$

Since stable uniform SW require $\alpha_r + \beta_r + \gamma_r < 0$, the condition for modulational instability is given by

$$(\alpha+\beta+\gamma)\bar{\delta}+(\bar{\alpha}+\bar{\beta}+\bar{\gamma})\delta > 0. \tag{48}$$

A similar calculation for the mixed parity state ASW gives the instability condition

$$(\alpha+\beta-\gamma)\bar{\delta}+(\bar{\alpha}+\bar{\beta}-\bar{\gamma})\delta > 0. \tag{49}$$

These conditions generalize the Benjamin-Feir criterion to pure and mixed parity states in a long box, and should be compared with the earlier conditions for TW instability (16) and standing wave instability (20) in an unbounded system. It should be noted that here the spatially uniform TW are stable provided $\alpha_r < 0$, $\alpha_r - \beta_r > 0$ and $|\alpha-\beta|^2-|\gamma|^2 > 0$, while the SW and ASW require that $\alpha_r+\beta_r\pm\gamma_r < 0$, $\alpha_r-\beta_r\mp 3\gamma_r < 0$ and $|\alpha-\beta\mp 3\gamma|^2 - |\alpha-\beta\pm\gamma|^2 > 0$. Since the presence of spatial modulation changes these conditions by $\mathcal{O}(q^2)$ only these conditions must be satisfied in order that the modulational instabilities identified above be observable. We do not discuss the modulational properties of the quasiperiodic states.

The inclusion of the group velocity v is straightforward. When $v = \mathcal{O}(1)$ the instability condition becomes, in the limit $q \to 0$,

$$\left(\delta_r \pm \frac{1}{4}v^2\frac{\alpha_r-\beta_r\mp\gamma_r}{(\alpha_r-\beta_r\mp\gamma_r)\gamma_r+(\alpha_i-\beta_i\mp\gamma_i)\gamma_i}\right)(\alpha_r+\beta_r\pm\gamma_r)$$
$$+\left(\delta_i+\frac{1}{2}v^2\frac{\gamma_i}{(\alpha_r-\beta_r\mp\gamma_r)\gamma_r+(\alpha_i-\beta_i\mp\gamma_i)\gamma_i}\right)(\alpha_i+\beta_i\pm\gamma_i) > 0. \tag{50}$$

Thus both δ_r and δ_i are "corrected" by finite group velocity effects. In particular with all the coefficients fixed changing the group velocity can lead to a change from stability to instability or vice versa.

Conclusions

In this paper we have pointed out that when stability computations are carried out on equations derived using an expansion in a small parameter one must watch out for potential problems. These problems arise because the stability calculations themselves use a small parameter (either an amplitude or a wavenumber) and the two limits in general do not commute. The real issue is of course which is the correct way to carry out such derivations. From a strict asymptotics point of view it is clear that one must derive the asymptotically exact equations first, before using them to compute stability. In practice, however, one may not be interested in the asymptotic limit. Instead one would like to have equations valid for small but finite values of the expansion parameter. Since asymptotic expansions do not in general converge there is no justification in using them at such parameter values. It is here that the recent theorems that have been proved justifying the use of asymptotic expansions at finite values of the small parameter are particularly important (e.g., Collet and Eckmann 1990, 1992; van Harten 1991; Schneider 1994a). In cases where such results are unavailable nonasymptotic techniques (e.g., reconstitution) will no doubt be used. These lead to simplified equations, usually containing explicitly the small parameter. In my opinion it is important to recognize that such equations are no more than models. *A priori* there is no way of telling whether such models are good or bad when used at finite values of the small parameter. The only way to do that is to perform explicit *quantitative* comparisons between the model equations and either careful experiments or careful numerical simulations of the underlying field equations. Regrettably such comparisons are very rare.

Acknowledgement: This work was supported by NSF under grant DMS-9406144. I am indebted to M.K. Hase and A.S. Landsberg for discussions, and to A. Doelman for a critical reading of the manuscript.

References

[1] M. Bestehorn, R. Friedrich and H. Haken, Z. Phys. B **77**, 151–155 (1989).

[2] T. Clune and E. Knobloch, Physica D **61**, 106–112 (1992).

[3] T. Clune and E. Knobloch, Physica D **74**, 151–176 (1994).

[4] T. Clune, M.C. Depassier and E. Knobloch, preprint (1994).

[5] P. Collet and J.-P. Eckmann, Commun. Math. Phys. **132**, 139–153 (1990).

[6] P. Collet and J.-P. Eckmann, Nonlinearity **5**, 1265–1302 (1992).

[7] J.D. Crawford, Rev. Mod. Phys. **63**, 991–1037 (1991).

[8] J.D. Crawford and E. Knobloch, Ann. Rev. Fluid Mech. **23**, 341–387 (1991).

[9] M.C. Cross, Phys. Rev. Lett. **57**, 2935–2938 (1986).

[10] M.C. Cross, Phys. Rev. A **38**, 3593–3600 (1988).

[11] M.C. Cross and P.C. Hohenberg, Rev. Mod. Phys. **65**, 851–1112 (1993).

[12] M.C. Cross and K. Kim, Phys. Rev. A **37**, 3909–3920 (1988).

[13] M.C. Cross and E.Y. Kuo, Physica D **59**, 90–120 (1992).

[14] P. Coullet, S. Fauve and E. Tirapegui, J. Physique Lett. **46**, L787–791 (1985).

[15] P. Coullet, R.E. Goldstein and G.H. Gunaratne, Phys. Rev. Lett. **63**, 1954–1957 (1989).

[16] G. Dangelmayr, preprint (1994).

[17] G. Dangelmayr and E. Knobloch, in *The Physics of Structure Formation*, W. Güttinger, G. Dangelmayr, eds., Springer-Verlag, Berlin, pp. 387–393 (1987).

[18] G. Dangelmayr and E. Knobloch, Nonlinearity 4, 399–427 (1991).

[19] G. Dangelmayr, E. Knobloch and M. Wegelin, Europhys. Lett. **16**, 723–729 (1991).

[20] R.E. Ecke, F. Zhong and E. Knobloch, Europhys. Lett. **19**, 177–182 (1992).

[21] W. Eckhaus, *Studies in Nonlinear Stability Theory*, Springer Tracts in Natural Philosophy **6**, Springer-Verlag (1965).

[22] S. van Gils and J. Mallet-Paret, Proc. Royal Soc. Edinburgh A **104**, 279–307 (1986).

[23] A. van Harten, J. Nonlinear Sci. **1**, 397–422 (1991).

[24] D. Jacqmin and J. Heminger, preprint (1993).

[25] P. Hirschberg and E. Knobloch, preprint (1994).

[26] E. Knobloch, in *Pattern Formation in Complex Dissipative Systems*, S. Kai, ed., World Scientific, Singapore, pp. 263–274 (1992).

[27] E. Knobloch, in *Lectures on Solar and Planetary Dynamos*, M.R.E. Proctor and A.D. Gilbert, eds., Cambridge University Press, pp. 331–372 (1994).

[28] E. Knobloch and J. De Luca, Nonlinearity **3**, 975–980 (1990).

[29] E. Knobloch and R. Pierce, in *Ordered and Turbulent Patterns in Taylor-Couette Flow*, C.D. Andereck and F. Hayot, eds., Plenum Press, pp. 83–90 (1992).

[30] P. Kolodner, Phys. Rev. E **47**, 1038–1048 (1993).

[31] P. Kolodner, C. M. Surko and H. Williams, Physica D **37**, 319–333 (1989).

[32] A.S. Landsberg and E. Knobloch, preprint (1994).

[33] A.S. Landsberg and E. Knobloch, preprint (1995).

[34] B.J. Matkowsky and V. Volpert, Physica D **54**, 203–219 (1992).

[35] W. Nagata, Proc. Royal Soc. Edinburgh A **117**, 1–20 (1991).

[36] R. Pierce, Phys. Fluids **4** , 2486–2494 (1992).

[37] R.D. Pierce and E. Knobloch, Phys. Fluids **6**, 1177–1190 (1994).

[38] R.D. Pierce and E. Knobloch, J. Fluid Mech. **264**, 137–163 (1994).

[39] R.D. Pierce and C.E. Wayne, preprint (1994).

[40] L. Pismen, Phys. Rev. A **38**, 2564–2572 (1988).

[41] H. Riecke, Phys. Rev. Lett. **68**, 301–304 (1992).

[42] G. Schneider, J. Nonlinear Sci. **4**, 23–34 (1994a).

[43] G. Schneider, preprint (1994b).

[44] W. Schöpf and W. Zimmermann, Europhys. Lett. **8**, 41–46 (1989).

[45] M. Silber and E. Knobloch, Nonlinearity **4**, 1063–1106 (1991).

[46] V. Steinberg, J. Fineberg, E. Moses and I. Rehberg, Physica D **37**, 359–383 (1989).

[47] K. Stewartson and J.T. Stuart, J. Fluid Mech. **48**, 529-545 (1971).

[48] T.S. Sullivan and G. Ahlers, Phys. Rev. A **38**, 3143–3146 (1988).

The mean flows driven by sandbar instabilities

N.L. KOMAROVA and A.C. NEWELL
Department of Mathematics, University of Arizona, Tucson, AZ 85721, U.S.A.

Abstract

A simple model describing the interaction between flow and bed sediment in a straight channel is considered. The model allows for a uniform current over a flat bottom, and this solution becomes unstable when the width-to-depth ratio exceeds some finite critical value. Linear theory gives a narrow spectrum of perturbations which grow at near-critical condition. A weakly nonlinear analysis is presented. It results in a system of three equations for three envelopes. One of them is the envelope of the group of marginally unstable alternate bars, and the two others are intensity of the sandbars. We discuss the applicability of these ideas to the sandwave - sandbank (tidal ridge) structures driven by tidal currents in the shallow ocean.

1 Introduction

Sandbars appear as the result of the interaction between river flow and an erodible bottom. The bed sediment in natural rivers usually consists of material which may be transported by currents. Sandbars are propagating waves in a downstream direction with an alternating transversal amplitude structure and typical wavelengths of the order of the several river widths. The nonlinear dynamics of these features was studied by M. Colombini, G. Seminara, M. Tubino [2] and R. Schielen, A. Doelman, H. E. de Swart [8]. The main goal of this paper is to augment an earlier analysis of R. Schielen et al who neglected mean flow effects, and to emphasize the importance of including such features, not only in the channel problem, but also in the problem of sandwave and sandbar generation in tidal seas.

To study the behavior of such bars, we consider a depth-averaged shallow water flow in a straight, infinitely long channel of width L, having a uniform mild slope, which we denote by i_0. For most rivers i_0 varies between 0.00008 and 0.0005. For instance, the Blue Nile has a length of 4,400 km and elevation 1,860 m, so its slope $i_0 = 0.0004$; for the White Nile or Mississippi $i_0 = 0.0001$. Note that the slope i_0 gives rise to an acceleration $i_0 g$ of fluid in the downstream direction.

Further, we consider the banks to be non-erodible, whereas the bottom consists of non-cohesive sediment, which we assume to be transported as bedload. The system then can be described by the following model:

$$\frac{\partial \mathbf{U}}{\partial t} + (\mathbf{U} \cdot \nabla)\mathbf{U} + g\nabla\zeta = \mathbf{F} + i_0 g\hat{x}, \tag{1}$$

$$\frac{\partial(\zeta + h - z_b)}{\partial t} + \nabla \cdot [\mathbf{U}(\zeta + h - z_b)] = 0, \tag{2}$$

$$\frac{\partial z_b}{\partial t} + \nabla \cdot \mathbf{S} = 0. \tag{3}$$

We use an orthogonal coordinate system, where we assume that the x- direction coincides with the streamwise direction ($\hat{x}$ is a unit vector in the x-direction). $\mathbf{U} = (u, v)$ is the depth-averaged velocity, ζ is the elevation of the disturbed free surface with respect to the undisturbed water-depth h, z_b is the disturbed bed-level with respect to the undisturbed water-depth h, and therefore $z = \zeta$ and $z = z_b - h$ are the positions of the surface and the bed respectively. Eqn. (1) is the equation of motion of the fluid flow. The vector $\mathbf{F}$ represents friction, which can be modeled by the expression [8]: $\mathbf{F} = \left(-C\frac{u\sqrt{u^2+v^2}}{\xi - z_b + h}, -C\frac{v\sqrt{u^2+v^2}}{\xi - z_b + h}\right)$. The coefficient C (the so-called drag coefficient) varies between 0.001 and 0.01. The fact that it is small means that the ratio of the water flow- sediment interaction and the weight of the fluid column is small. Equation (2) is conservation of water mass and equation (3) is conservation of sand. The volumetric sediment flux $\mathbf{S}$ is modeled as $\mathbf{S} = \sigma(|\mathbf{U}|^{\beta}\mathbf{U} - \gamma|\mathbf{U}|^{\beta+1}\nabla z_b)$, for some $\beta > 0$, $\gamma > 0$ and σ depending on the bed porosity and on the sediment properties. This empirical relationship is one of the most commonly used (see [7], for other models see [9]). It reflects the fact that transport increases with increasing velocities and that sediment is transported easier downhill than uphill (the second term involving the coefficient γ is a correction term for downslope effects). There are several ways to parametrize the coefficients β, C and γ depending on the real river properties (see, for instance, [1], [4] or [10]). However we will consider a simple conceptual model, which only contains the essential physical mechanisms, responsible for morphological instabilities, therefore we will take constant values for the parameters. For natural rivers β lies in the range 1 to 6, γ is of the order 1 and $\sigma \sim 10^{-4}\mathrm{m}^2\mathrm{s}^{-1}$.

We close the model by the boundary conditions $v = 0$, $S_y = 0$ for $y = 0$ and $y = L$. These boundary conditions represent the assumption that the walls are impermeable for water as well as sediment.

We scale the variables as follows:

$$(u, v) \to u^*(u, v); \tag{4}$$
$$(x, y) \to L(x, y); \tag{5}$$
$$z_b \to h z_b; \tag{6}$$
$$\zeta \to \frac{(u^*)^2}{g}\zeta; \tag{7}$$
$$t \to \frac{Lh}{\sigma(u^*)^{\beta+1}}t \equiv Tt, \tag{8}$$

where $u^* = \sqrt{\frac{i_0 g h}{C}}$ is velocity of uniform flow. The choice of scale for the free surface deformation is made in order to be consistent with the work of R. Schielen et al [8] and reflects the fact that the short wave amplitudes in ζ are of this order, but, as we shall see, the long wave (long compared to the sandbar wavelength) distortion of the free surface will turn out to be much larger and best scaled by h. The choice of time scale is made to balance the sediment transport (see Eqn. (3)). In these rescaled coordinates, the equations are:

$$\kappa\frac{\partial u}{\partial t}+u\frac{\partial u}{\partial x}+v\frac{\partial u}{\partial y}+\frac{\partial \zeta}{\partial x}=-R\left(-1+\frac{u\sqrt{u^2+v^2}}{\hat{f}\zeta+1-z_b}\right), \tag{9}$$

$$\kappa\frac{\partial v}{\partial t}+u\frac{\partial v}{\partial x}+v\frac{\partial v}{\partial y}+\frac{\partial \zeta}{\partial y}=-R\,\frac{v\sqrt{u^2+v^2}}{\hat{f}\zeta+1-z_b}, \tag{10}$$

$$\kappa\hat{f}\frac{\partial \zeta}{\partial t}-\kappa\frac{\partial z_b}{\partial t}+\frac{\partial}{\partial x}[u(\hat{f}\zeta+1-z_b)]+\frac{\partial}{\partial y}[v(\hat{f}\zeta+1-z_b)]=0, \tag{11}$$

$$\frac{\partial z_b}{\partial t}+\frac{\partial S_x}{\partial x}+\frac{\partial S_y}{\partial y}=0, \tag{12}$$

where $R=\frac{L}{h}C$ is the width-do-depth ratio of the channel (multiplied by the friction coefficient), $\hat{f}=\frac{(u^*)^2}{gh}=\frac{i_0}{C}$ is squared Froude number and $\kappa=\frac{L}{u^*T}$ is the ratio of the time scale of flow advection and the morphological time scale. Also it is convenient to introduce a new parameter $\delta\equiv\frac{\gamma C}{\beta}$. In dimensionless variables the sediment flux reads

$$\mathbf{S}=\left(|\mathbf{U}|^{\beta}u-|\mathbf{U}|^{\beta+1}\delta\frac{\beta}{R}\frac{\partial z_b}{\partial x},|\mathbf{U}|^{\beta}v-|\mathbf{U}|^{\beta+1}\delta\frac{\beta}{R}\frac{\partial z_b}{\partial y}\right). \tag{13}$$

We assume that the characteristic time of the bed motion is much larger than the one of the water flow, which implies that κ is very small. We can therefore neglect time derivatives in the flow equations which means that the flow instantaneously adapts to the evolution of the bed (quasi-stationary approach). Finally we obtain,

$$u\frac{\partial u}{\partial x}+v\frac{\partial u}{\partial y}+\frac{\partial \zeta}{\partial x}=-R\left(-1+\frac{u\sqrt{u^2+v^2}}{\hat{f}\zeta+1-z_b}\right), \tag{14}$$

$$u\frac{\partial v}{\partial x}+v\frac{\partial v}{\partial y}+\frac{\partial \zeta}{\partial y}=-R\,\frac{v\sqrt{u^2+v^2}}{\hat{f}\zeta+1-z_b}, \tag{15}$$

$$\frac{\partial}{\partial x}[u(\hat{f}\zeta+1-z_b)]+\frac{\partial}{\partial y}[v(\hat{f}\zeta+1-z_b)]=0, \tag{16}$$

$$\frac{\partial z_b}{\partial t}+\frac{\partial S_x}{\partial x}+\frac{\partial S_y}{\partial y}=0. \tag{17}$$

The important parameters of the system are:

1. $R=\frac{L}{h}C$ - the control parameter;
2. $\hat{f}=\frac{(u^*)^2}{gh}=\frac{i_0}{C}\ll 1$ (for typical slope $i_0=0.0001$ and the drag coefficient $C=0.007$, the squared Froude number is $\hat{f}=0.014$);
3. $\delta=\frac{\gamma C}{\beta}\ll 1$ (for $C=0.007$, $\gamma=1$ and $\beta=4$, this parameter has a value $\delta=0.0017$).

2 Rigid-lid approximation

In this section we reproduce some results obtained by R. Schielen, A. Doelman and H. E. de Swart in [8]. They studied system (9)- (12) with one more assumption, namely, the rigid-lid approximation. They considered the Froude number $\sqrt{\hat{f}}$ to be small and neglected the terms with premultiplier $\hat{f}$. It means that in their model the water height was (in dimension variables) $(h - z_b)$ instead of $(h + \zeta - z_b)$.

However, this approximation is not consistent with the dynamics. We will see shortly that downstream gradients in the intensity of the sandbar amplitude drive long waves on both the sand bed and the free surface. In fact, in order to conserve mass, the latter must deform. Nevertheless, the linear stability problem, the first step in the analysis of sandbar dynamics, is almost the same. The corrections are of order $\hat{f}$, which is small. However, the nonlinear corrections in the surface elevation, spoken of above, will scale with h and not $\frac{(u^*)^2}{g}$. We now review briefly both the linear and nonlinear analysis of Schielen et al (see [8]).

2.1 Linear theory

The linear stability of the state $\Phi_{steady} = (1, 0, 0, 0)^T$ was considered. Let

$$\Phi = (u, v, \zeta, z_b)^T = \Phi_{steady} + \Phi' \tag{18}$$

with $\Phi' = f(y)e^{ikx - i\omega t}$ (here k is real-valued wave number and ω is the complex frequency). The solution of the resulting eigenvalue problem is

$$\Phi' = Ae^{ikx - i\omega t} \begin{pmatrix} \alpha_u & \cos l\pi y \\ \alpha_v & \sin l\pi y \\ \alpha_\zeta & \cos l\pi y \\ & \cos l\pi y \end{pmatrix} \tag{19}$$

with $\omega = \omega(k)$ (see [8]). The stability of the basic state is determined by the imaginary part of ω. Calculations give us the expression for the neutral curve on the plane $(R^2, [k/l]^2) : \omega_i(R^2, [k/l]^2) = 0$, which can be approximated by a parabola near the minimum value of R. This curve separates the exponentially growing and decaying solutions of the linear problem. The minimum of this curve is achieved at the critical point (k_c, R_c). Since there is a natural small parameter δ in the system, it is very convenient to expand all the critical quantities as series in terms of δ. Here we show the order of magnitude the critical coefficients for the mode which gets excited the first $(l = 1)$:

$$k_c \sim \delta^{\frac{1}{4}}; \qquad R_c \sim \delta^{\frac{1}{2}}; \qquad \omega_c \sim k_c. \tag{20}$$

The curvature of the neutral stability curve at R_c is also of order $\delta^{\frac{1}{2}}$. It turns out that for width-to-depth ratios larger than the critical value R_c the uniform flow over a flat bed is unstable and a spectrum of wave-like perturbations with exponentially growing amplitudes will develop.

To explain the physical mechanism of the instability, let us write the linearized continuity equation for the sand in unscaled variables:

$$\frac{1}{\sigma}\partial_t z_b = -(u^*)^\beta \Big[\partial_x u(\beta+1) + \partial_y v\Big] + \gamma (u^*)^{\beta+1}(\partial_x^2 + \partial_y^2) z_b, \tag{21}$$

where $u^* = \sqrt{\frac{i_0 g h}{C}}$ as before. We will refer to the first term of the right hand side as the flow component of sediment flux and to the second one as the down-slope correction. Note that the former term makes the sand bumps grow (see Fig. [1]) whereas the latter one always causes damping of the perturbations. For a fixed river width L both terms grow as h increases, but the flow term grows as $h^{\frac{\beta}{2}}$, and the down-slope correction grows as $h^{\frac{\beta+1}{2}}$, i.e. faster. Assume that $z_b \sim e^{-i(\omega_r + i\omega_i)t}$ and determine the sign of growth rate ω_i. If the ratio $\frac{L}{h} \propto R$ is increased, first the absolute value of the contribution to ω_i which comes from down-slope correction is larger than the one from

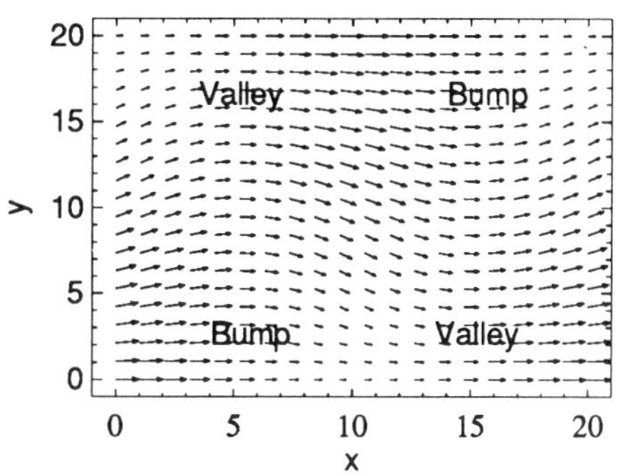

Figure 1: The velocity field in the laboratory frame (qualitatively, at a fixed time). Steady state flow and the first correction $\Phi_{steady} + e^{ik_c x - i\omega_c t}\Phi_{11}$ are shown. It is now clear how the stream flow creates the sand bumps: the velocity is large when water comes to the bump and small from the downslope side, so the bump grows because sediment flux is proportional to water velocity.

the stream term (therefore the growth rate is negative), but as soon as the width-to depth ratio exceeds the value corresponding to R_c, the contribution from the stream term becomes larger and the total contribution is positive (see Fig. [2]). We can see that the competition of the diffusive term and the flow term in the expression for the sediment flux gradient gives rise to a finite value of R_c which balances the negative and positive contributions to the growth rate ω_i.

2.2 Nonlinear theory

Of course, the linear description is valid only when wave amplitudes are infinitesimally small. For slightly supercritical values of the control parameter, a nonlinear theory

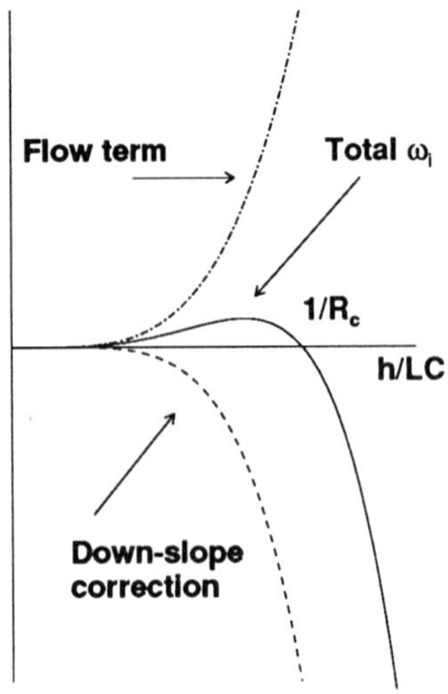

Figure 2: Competition of the flow term of sediment flux and the down-slope correction leads to finite value of R_c (qualitatively).

must be developed. For R near threshold, namely,

$$R = R_c(1 + \epsilon^2 r) \quad r = O(1), \quad \epsilon^2 \ll 1, \tag{22}$$

a perturbation expansion of each of the fields can be carried out and the solvability conditions for the various iterates used to write down evolution equations for the envelopes of the active modes, namely the envelope of the unstable wave and the long wave distortions of the sandbed and the free surface. Ordinarily, this will be an expansion in the small parameter ϵ. However, since δ is also small in the many situations of interest, it turns out that ϵ does not have to be very small because it is sufficient that ϵ be bounded by $\delta^{\frac{1}{4}}$, which is about 0.2. The reason for this is that the neutral stability curve (see Fig. [3]) near k_c can be written as

$$R = R_c + \delta^{\frac{1}{2}}\alpha_1(k - k_c)^2 + \delta^{\frac{1}{4}}\alpha_2(k - k_c)^3 + \alpha_3(k - k_c)^4 + \ldots; \quad \alpha_i = O(1), \tag{23}$$

and therefore for $|R - R_c|$ of order $\epsilon^2\delta^{\frac{1}{2}}$, the second term in the expansion is only greater then each of the third and fourth when ϵ is less than $\delta^{\frac{1}{4}}$. For ϵ of the same order of magnitude or greater than $\delta^{\frac{1}{4}}$, a new fully nonlinear analysis will be required because then one cannot simply divide the active modes into an envelope about k_c and mean components. This will involve similar ideas to the far from onset theory of Cross-Newell [3], [6]. Despite the fact that the structure of the solution is less clear,

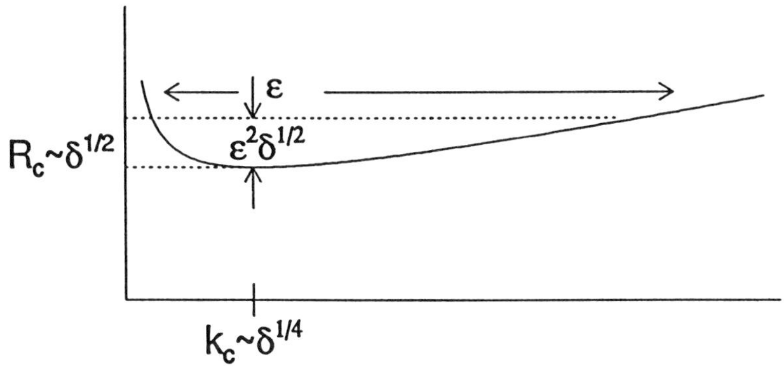

Figure 3: The neutral curve (qualitatively).

there is no reason to expect a qualitative change in behavior as ϵ increases, other than the mixing of the wavy mode with the mean. Therefore, a careful analysis of the small ϵ case is very important. Moreover, one can still use the results of the weakly nonlinear theory to get a qualitative idea of how much mean flow is produced for ϵ up to about 0.5.

The Taylor expansion for the complex frequency $\omega = \omega_r + i\omega_i$ near (k_c, R_c) reads

$$\begin{aligned} \omega &= \omega_c + \left(\frac{\partial \omega_r}{\partial k}\right)_c (k - k_c) + i\left(\frac{\partial \omega_i}{\partial R}\right)_c (R - R_c) \\ &+ \frac{1}{2}\left[\left(\frac{\partial^2 \omega_r}{\partial k^2}\right)_c + i\left(\frac{\partial^2 \omega_i}{\partial k^2}\right)_c\right](k - k_c)^2 + \cdots. \end{aligned} \tag{24}$$

We will use the following notations:

$$\left(\frac{\partial \omega_r}{\partial k}\right)_c = \omega_c' - \text{ the group velocity;} \tag{25}$$

$$\left(\frac{\partial \omega_i}{\partial R}\right)_c = \sigma_r; \qquad -\left(\frac{\partial \omega_r}{\partial R}\right)_c = \sigma_i; \tag{26}$$

$$-\frac{1}{2}\left(\frac{\partial^2 \omega_i}{\partial k^2}\right)_c = \gamma_r; \quad \frac{1}{2}\left(\frac{\partial^2 \omega_r}{\partial k^2}\right)_c = \gamma_i. \tag{27}$$

The coefficients ω_c' and σ_r are of order one; $\gamma_r = O(\delta^{\frac{1}{2}})$, σ_i and γ_i are both $O(\delta^{\frac{1}{4}})$.

Again the expansion is only valid for $|k - k_c|$ which is of order ϵ small. It has to be sufficiently small so that the quadratic terms dominate and this again requires $\epsilon < \delta^{\frac{1}{4}}$.

To describe the nonlinear dynamics of the system, Schielen et al used the following expansion for Φ:

$$\begin{aligned} \Phi = \Phi_{steady} \quad & + \epsilon^2 \Phi_{02} + \epsilon^3 \Phi_{03} + \cdots \\ + e^{i(k_c x - \omega_c t)} & [\epsilon \Phi_{11} + \epsilon^2 \Phi_{12} + \epsilon^3 \Phi_{13} + \cdots] + \text{c.c} \end{aligned} \tag{28}$$

$$+e^{2i(k_cx-\omega_ct)}[\epsilon^2\Phi_{22}+\epsilon^3\Phi_{23}+\quad\cdots\quad]+\text{c.c}$$
$$+e^{3i(k_cx-\omega_ct)}[\epsilon^3\Phi_{33}+\quad\cdots\quad]+\text{c.c}+\cdots,$$

where Φ_{11} coincides with (19), but now the amplitude $A = A(\xi,\tau)$ is no longer a constant but some function of slow variables $\tau=\epsilon^2 t$ and $\xi=\epsilon(x-\omega_c' t)$, where ω_c' is the group velocity of the most excited wave R_c. The analysis of nonlinear interactions between different modes gave expressions for Φ_{pq} (here q stands for the order of the correction and p for the order in the corresponding critical exponent $e^{p(ik_cx-i\omega_ct)}$). It was also found that the evolution of A is described by Ginzburg-Landau equation (for the values of the coefficients we again refer to [8]):

$$\frac{\partial A}{\partial\tau}=r\,R_c(\sigma_r+i\sigma_i)A+(\gamma_r+i\gamma_i)\frac{\partial^2 A}{\partial\xi^2}+(c_r+ic_i)|A|^2A. \tag{29}$$

3 What is wrong with the above analysis

Now let us discuss the expansion (28), proposed in [8]. Substitute it in system (14)-(15) with $\hat{f}=0$ and consider the mode with $p=0,\ \ q=3$. In this order the problem reads

$$\mathcal{L}_0\Phi_{03}=f_{03}, \tag{30}$$

where $\mathcal{L}_{03}$ is the linear matrix

$$\mathcal{L}_0=\begin{pmatrix} 2R & 0 & 0 & R\\ 0 & R & \partial_y & 0\\ 0 & \partial_y & 0 & 0\\ 0 & \partial_y & 0 & -\frac{\delta\beta}{R}\partial_y^2\end{pmatrix}. \tag{31}$$

Since the matrix is singular, the inhomogeneous problem does not always have non-trivial solutions. The Fredholm alternative theorem gives the condition on the right hand side (function f_{03}) which assures that the problem does have non-trivial solutions. In our case it reads

$$\langle\phi,f_{03}\rangle=0, \tag{32}$$

where ϕ stands for any vector of the null-space of the adjoint linear operator of the problem. It is easy to check that the vectors $\phi_{\nu\mu}=\begin{pmatrix}0&0&\nu&\mu\end{pmatrix}$ belong to the null-space of the adjoint problem (here ν and μ are arbitrary constants). If we represent the right hand side of the equation for the (03) mode by $f=\begin{pmatrix}f_u&f_v&f_\zeta&f_z\end{pmatrix}^T$, equation (32) with $\phi=\phi_{\nu\mu}$ gives

$$\nu\int_0^1 f_\zeta dy+\mu\int_0^1 f_z dy=0. \tag{33}$$

The problem has a solution only if the relation (33) holds for *any* ν and μ. This results in the equation $\frac{\partial|A|^2}{\partial\xi}=0$, which suggests that the initial expansion (28) was not complete. It is not even correct when $\frac{\partial |A|^2}{\partial \xi} = 0$ initially, because, for certain parameter ranges, this state is unstable. Physically it means that nontrivial solutions fail to satisfy the mass conservation law, because the first of the Eqns. (33) is merely the continuity equation (16) averaged over y (with $\hat{f} = 0$). Namely, the operation of averaging gives the expression $\partial_x \langle u(1 - z_b) \rangle + v(1 - z_b)|_{y=0}^{y=1} = 0$ which for the mode (03) reads

$$\frac{\partial}{\partial \xi}\langle u_{02} - u_{steady} z_{02} - u_{11} z_{11}^* - u_{11}^* z_{11} \rangle = \frac{\partial |A|^2}{\partial \xi} \frac{u_{02s} + u_{02c} - 2Re(\alpha_u)}{3} = 0. \quad (34)$$

Similarly, the second of the Eqns. (33) is the averaged continuity equation for sand. Note that for the mode (02) condition (33) can be rewritten as $\nu(v_{11} z_{11}^* + c.c.) + \mu \left[\beta(v_{11} u_{11}^* + c.c.) + \frac{\gamma \delta (\beta+1)}{R} (u_{11} \partial_y z_{11}^* + c.c.) \right] = 0$, which holds automatically. Moreover, a stronger statement can be proved: for the (02) mode the Fredholm condition (32) (with f_{02} instead of f_{03}) holds for any vector ϕ from the null-space of the adjoint matrix $\mathcal{L}_0^\dagger$. The above shows that the system first becomes inconsistent at level (03).

From this simple consideration we conclude that some mode has been overlooked, namely, so-called mean flow has not been taken into account. The mean flow plays an important role in the dynamics of the system, and the rest of the paper will be devoted to an introduction of the mean flow and the analysis of its influence on the system.

An intuitive thing to do would be to add a new term (the mean flow) in the expansion (28) and derive equations for the mean flow envelopes, following the procedure described below (but with $\hat{f} = 0$). However, if one tries to introduce a mean flow, staying within the framework of the rigid-lid model, difficulties are encountered. The mean flow then comes in two components (see expression (39)), and the application of the Fredholm alternative leads to the system of three equations (see Eqns. (54)-(56)) for the two envelopes A and B (the premultiplier in front of C is zero). This system is overdetermined, and it means that the initial model is inconsistent (see also the comment on system (57)-(58) on page 14). To introduce a mean flow one must drop the rigid-lid approximation and consider a different model, which is discussed below.

4 Linear analysis of model without the rigid - lid approximation

It will be clear from our analysis that the rigid-lid approximation which has been implemented in [8] is not valid. Namely although $\hat{f}$ (the squared Froude number) in Eqns. (9)-(12) can be small, we cannot take it to be zero. The reason for this is that in system (9)-(12) $\hat{f}$ always appears as a premultiplier of the variable ζ (elevation of the free surface) and it turns out that the long wave component of ζ is proportional to $\frac{1}{\hat{f}}$ which means $\lim_{\hat{f} \to 0} \hat{f} \zeta \neq 0$ because ζ itself depends on $\hat{f}$. So to continue the analysis we drop the rigid-lid approximation and consider system (14)-(17).

The linear analysis of this system looks very much like the one for the rigid-lid model. The linear matrix of the problem is now

$$\mathcal{L} = \begin{pmatrix} \frac{\partial}{\partial x} + 2R & 0 & \frac{\partial}{\partial x} - R\hat{f} & R \\ 0 & \frac{\partial}{\partial x} + R & \frac{\partial}{\partial y} & 0 \\ \frac{\partial}{\partial x} & \frac{\partial}{\partial y} & \hat{f}\frac{\partial}{\partial x} & -\frac{\partial}{\partial x} \\ (\beta+1)\frac{\partial}{\partial x} & \frac{\partial}{\partial y} & 0 & \frac{\partial}{\partial t} - \frac{\delta\beta}{R}\left(\frac{\partial^2}{\partial x^2} + \frac{\partial^2}{\partial y^2}\right) \end{pmatrix}. \tag{35}$$

We find that (19) is still the eigenvector of the eigenvalue problem. The new values for the coefficients α_i now depend on $\hat{f}$ and given in Appendix A. The dispersion relation becomes more complicated and the complex frequency ω is now related to the wavenumber k as

$$\omega = \frac{l^4\pi^4 b_3 - l^2\pi^2 b_2 + b_1}{l^2\pi^2 a_2 - a_1}, \tag{36}$$

where l has the same meaning as in (19) and the coefficients

$$\begin{aligned} b_3 &= -\frac{\delta\beta\,(i\,k + 2\,R)}{R}, \\ b_2 &= [-1 + \delta\beta\left(3 - 4\,\hat{f}\right)]\,k^2 + \frac{i\left(2 - \hat{f}\right)\,\delta\beta\,k^3}{R} + i\,\frac{\delta\beta}{\gamma}\left(2 - \beta + 3\,\delta\beta\,\hat{f}\right)\,k\,R, \\ b_1 &= [1 + \beta - \delta\beta\left(1 - 4\hat{f}\right)]k^4 - i(1 - \hat{f})k^5\frac{\delta\beta}{R} - i\,\frac{\delta\,\beta}{\gamma}\left(1 + \beta + 3\,\delta\,\beta\,\hat{f}\right)\,k^3\,R, \\ a_2 &= -k + 2iR, \quad a_1 = \left(1 - \hat{f}\right)\,k^3 - i\left(1 - 4\,\hat{f}\right)\,k^2\,R + 3\,\hat{f}\,k\,R^2, \end{aligned}$$

have as a limit of $\hat{f} \to 0$ the corresponding values from [8]. The analysis of the neutral curve gives us the new values of the critical parameters (note that all of them depend now on $\hat{f}$ as well as on the other physical parameters of the system). Calculations show that the most unstable mode again corresponds to $l = 1$, and the order of the magnitude of k_c, R_c and ω_c remains the same (see Appendix A).

However we would like to emphasize that there is one point missing in this analysis of expression (36). The neutral curve which was described in Section 2 as some function $R^2 = R^2([k/l]^2)$ is not the only set of solutions of the equation $\omega_r(R,k) = 0$. If we take $l = 0$ in Eqn. (36), we will find another solution, namely $k = 0$ for arbitrary R. It corresponds to the family of uniform in y-direction solutions which are always neutral (neither grow nor decay) in the linear problem. These solutions are the eigenvectors of linear matrix (35) with $\partial_x = 0$, $\partial_y = 0$ and the correspondent eigenvalue $\omega = 0$. We find two linearly independent eigenvectors:

$$B\begin{pmatrix} -\frac{1}{2} \\ 0 \\ 0 \\ 1 \end{pmatrix} \quad \text{and} \quad C\begin{pmatrix} \frac{\hat{f}}{2} \\ 0 \\ 1 \\ 0 \end{pmatrix}. \tag{37}$$

It seems to be very natural that y-dependent structures get excited first and the uniform in y solutions are always damped. The latter if being developed would make the water go up and down following x-structure of the sandbars. The most active y-dependent mode (see z_b-component of vector (19) with $l = 1$) makes the water meander around the maxima of $cos\pi y$ keeping the potential energy of the water lowest. Modes (37) are obviously neutral because they just give corrections to the undisturbed river depth.

To see that vectors (37) are additional symmetries of the original equations (9)-(12), simply observe that $u = \sqrt{1 + \hat{f}\zeta - z_b}$, $v = 0$, $\zeta = \zeta$, $z_b = z_b$ is an exact solution for any values of constants ζ and z_b. Thus the modes (37) will be solutions of these equations linearized about $u = 1$, $v = \zeta = z_b = 0$. They are Goldstone modes. They are neutral. But they can be driven by gradients of the intensity of the short waves.

5 Nonlinear analysis of the model without the rigid-lid approximation

The purpose of this section is to develop a self-consistent nonlinear theory for the model. We start with the same assumptions as in the analysis for the rigid-lid model, namely, R lying within a narrow interval close to the threshold and the amplitude A in (19) being a slowly varying function so that $A = A(X, T, \tau)$ with the variables $X = \epsilon x$, $T = \epsilon t$ and $\tau = \epsilon^2 t$. However we propose a different expansion for Φ:

$$\begin{aligned}\Phi = \Phi_{steady} \quad & + \epsilon^2\Phi_{02} + \epsilon^3\Phi_{03} + \cdots \\ +e^{i(k_c x - \omega_c t)}[\epsilon\Phi_{11} & + \epsilon^2\Phi_{12} + \epsilon^3\Phi_{13} + \cdots] + \text{c.c} \\ +e^{2i(k_c x - \omega_c t)}[\epsilon^2\Phi_{22} & + \epsilon^3\Phi_{23} + \cdots] + \text{c.c} \\ +e^{3i(k_c x - \omega_c t)}[\epsilon^3\Phi_{33} & + \cdots] + \text{c.c} \\ +\epsilon^2\Phi_{mean} & + \cdots ,\end{aligned} \tag{38}$$

where Φ_{mean} is the mean flow modes which we introduce as

$$\begin{pmatrix} -\frac{1}{2} \\ 0 \\ 0 \\ 1 \end{pmatrix} B(X,T) + \begin{pmatrix} \frac{\hat{f}}{2} \\ 0 \\ 1 \\ 0 \end{pmatrix} C(X,T) \tag{39}$$

The reason for our choice of the form for the mean flow is that the set (37) for arbitrary constants B and C spans the (two-dimensional) eigenspace of the linear matrix of the problem, corresponding to the eigenvalue $\partial_t = 0$. B and C are slow varying in space and time. As we mentioned in the previous section, modes (37) do not grow linearly, but they participate in the nonlinear dynamics of the system. The physical meaning of the expansion (38) is the following: the mean flow B and C describe large scale waves on the sand and the surface respectively, whereas A describes the amplitude of the sandbars.

Rewrite system (14)-(17) as

$$\mathcal{L}\Phi = f; \quad f = f^{lin} + f^{nonlin}, \tag{40}$$

where $\mathcal{L}$ is given in (35) and Φ is given in (38). The right hand side of Eqn. (40), $f = \begin{pmatrix} f_u & f_v & f_\zeta & f_z \end{pmatrix}^T$, consists of the linear part f^{lin} and the nonlinear part f^{nonlin}. The former term comes from the "slow" derivatives in the problem and the deviation of the control parameter R from its critical value, R_c: $R = R_c + \epsilon^2 r$. The latter term contains nonlinearities, and since we will need only the three first orders in ϵ, f^{nonlin} can be expanded into a series about the solution Φ_{steady} up to the square terms:

$$f^{nonlin} = \begin{pmatrix} -R(u^2 + \frac{v^2}{2} + z^2 + 2uz_b - 2\hat{f}u\zeta - 2f\zeta z_b + \hat{f}^4\zeta^2) - u\partial_x u - v\partial_y u \\ -R(uv + vz_b - \hat{f}v\zeta) - u\partial_x v - v\partial_y v \\ \partial_x(uz_b - \hat{f}u\zeta) + \partial_y(vz_b - \hat{f}v\zeta) \\ -(\beta+1)\beta u\partial_x u - \beta v\partial_x v + \frac{\gamma\delta\beta(\beta+1)}{R}\partial_x(u\partial_x z_b) - \beta\partial_y(uv) + \frac{\gamma\delta\beta(\beta+1)}{R}\partial_y(u\partial_y z_b) \end{pmatrix} \tag{41}$$

Now we can equate terms in Eqn. (40) at different orders of ϵ. At second order, we get that all of the Φ_{p2} have the same form as the ones obtained for the rigid-lid model. The updated expressions for the coefficients are given in Appendix A.

At third order we obtain the equations governing the envelopes A, B and C. First consider the mode with $p = 1$ and $q = 3$. By applying the Fredholm condition (32), we reproduce the derivation of R. Schielen et al [8] and obtain the complex equation for A, which is equivalent to Eqn. (29) but has two additional terms:

$$\frac{\partial A}{\partial \tau} = r\,R_c(\sigma_r + i\sigma_i)A + (\gamma_r + i\gamma_i)\frac{\partial^2 A}{\partial \xi^2} + (c_r + ic_i)|A|^2 A + (m_r + im_i)AB + \hat{f}(l_r + il_i)AC \tag{42}$$

(here $\xi = X - \omega_c' T$). Then we consider the mode with $p = 0$ and $q = 3$. To apply the Fredholm alternative in this case, note that the left-eigenspace of the adjoint problem corresponding to $\partial_x = \partial_y = \partial_t = 0$ is

$$\phi = \begin{pmatrix} 0 & 0 & 1 & 0 \end{pmatrix}\lambda + \begin{pmatrix} 0 & 0 & 0 & 1 \end{pmatrix}\mu \tag{43}$$

for arbitrary λ and μ. One can substitute this into (32) and get

$$\int_0^1 f_\zeta dy = 0 \text{ and } \int_0^1 f_z dy = 0. \tag{44}$$

Again, these equations are the same as if we had averaged Eqns. (16) (conservation of water mass) and (17) (conservation of sand) in y. For this mode, the linear term in the expression for f becomes

$$f^{lin} = \begin{pmatrix} \partial_X & 0 & \partial_X & 0 \\ 0 & \partial_X & 0 & 0 \\ \partial_X & 0 & \hat{f}\partial_X & -\partial_X \\ (\beta+1)\partial_X & 0 & 0 & \partial_T \end{pmatrix} (\Phi_{02} + \Phi_{mean}). \tag{45}$$

The explicit expressions for f_ζ^{nonlin} and f_z^{nonlin} are:

$$f_\zeta^{nonlin} = \partial_X(u_{11}z_{11}^* - \hat{f}u_{11}\zeta_{11}^* + c.c.) + \partial_y(v_{12}z_{11}^* + v_{11}^*z_{12} - \hat{f}v_{12}\zeta_{11}^* - \hat{f}v_{11}^*\zeta_{12}); \quad (46)$$

$$\begin{aligned} f_z^{nonlin} &= -\beta(\beta+1)\partial_X(u_{11}u_{11}^*) - \beta\partial_X(v_{11}v_{11}^*) - \frac{\gamma\delta\beta(\beta+1)}{R}ik\partial_X(u_{11}z_{11}^* - u_{11}^*z_{11}) \\ &- \beta\partial_y(u_{12}v_{11}^* + u_{11}^*v_{12}) + \frac{\gamma\delta\beta(\beta+1)}{R}\partial_y(u_{12}\partial_y z_{11}^* + u_{11}^*\partial_y z_{12}). \end{aligned} \quad (47)$$

Evaluating (44) leads to two real equations for $|A|^2$, B and C:

$$\frac{\partial B}{\partial T} = \frac{\beta+1}{2}\frac{\partial B}{\partial X} - \frac{\beta+1}{2}\hat{f}\frac{\partial C}{\partial X} + n_1\frac{\partial |A|^2}{\partial X}, \quad (48)$$

$$0 = n_2\frac{\partial |A|^2}{\partial X} + \frac{3}{2}\frac{\partial B}{\partial X} - \frac{3\hat{f}}{2}\frac{\partial C}{\partial X}. \quad (49)$$

We introduced the new coefficients

$$\begin{aligned} n_1 &= -\frac{\beta+1}{2}(u_{02s} + u_{02c}) - \frac{(\beta+1)\beta}{2}|\alpha_u|^2 - \frac{\beta}{2}|\alpha_v|^2 \\ &+ \frac{\gamma\delta\beta(\beta+1)}{R}k_c Im(\alpha_u) \sim O(\delta^{\frac{1}{2}}, \hat{f}); \end{aligned} \quad (50)$$

$$n_2 = \frac{1}{2}\Big[2Re(\alpha_u) - (u_{02_s} + u_{02_c}) - \hat{f}(\alpha_u\alpha_\zeta^* + \alpha_u^*\alpha_\zeta)\Big] \sim O(\delta^{\frac{1}{2}}, \hat{f}). \quad (51)$$

To write down all of the obtained equations in terms of the original space and time variables, one has to take into account the simple relations:

$$\partial_x A = \epsilon\partial_\xi A = \epsilon\partial_X A; \;\; \partial_x B = \epsilon\partial_X B; \;\; \partial_x C = \epsilon\partial_X C; \quad (52)$$

$$\partial_t A = -\epsilon\omega_c'\partial_\xi A + \epsilon^2\partial_\tau A; \;\; \partial_t B = \epsilon\partial_T B + \epsilon^2\partial_\tau B. \quad (53)$$

We now substitute expressions (42) and (48) into Eqns. (53) and apply formulae (52). Then we multiply the obtained equation for $\partial_t A$ by ϵ and the equations for B and C by ϵ^2. Next, we rescale $\tilde{A} = \epsilon A$ and $(\tilde{B}, \tilde{C}) = \epsilon^2(B, C)$. Finally, we achieve a closed system for the three envelopes:

$$\begin{aligned} \frac{\partial \tilde{A}}{\partial t} &= -\omega_c'\frac{\partial \tilde{A}}{\partial x} + (R - R_c)(\sigma_r + i\sigma_i)\tilde{A} + (\gamma_r + i\gamma_i)\frac{\partial^2 \tilde{A}}{\partial x^2} + (c_r + ic_i)|\tilde{A}|^2\tilde{A} \\ &+ (m_r + im_i)\tilde{A}\tilde{B} + \hat{f}(l_r + il_i)\tilde{A}\tilde{C}, \end{aligned} \quad (54)$$

$$\frac{\partial \tilde{B}}{\partial t} = \frac{\beta+1}{2}\frac{\partial \tilde{B}}{\partial x} - \frac{\beta+1}{2}\hat{f}\frac{\partial \tilde{C}}{\partial x} + n_1\frac{\partial |\tilde{A}|^2}{\partial x}, \quad (55)$$

$$0 = \frac{3}{2}\frac{\partial \tilde{B}}{\partial x} - \frac{3\hat{f}}{2}\frac{\partial \tilde{C}}{\partial x} + n_2\frac{\partial |\tilde{A}|^2}{\partial x}. \quad (56)$$

Note that the group velocity $\omega' \approx 1$. All the coefficients are calculated as series in terms of $\delta^{\frac{1}{4}}$ and $\hat{f}$ and given in Appendix A. The biggest contributions to both n_1

and n_2 are of order $\delta^{\frac{1}{2}}$ and $\hat{f}$. Rigorously speaking, to determine all the quantities up to some given accuracy, we have to relate parameters δ and $\hat{f}$. But since we do not consider the finite amplitude behavior of perturbations, it is enough just to keep in mind that typical values of $\hat{f}$ are less than $\delta^{\frac{1}{2}}$. The expansions in Appendix A reflect the limiting behavior of the coefficients if $\hat{f}$ goes to zero faster than $\delta^{\frac{1}{4}}$.

6 Analysis of the results

Now let us take a close look at system (54)-(56). The first of the three equations is a complex one and describes the time-evolution of the complex envelope $\hat{A}$, whereas equations (54) and (55) have real coefficients and describe the behavior of the two real mean flows $\hat{B}$ and $\hat{C}$. How important are the mean flows and why do we have to take them into account in our analysis?

First let us notice that A in equation (42) is a function of $\xi = X - \omega_c' T$ and τ. It forces the mean flow envelopes B and C to be functions of ξ and τ as well. Now one can substitute $B(\xi,\tau)$ and $C(\xi,\tau)$ into equations (48)-(49) and note that $\frac{\partial B}{\partial T} = -\omega_c' \frac{\partial B}{\partial \xi}$. It results in the following system of linear equations for $\partial_\xi B$ and $\partial_\xi C$:

$$\left(\frac{\beta+1}{2} + \omega_c'\right) \partial_\xi B - \frac{\beta+1}{2} \hat{f}\, \partial_\xi C = -n_1\, \partial_\xi |A|^2; \tag{57}$$

$$\frac{3}{2}\, \partial_\xi B - \frac{3}{2} \hat{f}\, \partial_\xi C = -n_2\, \partial_\xi |A|^2. \tag{58}$$

It is obvious that system (57)-(58) is inconsistent if $\hat{f} = 0$. System (54)-(56) also becomes overdetermined when $\hat{f} = 0$, and any couple of solutions (A, B) violates the mass conservation law. If $\hat{f} \neq 0$, system (57)-(58) can be solved and the solution reads:

$$\partial_\xi B = \Lambda_1\, \partial_\xi |A|^2 \text{ and } \partial_\xi C = \Lambda_2\, \partial_\xi |A|^2, \tag{59}$$

where Λ_1 and Λ_2 are constants. The order of magnitude of these coefficients turns out to be

$$\Lambda_1 \sim \delta^{\frac{1}{2}}; \quad \Lambda_2 \sim \frac{\delta^{\frac{1}{2}}}{\hat{f}} \tag{60}$$

(the expansion of these quantities in terms of δ and f are given in Appendix A). We justified the statement that the limit $\hat{f} \to 0$ in the system of equations (14)-(17) is singular.

Now we can estimate quantitatively the role of the mean flows and compare the contribution of the mean flows with the contribution of the basic mode. Write down the expansion for Φ symbolically, i.e. keeping only the order of different components:

$$\Phi = \begin{pmatrix} 1 \\ 0 \\ 0 \\ 0 \end{pmatrix} + A\epsilon \begin{pmatrix} \delta^{\frac{1}{4}} \\ \delta^{\frac{1}{4}} \\ \delta^{\frac{1}{2}} \\ 1 \end{pmatrix} + |A|^2\epsilon^2 \begin{pmatrix} 1 \\ 1 \\ \delta^{\frac{1}{2}} \\ 1 \end{pmatrix} + A^2\epsilon^2 \begin{pmatrix} 1 \\ 1 \\ \delta \\ \delta^{-\frac{1}{4}} \end{pmatrix}$$

$$+|A|^2\epsilon^2\delta^{\frac{1}{2}}\begin{pmatrix} -\frac{1}{2} \\ 0 \\ 0 \\ 1 \end{pmatrix} + |A|^2\epsilon^2\frac{\delta^{\frac{1}{2}}}{\hat{f}}\begin{pmatrix} \frac{\hat{f}}{2} \\ 0 \\ 1 \\ 0 \end{pmatrix} + \cdots. \tag{61}$$

In this formula the first term corresponds to Φ_{steady}, the second stands for Φ_{11}, the third and fourth are Φ_{02} and Φ_{22} respectively and finally the last two terms denote the mean flows B and C. We can see that in the second order of ϵ the last terms give just a small corrections to the components u, v and z_b. However the correction to the elevation of the water which comes from the C mean flow term is proportional to $\frac{\delta^{\frac{1}{2}}}{\hat{f}}$ (it also follows from the expression for Λ_2). The ratio $\frac{\delta^{\frac{1}{2}}}{\hat{f}}$ can be big depending on the concrete values of δ and $\hat{f}$. For instance, for the values of the constants given in the first section, this ratio is 2.9, and for milder river slopes (i.e. smaller i_0) and larger drag coefficients this ratio can be made of order 10. If we estimate the ratio of the ζ-component of the last term in expression (61) and the ζ-component of Φ_{11} we will see that it is of the order $\frac{\epsilon}{\hat{f}}$. As it was mentioned in Section 3, ϵ must be less than $\delta^{\frac{1}{4}}$. If $\delta^{\frac{1}{4}} = 0.2$, ϵ can be 0.02 and the ratio $\frac{\epsilon}{\hat{f}} = 1.43$. It also can be shown that within physically interesting domain of parameters this ratio can be close to 4.

One more remark has to be made concerning the size of the mean flows and the limit $\hat{f} \to 0$. Expansion (61) could be a reason for confusion, because one might conclude that if $\hat{f}$ is very small, the mean flow seems to come in at the first order of ϵ and not at the second. However it does not mean that expansion (38) is not valid. If we consider only the terms related to the sandbars (envelope A), each term of the expansion is obviously less than the previous one in the series in ϵ. The mean flows are described with independent envelope variables B and C. We could introduce them in any order of the expansion for A, and after the analysis is done, the result shows explicitly in what order of small parameters the mean flows make their contribution. One has to remember that the singularity in the limit $\hat{f} \to 0$ is not a true one. It can be excluded by the right scaling (we discussed this in the introduction). The scaling we chose for ζ is motivated by the paper of R. Schielen et al [8], but it follows from the analysis that the right order of magnitude for the long-wave distortions of the surface is h. When we go back to the original variables, we can see that there is no inconsistency in orders.

Let us consider a real physical situation, when the depth of the river is 2m, the slope $i_0 = 0.0001$, the drag coefficient $C = 0.007$ and $\delta = 0.0017$. In this case the velocity of the steady-state flow $u^* = 53$cm/s. If we take the value of the control parameter above threshhold $|R - R_c| \approx 0.01$ (which corresponds to $\epsilon = 0.1$), the dimensional value for the unstable mode is:

$$\begin{pmatrix} u \\ v \\ \zeta \\ z_b \end{pmatrix} \approx \begin{pmatrix} 0.4\text{cm/s} \\ 0.4\text{cm/s} \\ 0.03\text{mm} \\ 6\text{cm} \end{pmatrix}. \tag{62}$$

The values for the bottom and surface mean flow modes will read respectively:

$$\Phi_B \approx \begin{pmatrix} -0.004\text{mm/s} \\ 0 \\ 0 \\ 0.03\text{mm} \end{pmatrix}; \quad \Phi_C \approx \begin{pmatrix} 0.004\text{mm/s} \\ 0 \\ 0.03\text{mm} \\ 0 \end{pmatrix}. \tag{63}$$

From the analysis performed we can conclude that the amplitude of the surface distortion mean flow can be of order or even bigger than the amplitude of the short-scale structures and it is clear that one can not neglect this effect in the analysis. Moreover, this effect can be used in practice, namely, it allows to detect sandbars by means of observation the surface deformation. It can be understood in the following way. As it has been shown,

$$\zeta_{mean} = \frac{\epsilon^2}{\hat{f}}\,\delta^{\frac{1}{2}}\,\Lambda_2\,|A|^2, \text{ and} \tag{64}$$

$$\zeta_{basic} = \epsilon\,\alpha_\zeta\,A\,e^{i(kx-\omega t)} + c.c. \quad \propto \quad \epsilon\,\delta^{\frac{1}{2}}\,|A|\,\cos\left(kx - \omega t + \phi\right), \tag{65}$$

so one can predict the large scale bottom features just by looking at the surface of the opaque water.

Now return to the analysis of equations (54)-(56). It can be seen right away that the first two equations contain the time derivatives of A and B respectively. The last equation has 0 in the right hand side. The reason for the absence of $\partial_t C$ in the system can be explained physically, if we look at system of equations (9)-(12). We know that the both A and B variables describe the motion of sand, whereas the envelope C is associated with the motion of the fluid. In other words, the time derivatives in equations (54) and (55) come from the term $\frac{\partial z_b}{\partial t}$ in Eqn.(12). Equations (9)-(11) contain time derivatives of u and ζ (that is where $\partial_t C$ could come), but they all contain small premultiplier. In our model we neglect it, which, as we have already mentioned in Section I, means that we assume instantaneous adaptation of the flow to the bed evolution.

The next thing to check is if the stability criterion for periodic solutions has been modified in comparison with the one for single equation (29).

Suppose our periodic solution has a form $A(\xi,\tau) = G\,e^{i(K\xi + W\tau)}$ with some real parameters G, K and W. In our problem it corresponds to the so-called alternate bars (bifurcating unstable waves). It is well-known fact that the wave number K obeys the following relation $K \leq h\left(|c_r|\,|\tau_{k^2}| + c_i\,\nu_{k^2}\right)$, where h stands for some known positively determined function of the coefficients of the equation of the type (29). This condition on K determines a region on the phase plane of a system where periodic solutions are stable with respect to general perturbations. Substitute B and C in the form (59) into A and get the equation of the form (29) with the corrected coefficients c_r and c_i: $c_r^{total} = c_r + c_r'$; $c_i^{total} = c_i + c_i'$, where

$$c_r \sim 1; \quad c_i \sim \delta^{\frac{1}{4}}; \quad c_r' = m_r\Lambda_1 + \hat{f}l_r\Lambda_2 \sim \delta; \quad c_i' = m_i\Lambda_1 + \hat{f}l_i\Lambda_2 \sim \delta^{\frac{3}{4}}. \quad (66)$$

If the quantity $\chi = |c_r||\gamma_r| + c_i\gamma_i$ is less than zero, the system is nonlinearly unstable. As we can see the mean flow gives the correction to the region of nonlinear stability only in higher orders of $\delta^{\frac{1}{2}}$ and $\hat{f}$.

However even there is only a little modification of the stability criterion, the presence of an instability leads to a nontrivial $\frac{\partial |A|^2}{\partial x}$, namely, if $|c_r||\gamma_r| + c_i\gamma_i < 0$, there is no x-independent solution $|A|^2$ satisfying Eqn. (54). Therefore, for the parameters within the instability domain, the intensity (which is $\sim |A|^2$) varies in space, and this is the necessary condition for the mean modes forming, independent of whether the mean flow causes this instability.

Finally, as it has been mentioned before, application of the Fredholm condition yields equations which are identical to the ones obtained by the spatial averaging of continuity equations (16)-(17), namely, Eqns. (55)-(56). One can compare different terms in expressions (50) and (51) and keep track of where the biggest terms come from. We will see that among the terms of equations (16) and (17) the largest contributions to the coefficients n_1 and n_2 are made by:

1. The flow term of the sediment flux $(\beta+1)\partial_x u + (\beta+1)\beta u\partial_x u + \beta v\partial_x v \rightarrow$ to n_1;

2. The down-slope component of the sediment motion $\partial_x(1 - uz_b) \rightarrow$ to n_2.

7 General discussion

There are a lot of interesting consequences due to the presence of the mean flows in nonlinear systems. For example, the mean flows participate in the dynamics of the system and there can be a situation in which the mean field quenches the short-scale instability. Imagine that the gradient of $|A|^2$ leads to a local increase in B and/or C as x increases (or decreases). In Eqn. (54) the linear growth (the time derivative on the right hand side) can be balanced or even damped if the mean flow amplitudes are large enough and if the expression $m_rB + l_r\hat{f}C$ has the proper sign. As soon as the values B and C are large enough, the basic wave gets quenched and only the long-wave mean flow is present. Of course, immediately after this point, there is nothing to drive the mean fields any more, so B and C will decrease and the wave can return, which starts the process over again.

We have solved the problem of mean flows driven by sandbar instabilities for straight channels. The main ideas can be applied for a wider range of problems. One such problem could be the dynamics of tidally induced sandwaves in the shallow ocean (see [5]). There exists a variety of bottom features of different scales in tidal basins. The largest structures are called sandbanks or tidal ridges and have characteristic wave-lengths of 100 times the undisturbed water depth. This phenomenon is somewhat similar to the sandbars considered above in the case of a river, because the both features can be excited by a depth averaged fluid. Other features are called sand waves and have typical wave-lengths of ten times the local water depth. As it was shown in [5], for their excitation the vertical structure of the fluid is important. S. H. Hulscher et al [5] solved the linear problem for the largest scale features, namely, the tidal ridges, and found the period of their linear excitation to be about 200 years. It seems to be more realistic to think that the presence of the tidal ridges relies on the nonlinear effects that couple short and long waves, rather than on linear mechanisms. In other words, our hypothesis is that the tidal ridges are formed as the mean flow driven by the sandwave instabilities, just as in our case the mean field is driven by the sandbar instabilities.

To check this hypothesis we will have to abandon the depth-averaged model and consider the whole three-dimensional problem, or at least introduce some effective sediment flux which would include the influence of z-dependence of the problem (as the depth-averaged model is unable to predict the growth of sand waves). This problem is quite complicated but challenging and also has some practical applications. The predictions of the dynamics of the bottom features could be used in the offshore industry and in the maintenance of navigation channels.

References

[1] Bagnold, R A. (1956) The flow of cohesionless grains in fluids. *Proc. R. Soc. Lond. A* **249**, 235-279.

[2] Colombini, M, Seminara, G and Tubino, M. (1987) Finite-amplitude alternate bars. *J. Fluid Mech.* **181**, 213-232.

[3] Cross, M C and Newell, A C. (1984) Convection patterns in large aspect ratio systems. *Physica* **10D**, 299-328.

[4] Engelund, F. (1974) Flow and bed topography in channel bends. *J. Hydr. Div., ASCE* **100**(HY11), 1631-1648.

[5] Hulscher, S J, Swart, H E de and Vriend, H J de. (1993) The generation of offshore tidal sand banks and sand waves.*Continental Shelf Research* **13** (11), 1183-1204.

[6] Passot, T and Newell, A C. (1994) Towards a universal theory for natural patterns. *Physica D* **74**, 301-352.

[7] Rijn, L C van. (1989) *Handbook of sediment transport by currents and waves.* Delft Hydraulics, delft, The Netherlands.

[8] Schielen, R, Doelman, A and Swart, H E de. (1993) On the nonlinear dynamics of free bars in straight channels. *J. Fluid Mech.* **252**, 325-356.

[9] Sedimentation Engineering. *ASCE-manuals and reports on engineering practice* (54).

[10] Sekine, M, Parker, G. (1992) Bed-load transport on transverse slope. *J. Hydraul. Engng ASCE* **118**(4), 513-535.

Appendix A

Here we show the asymptotic results for all the coefficients which appear in the above formulae. All the quantities are given as series in terms of $\delta^{\frac{1}{4}}$ up to $\delta^{\frac{1}{2}}$ and in terms of $\hat{f}$ up to first order (first we expanded them in terms of $\delta^{\frac{1}{4}}$, and then in terms of $\hat{f}$, using the fact that $\hat{f} < \delta^{\frac{1}{2}}$). These are the coefficients for the eigenvalue (19) of the linear problem:

$$\begin{aligned}
\alpha_u &= \delta^{\frac{1}{2}}\left(1+\frac{3\hat{f}}{2}-\frac{3\hat{f}}{\beta}+\left(-5-22\hat{f}+\frac{4\hat{f}}{\beta}\right)\delta^{\frac{1}{2}}\right)+\frac{i\,\delta^{\frac{1}{4}}}{\sqrt{2}}\left(1-\frac{\hat{f}}{2}+\frac{\hat{f}}{4\beta}+\frac{3\hat{f}\delta^{\frac{1}{2}}}{2}\right)+\cdots;\\
\alpha_v &= \delta^{\frac{1}{2}}\left(1+6\hat{f}\delta^{\frac{1}{2}}\right)+i\sqrt{2}\,\delta^{\frac{1}{4}}\left(\left(1+\left(\frac{1}{4}-\frac{1}{2\beta}\right)\hat{f}\right)-\left(1+\frac{(14+15\beta)\hat{f}}{4\beta}\right)\delta^{\frac{1}{2}}\right)+\cdots;\\
\alpha_\zeta &= \delta^{\frac{1}{2}}\left(-2-\hat{f}+\frac{2\hat{f}}{\beta}+\left(3+16\hat{f}-\frac{8\hat{f}}{\beta}\right)\delta^{\frac{1}{2}}\right)+\\
&\quad i\sqrt{2}\,\delta^{\frac{3}{4}}\left(2+\frac{\hat{f}}{2}-\frac{\hat{f}}{\beta}+\left(1-\frac{9\hat{f}}{4}+\frac{9\hat{f}}{2\beta}\right)\delta^{\frac{1}{2}}\right)+\cdots.
\end{aligned}$$

The critical wave number, the control parameter and the critical frequency are:

$$\begin{aligned}
k_c &= \pi\sqrt{2}\delta^{\frac{1}{4}}\left(1+\frac{\hat{f}}{4}-\frac{\hat{f}}{2\beta}+4\hat{f}\delta^{\frac{1}{2}}+\left(\frac{19}{4}+\frac{741\hat{f}}{16}+\frac{233}{8\beta}\right)\delta\right)+\cdots;\\
R_c &= \pi\delta^{\frac{1}{2}}\left(1+\left(2-\hat{f}+\frac{2\hat{f}}{\beta}\right)\delta^{\frac{1}{2}}+\left(4-\frac{\hat{f}}{2}-\frac{17\hat{f}}{\beta}\right)\delta\right)+\cdots;\\
\omega_c &= k_c\left(1+\delta^{\frac{1}{2}}\left(\beta-\hat{f}+\frac{3\beta\hat{f}}{2}\right)-\delta(5\beta-\hat{f}+22\beta\hat{f})\right)+\cdots.
\end{aligned}$$

The following coefficients refer to the formula (38).

$$\Phi_{02} = \epsilon^2|A|^2\begin{pmatrix} u_{02_s}\sin^2\pi y+u_{02_c}\cos^2\pi y\\ v_{02}\sin 2\pi y\\ \zeta_{02}\cos 2\pi y\\ z_{02}\cos 2\pi y\end{pmatrix};\quad \Phi_{22}=\epsilon^2A^2\begin{pmatrix} u_{22_s}\sin^2\pi y+u_{22_c}\cos^2\pi y\\ v_{22}\sin 2\pi y\\ \zeta_{22_s}\sin^2\pi y+\zeta_{22_c}\cos^2\pi y\\ z_{22_s}\sin^2\pi y+z_{22_c}\cos^2\pi y\end{pmatrix};$$

$$\Phi_{12} = \epsilon^2\left(-i\frac{\partial A}{\partial\xi}\right)\begin{pmatrix}\beta_u\cos\pi y\\ \beta_v\sin\pi y\\ \beta_\zeta\cos\pi y\\ 0\end{pmatrix}+\epsilon^2\tilde{A}\begin{pmatrix}\alpha_u & \cos\pi y\\ \alpha_v & \sin\pi y\\ \alpha_\zeta & \cos\pi y\\ & \cos\pi y\end{pmatrix},$$

$$\begin{aligned}
u_{02s} &= \frac{3\beta-1}{4\beta}-\frac{3\hat{f}}{16\beta}(\beta+1)+\\
&\quad\left(-3-\frac{(\beta+1)(\beta+2)}{4\beta}-\frac{\hat{f}(6\beta^3+7\beta^2+11\beta+8)}{16\beta^2}\right)\delta^{\frac{1}{2}}+\cdots;\\
u_{02c} &= -\frac{3\beta-1}{4\beta}-\frac{\hat{f}}{16\beta}(\beta+1)+
\end{aligned}$$

$$\left(-\frac{5}{2}+\frac{(\beta+1)(\beta+2)}{4\beta}+\frac{\hat{f}(6\beta^3-85\beta^2+99\beta+8)}{16\beta^2}\right)\delta^{\frac{1}{2}}+\cdots;$$

$$
\begin{aligned}
v_{02} &= \delta^{\frac{1}{2}}+4\hat{f}\delta+\cdots;\\
\zeta_{02} &= \left(4+2\hat{f}-\frac{4\hat{f}}{\beta}\right)\delta^{\frac{1}{2}}+\left(-1-\frac{65\hat{f}}{2}+\frac{63\hat{f}}{2\beta}\right)\delta+\cdots;\\
z_{02} &= \frac{(1+\beta)(-4+\hat{f})}{8\beta}+\left(-\frac{(\beta+1)(\beta+2)}{2\beta}\frac{\hat{f}(6\beta^3-13\beta^2+7\beta+8)}{8\beta^2}\right)\delta^{\frac{1}{2}}+\cdots.\\
u_{22s} &= -\frac{1}{6\beta}+\frac{i\sqrt{2}\delta^{\frac{1}{4}}}{12\beta}[5-\beta+\\
&\quad \frac{\hat{f}}{2\beta}(20\beta-19)]+\frac{\delta^{\frac{1}{2}}}{12\beta}[\beta^2+18\beta-3+\frac{\hat{f}}{4\beta}(-5\beta^3-55\beta^2+34\beta+42)]+\cdots\\
u_{22c} &= \frac{1}{6\beta}+\frac{i\sqrt{2}\delta^{\frac{1}{4}}}{12\beta}[\beta-11-\frac{\hat{f}}{\beta}(\beta-14)]+\\
&\quad \frac{\delta^{\frac{1}{2}}}{12\beta}[2\beta^2+37\beta-10+\frac{\hat{f}}{4\beta}(-7\beta^3+61\beta^2-172\beta+186)]+\cdots\\
v_{22} &= \frac{2}{3\beta}[1+\frac{\hat{f}}{\beta}(\beta-2)]+\frac{i\sqrt{2}\delta^{\frac{1}{4}}}{6\beta}[5\beta-3+\frac{\hat{f}}{\beta}(\beta^2-4\beta+1)]+\\
&\quad \frac{\delta^{\frac{1}{2}}}{6\beta}[\beta^2+11\beta-8+\frac{\hat{f}}{4\beta}(81\beta^2-210\beta+125)]+\cdots;\\
\zeta_{22s} &= -\frac{i\sqrt{2}\delta^{\frac{1}{4}}}{12\beta}[5-3\beta+\frac{\hat{f}}{8\beta}(9\beta^2+75\beta-78)]\\
&\quad +\frac{\delta^{\frac{1}{2}}}{24\beta}[-3\beta^2+46\beta-11+\frac{\hat{f}}{4\beta}(12\beta^3-287\beta^2-125\beta+114)]+\cdots;\\
\zeta_{22c} &= -\frac{i\sqrt{2}\delta^{\frac{1}{4}}}{12\beta}[-3\beta-11+\frac{\hat{f}}{8\beta}(9\beta^2-21\beta+114)]\\
&\quad +\frac{\delta^{\frac{1}{2}}}{24\beta}[-3\beta^2-58\beta+53+\frac{\hat{f}}{4\beta}(12\beta^3-78\beta^2+305\beta-349)]+\cdots;\\
z_{22s} &= \frac{i\sqrt{2}\delta^{-\frac{1}{4}}}{3\beta}[1+\frac{\hat{f}}{4\beta}(\beta-2)]+\frac{1}{3\beta}[1-\beta+\frac{\hat{f}}{8\beta}(\beta^2+3\beta+2)]+\frac{i\sqrt{2}\delta^{\frac{1}{4}}}{12\beta}[\beta^2+\\
&\quad 18\beta-3+\frac{\hat{f}}{4\beta}(\beta^3+76\beta^2-80\beta+71)]+\frac{\delta^{\frac{1}{2}}}{24\beta}(5\beta^2+63\beta-4+\cdots)+\cdots;\\
z_{22c} &= \frac{i\sqrt{2}\delta^{-\frac{1}{4}}}{3\beta}[1+\frac{\hat{f}}{4\beta}(\beta-2)]+\frac{1}{3\beta}[1-\beta-\frac{\hat{f}}{8\beta}(\beta^2+3\beta+2)]+\frac{i\sqrt{2}\delta^{\frac{1}{4}}}{12\beta}[-(\beta^2+\\
&\quad 24\beta+3)+\frac{\hat{f}}{4\beta}(\beta^3-40\beta^2+134\beta-53)]+\frac{\delta^{\frac{1}{2}}}{24\beta}(\beta^2-57\beta-20+\cdots)+\cdots.
\end{aligned}
$$

The following coefficients appear in Eqn. (54):

$$\frac{\partial\tilde{A}}{\partial t} = -\omega_c'\frac{\partial\tilde{A}}{\partial x}+(R-R_c)(\sigma_r+i\sigma_i)\tilde{A}+(\gamma_r+i\gamma_i)\frac{\partial^2\tilde{A}}{\partial x^2}+(c_r+ic_i)|\tilde{A}|^2\tilde{A}$$

$$+(m_r + im_i)\tilde{A}\tilde{B} + \hat{f}(l_r + il_i)\tilde{A}\tilde{C}.$$

$$\begin{aligned}
\gamma_r &= \frac{4\delta^{\frac{1}{2}}}{\pi}(2\hat{f} + \beta - \hat{f}\beta) + \cdots; \\
\gamma_i &= \frac{\sqrt{2}\delta^{\frac{1}{4}}}{8\pi}(20\beta + 6\hat{f} + 9\beta\hat{f}) + \frac{2\sqrt{2}\delta^{\frac{3}{4}}}{\pi}(-8\beta + 10\hat{f} - 9\beta\hat{f}) + \cdots; \\
\sigma_r &= 2\beta + \cdots; \qquad \sigma_i = \frac{\delta^{\frac{1}{4}}}{\sqrt{2}}(4\beta + 2\hat{f} - \beta\hat{f}) + \cdots; \\
c_r &= -\frac{2\pi}{3\beta}[1 + \frac{\hat{f}}{\beta}(\beta - 2)] + \frac{\pi\delta^{\frac{1}{2}}}{12\beta}[9\beta^3 - 8\beta^2 - 35\beta + 26 + \\
&\quad \frac{\hat{f}}{8\beta}(15\beta^4 + 54\beta^3 - 39\beta^2 + 732\beta - 758)] + \cdots; \\
c_i &= \frac{\sqrt{2}\pi\delta^{\frac{1}{4}}}{24\beta}\left(9\beta^2 - 12\beta - 1 + \frac{\hat{f}}{2\beta}(9\beta^3 - 5\beta^2 + 19\beta + 3)\right) + \cdots; \\
m_r &= \delta^{\frac{1}{2}}\pi\left(\beta + \frac{\hat{f}}{8\beta}(\beta + 1)(\beta^2 - 2)\right) + \cdots; \\
m_i &= \frac{\pi\delta^{\frac{1}{4}}}{\sqrt{2}}\left(\beta - 1 + \frac{\hat{f}}{4\beta}(\beta - 1)(\beta - 2)\right) + \cdots; \\
l_r &= -2\beta\pi\delta^{\frac{1}{2}} + \cdots; \qquad l_i = \frac{\pi\delta^{\frac{1}{4}}}{\sqrt{2}}(1 - \beta) + \cdots.
\end{aligned}$$

The coefficients in equations (55) and (56) are:

$$\frac{\partial B}{\partial t} = \frac{\beta + 1}{2}\frac{\partial B}{\partial x} - \frac{\beta + 1}{2}\hat{f}\frac{\partial C}{\partial x} + n_1\frac{\partial |A|^2}{\partial x}, \qquad 0 = n_2\frac{\partial |A|^2}{\partial x} + \frac{3}{2}\frac{\partial B}{\partial x} - \frac{3\hat{f}}{2}\frac{\partial C}{\partial x}.$$

$$\begin{aligned}
n_1 &= \frac{\hat{f}(\beta + 1)^2}{8\beta} + \frac{\delta^{\frac{1}{2}}}{4}\left(-\beta^2 + 6\beta + 11 + \frac{\hat{f}}{8\beta}(8\beta^3 + 35\beta^2 - 13\beta - 88)\right) + \cdots; \\
n_2 &= \frac{\hat{f}}{8\beta}(\beta + 1) + \delta^{\frac{1}{2}}\left(\frac{15}{4} + \frac{\hat{f}}{8\beta}(95\beta - 184)\right) + \cdots.
\end{aligned}$$

The coefficients in (59) are:

$$B = \Lambda_1|A|^2; \qquad C = \Lambda_2|A|^2.$$

$$\begin{aligned}
\Lambda_1 &= -\frac{5\hat{f}}{48\beta}(\beta + 1)^2 + \frac{\delta^{\frac{1}{2}}}{8}\left(2\beta^2 - 7\beta - 17 + \frac{\hat{f}}{12\beta}(46\beta^3 + 435\beta^2 - 69\beta - 104)\right) + \cdots; \\
\Lambda_2 &= -\frac{5\beta^2 + 8\beta + 3}{48\beta} + \frac{\delta^{\frac{1}{2}}}{8\hat{f}}(2\beta^2 - 7\beta - 7) + \frac{\delta^{\frac{1}{2}}}{96\beta}(46\beta^3 + 435\beta^2 + 311\beta - 840) + \cdots.
\end{aligned}$$

And finally the expression for the group velocity is

$$\omega_c' = 1 + \left(7\beta + \hat{f} + \frac{5}{2}\beta\hat{f}\right)\delta^{\frac{1}{2}} + \cdots.$$

Dynamical systems, temporal vs. spatio-temporal chaos, and climate

P. MANNEVILLE

Laboratoire d'Hydrodynamique, Ecole Politechnique, F-91128 Palaiseau, France

Abstract

In this contribution, we review the current understanding of chaos and spatio-temporal chaos within the framework of dissipative dynamical systems theory. Chaotic behavior is interpreted through a heuristic introduction to instability theory and the reduction to center manifold. Subsequent developments linked with the spatial extension of physical systems and confinement are discussed in terms of envelope equations, phase instability, and phase turbulence, providing hints on the nature of spatio-temporal chaos. Then we attempt to discuss climate modeling from the viewpoint of dynamical systems. Characterization of paleoclimatic variations is examined through phase-space reconstruction from time-series, and the design of low-dimensional models of climate variability is illustrated by a "box model" example of thermohaline circulation coupled to an ice sheet. We conclude by stressing the virtues and limits of the approach in terms of dynamical systems with respect to the general methodology followed in climate modeling.

Dynamical systems, nonlinearities, and chaos

The natural environment is the site of processes developing in space and time over a wide range of scales. Understanding its past evolution and attempting to predict its future requires to consider it as a *system* sampling a set of *states* and governed by *rules*. The archetype of such systems is offered by ordinary point mechanics where Newton's law entirely accounts for the motion of *degrees of freedom* in their *phase space*, yielding an initial value problem for the corresponding Hamilton equations. The generalization of this "mechanical" description directly leads to the notion of differential dynamical systems, formally written as

$$\frac{d}{dt}\mathbf{X} = \mathbf{F}_r(\mathbf{X};t) \qquad \text{with} \quad \mathbf{X} = \{X_1, \ldots, X_n\}\,, \tag{1}$$

where $\mathbf{X}$ denotes the instantaneous state of the system, $\mathbf{F}_r$ a vector field on the state space, and r some set of control parameters. The independent variable t in (1) is the usual time, tacitly assumed continuous. A continuous-time flow is then defined in the state space, and an *orbit* $\mathbf{X}(t)$ emerges from a given initial condition $\mathbf{X}_0$. Some processes are preferably described by a discrete time version of the evolution rule, namely an iteration

$$\mathbf{X}_{k+1} = \mathbf{G}_r\left(\mathbf{X}_k; k\right). \tag{2}$$

From a general standpoint, this is the case after a *Poincaré section* of the flow has been taken, a useful procedure that generalizes stroboscopic analysis of periodically-forced systems to more general cases. While degrees of freedom are discrete at the most microscopic scale, our current experience is with *continuous media* governed by partial differential equations (PDE), vis., $\mathbf{H}_r(\mathbf{X}; \partial_t, t; \partial_\mathbf{x}, \mathbf{x}) = 0$, e.g., the Navier–Stokes equations (NSE) for concrete realizations of incompressible fluid flow. However most results have been obtained in the framework of *low-dimensional* systems, i.e., iterations (2) or ordinary differential sets (1) involving a small number of variables, mainly focusing on the notion of chaos and on the scenarios of transition to turbulence, which we now briefly recall [1, 2].

The asymptotic regime, as t tends to infinity, of *deterministic*, *dissipative*, *autonomous* dynamical systems is accounted for by *attractors*. Dissipation is associated with a persistent reduction of the number of accessible states. Accordingly, if a given system is not driven sufficiently away from equilibrium, it relaxes towards a time-independent state corresponding to a *fixed point* in state space. By driving the system further away, one can obtain less trivial attractors. Control parameters can be varied to make it bifurcate toward periodic regimes (*limit cycle* in state space), or multi-periodic regimes (the attractor is then a *torus*). All these behaviors are characterized by the regular "recurrence" of states belonging to the corresponding attractors. But there is also the possibility of an aperiodic behavior called *chaos* which turns out to be generic for systems with sufficient degrees of freedom: let n be the number of degrees of freedom, i.e., the number of components of $\mathbf{X}$ in (1) or (2), then chaos can be observed in *non-invertible iterations* with $n \geq 1$, e.g., the dyadic map: $X_{k+1} = 2X_k \pmod 1$; in *invertible iterations* with $n \geq 2$, e.g., the Hénon map [3]; in *differential systems* with $n \geq 3$, e.g., the Lorenz model [4] or the Rössler system [5].

Chaos is characterized by the instability of individual trajectories on the attractor, i.e., the indefinite amplification of infinitesimal perturbations which makes two initially close trajectories diverge inexorably [1]. This property called the *sensitivity to initial conditions* [6] is easily illustrated using the dyadic map introduced above: let two trajectories be initiated at X_0 and $X_0 + \delta X_0$ then, after k iterations, the distance between them is $\delta X_k = 2^k \delta X_0$. When $\delta X_k \sim 1$, i.e., for $k \sim \log_2(1/\delta X_0)$, the trajectories are so distant that their subsequent evolutions seem independent.

Strange attractors [7] associated with chaotic behavior most often display a *fractal transverse structure* able to handle both stretching due to instability and contraction imposed by dissipation. This is well illustrated by the celebrated Hénon attractor issued from the iteration mentioned above with $a = 1.4$ and $b = 0.3$ (see pictures in [3]).

The notion of chaos as accounted for by strange attractors is then associated with long-term *unpredictability* in spite of short term *determinism*, which makes it compatible with apparent *stochasticity* [1]. However, some supplementary work is needed to understand how this can be applied to physical (or more general [8]) systems which are *a priori* not explicitly written as low-dimensional dynamical systems (e.g., PDEs).

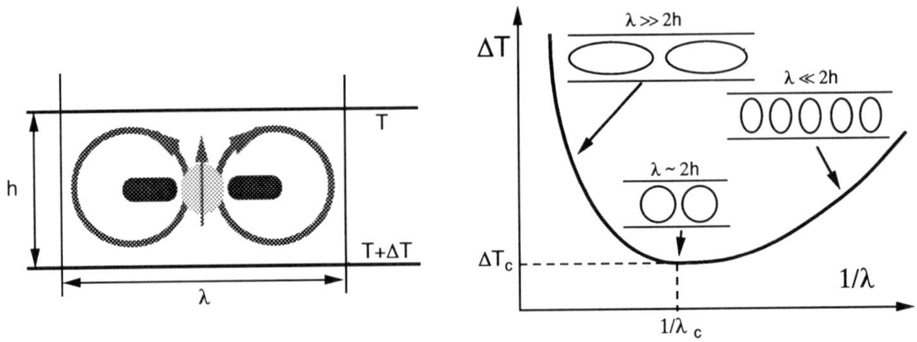

Figure 1: Rayleigh–Bénard convection. Left: a small blob of fluid slightly hotter (and lighter) than its surroundings tends to rise in an ever colder (and denser) environment, hence an instability, at least if heating is great enough; the closing of low lines yields a cellular pattern. Right: Patterns with too small or too large wavelengths are destabilized at larger values of ΔT than a pattern with $\lambda \simeq 2h$, hence a marginal stability curve, the minimum of which defines the threshold and the critical wavelength.

Dissipative structures, coherence, and slaving

A given regime can be assumed by the system under study only if it is able to withstand perturbations, which can be determined on the basis of conventional *stability analysis*. The *modes* which emerge from such an analysis express the space-time *coherence* of couplings between primitive variables (fields) on *specific scales* [2]. For example, the natural convection mode in a fluid layer of height h heated from below (see Fig. 1, left) is the result of the co-operative interaction of velocity and temperature fluctuations at a scale of the order of the layer thickness. This ends in the formation of stationary convection rolls with wavelength λ provided that the temperature difference ΔT between bottom and top is large enough so that the gain of gravitational energy by overturning the layer is not completely outweighted by heat diffusion and viscous damping (the stabilizing dissipative processes). But the latter depend on the scale of the structure, so that a well-defined wavelength $\lambda_c \sim 2h$ is favored. The corresponding pattern is therefore unstable at some critical value ΔT_c, whereas patterns with smaller or larger wavelengths are still damped at the same value of the control parameter (see Fig. 1, right).

The instability mechanism at work in a given system operates a severe selection among physical fluctuations, which in turn implies an effective reduction of the number of degrees of freedom since, out of (infinite dimensional) field solutions of the primitive equations (e.g., the NSE), the system extracts a "macroscopic mode" that can be written as

$$\mathbf{X}(\mathbf{x}, t) = A(t)\overline{\mathbf{X}}(\mathbf{x}) \tag{3}$$

where $\overline{\mathbf{X}}(\mathbf{x})$ defines the (essentially frozen) spatial structure of the mode and $A(t)$ its *amplitude* slowly function of time.

On general grounds, a whole basis of eigen-modes is obtained, on which the dynamics can be projected yielding

$$\frac{dA_m}{dt} = L_r A_m - \mathcal{N}(\{A_p; p = 1, \ldots\}) , \qquad m = 1, \ldots , \tag{4}$$

where L_r is the linear stability operator, which, when acting up on (generically non-degenerate) eigen-mode m, simply gives

$$L_r A_m = s_m A_m \qquad \text{with} \qquad s_m = \sigma_m - i\omega_m .$$

$\mathcal{N}$ gathers all nonlinear contributions to the dynamics. In this way, the *degrees of freedom* of the problem are now the amplitudes A_m; this constitutes significant progress since the coherence implied by the existence of the instability mechanism is better taken into account than by using an arbitrary basis. Furthermore, these degrees of freedom can be divided into two sets according to the real part σ of their evolution rate s [9]: *nearly neutral* ("nn") modes with $\sigma_m \equiv \sigma_{\rm nn} \simeq 0$, either unstable or stable ($\sigma_{\rm nn} > 0$ or < 0 but $|\sigma_{\rm nn}|$ "small") are slowly evolving; they are often called the *central modes*. Meanwhile, *strongly stable* ("ss") modes with $\sigma_m \equiv \sigma_{\rm ss}$ negative and "large" are rapidly relaxing. As a matter of fact, the effective number of degrees of freedom can be further reduced to just the number $n_{\rm nn}$ of central modes by "adiabatic elimination" of the strongly stable modes, also called *enslaved modes* [10]. This yields an effective dynamics involving only the amplitudes $A_{\rm nn}$ of the nearly neutral modes

$$\frac{dA_{\rm nn}}{dt} = s_{\rm nn} A_{\rm nn} - \widetilde{\mathcal{N}}_{\rm nn}(A_{\rm nn}) \tag{5}$$

valid for t large enough (see Fig. 2), provided that central modes and enslaved modes have widely different time scales (i.e., a wide gap exists in the spectrum). This assumption will become problematic in particular for the extended systems we will now discuss.

Space-time effects and envelopes

In the presentation above, amplitudes are implicitly assumed to form a discrete set of variables associated with distinct physical modes. In practice this is only valid when confinement effects are strong enough, which can be appreciated from the definition of the system's *aspect ratio*:

$$\Gamma = \ell / \lambda_{\rm c} , \tag{6}$$

measuring the size ℓ of the region affected by the instability in units of the length scale $\lambda_{\rm c}$ generated by the mechanism. When Γ is small, typically of order unity, the geometry is said to be *confined* and, as a result, the situation sketched above holds without restriction. But this is no longer the case when $\ell \gg \lambda_{\rm c}$, i.e., $\Gamma \gg 1$. The system, then called *extended*, can be considered as a large assembly of individual "cells" of size $\sim \lambda_{\rm c}$. The specificities of this case arise from the fact that the strong local coherence implied

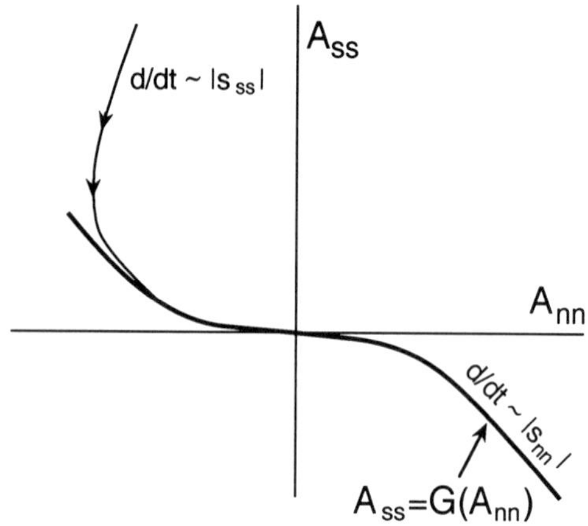

Figure 2: Adiabatic elimination of strongly stable modes: except during a brief transient, the time scale is fixed by the slow modes, so that $|dA_{\mathrm{ss}}/dt| \ll |s_{\mathrm{ss}}A_{\mathrm{ss}}|$; the evolution of the enslaved modes initially given by: $dA_{\mathrm{ss}}/dt = s_{\mathrm{ss}}A_{\mathrm{ss}} - \mathcal{N}_{\mathrm{ss}}(A_{\mathrm{nn}}, A_{\mathrm{ss}})$ can thus be solved as: $A_{\mathrm{ss}} \simeq s_{\mathrm{ss}}^{-1}\mathcal{N}_{\mathrm{ss}}(A_{\mathrm{nn}}, A_{\mathrm{ss}})$ yielding $A_{\mathrm{ss}} = G(A_{\mathrm{nn}})$ and by substitution the effective dynamical system for the A_{nn}.

by the instability mechanism within cells is not incompatible with local fluctuations of the amplitude or/and position of the cells [2], which determines the emergence of new *collective modes*. A large number of modes with similar spatial structures (except for the precise value of the wavelength) are excited even close to the instability threshold and a mere superposition of such modes within wave-packets allows for long wavelength modulations of the emerging structure. Close to the threshold, a single branch of modes is generically concerned, which results in an effective equation (5) for a single amplitude A rewritten here for convenience as

$$\tau_0 \frac{dA}{dt} = \delta r A - \widetilde{\mathcal{N}}(A)\,, \tag{7}$$

where τ_0 is a characteristic time and δr a dimensionless measure of the distance to the threshold. In the previous section, the amplitudes, here simply A, were just scalar functions of time. Now, owing to the possibility of modulations, A has to be considered as a field dependent on space and time and governed by an appropriate generalization of (7). At lowest order, modulations are expected to behave in a diffusive way, which transforms (7) into a partial differential equation:

$$\tau_0 \partial_t A = \delta r A + \xi_0^2 \partial_{xx} A - \widetilde{\mathcal{N}}(A) \tag{8}$$

known under the generic name *Ginzburg–Landau equation*.[1] Such a formulation, here restricted to one space dimension, can be extended to deal with two-dimensional pat-

[1]The explicit form of this equation depends on the nature, stationary or oscillatory, of the instability mode, and on the character, super-critical or sub-critical, of the bifurcation.

terns. It introduces itself naturally in many different contexts [13] to model processes with specific symmetries, often just at a phenomenological level.

From modulations to spatio-temporal chaos

In extended systems, the origin of modulations in patterns generated by instabilities is twofold: either *extrinsic*, i.e., due to confinement effects inducing boundary conditions on the envelope, e.g., $A(x = \pm\ell/2) = 0$; or *intrinsic*, that is, due to secondary instabilities and/or the presence of topological defects (for example, dislocations in a roll pattern or spirals in chemical waves). Here we shall consider, owing to their universal meaning [14, 2], only effects related to long wavelength secondary instabilities of envelope modes.

To understand the origin of these instabilities, let us return to expression (3) of an unstable mode accounting for the emergence of a periodic structure with wavelength $\lambda \sim \lambda_c$ in a laterally unbounded system ($\Gamma \to \infty$). For such a mode, the space-dependent part $\overline{X}(x)$ is a 2π-periodic function of a "phase" $\theta = kx$, where $k = 2\pi/\lambda$ is the local wavevector and λ the local wavelength. Above threshold ($\delta r > 0$), the modulus of the amplitude A generically saturates to a value $A_{\mathrm{sat}} \simeq \sqrt{\delta r}$, but the phase of the pattern is determined only up to a constant accounting for *translational invariance*. A neutral mode is associated with this invariance property since, if $A_{\mathrm{sat}}\overline{X}(x)$ is a solution, then $A_{\mathrm{sat}}\overline{X}(x - x_0) \simeq A_{\mathrm{sat}}\left(\overline{X}(x) - x_0\partial_x\overline{X}(x)\right)$ is also a solution for x_0 infinitesimal, so that the "perturbation" proportional to $\partial_x\overline{X}(x)$ does not relax. Accordingly, long wavelength phase fluctuations $\phi = -kx_0$ are *nearly neutral*. In the simplest cases, a simple diffusion equation obtains [15]:

$$\partial_t\phi = D\partial_{xx}\phi\,. \tag{9}$$

The effective diffusion coefficient D depends on the control parameter δr and the value of the underlying wavelength λ of the pattern. The latter dependence accounts for the possibility of *side-band* instabilities, i.e., instabilities of a pattern with given λ against the growth of a perturbation corresponding to a neighboring wavelength $\lambda+\delta\lambda$. Indeed, as long as D is positive, phase fluctuations relax and the pattern is stable; but when D becomes negative, phase fluctuations diverge at the linear stage described by (9). In some cases, the magnitude of phase fluctuations arising from this phase instability remains bounded while the modulus of the amplitude stays enslaved to the phase gradient. A regime of *phase turbulence* may then develop [16], as exemplified in Fig. 3. Otherwise, defects may be created, resulting in a different regime called *defect mediated turbulence* [17], where enslaving of the modulus no longer holds everywhere. Where relevant, an equation governing phase turbulence can be obtained explicitly by a long wavelength expansion for phase fluctuations. At lowest significant order this yields the Kuramoto–Sivashinsky equation (KSE, [19]) here written as:

$$\partial_t\phi = -\partial_{xx}\phi - \partial_{xxxx}\phi - \tfrac{1}{2}\left(\partial_x\phi\right)^2\,, \tag{10}$$

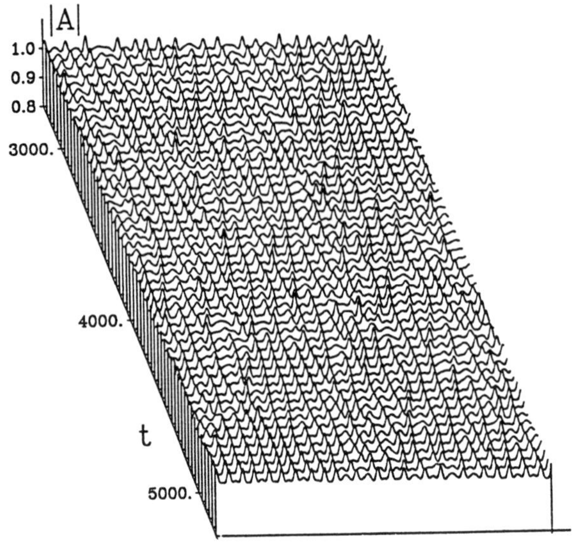

Figure 3: Phase turbulence in the one-dimensional complex Ginzburg–Landau equation: $\partial_t A = A + (1 - i\alpha)\partial_{xx} A - (1 + i\beta)|A|^2 A$ for $\alpha = 2.0$ and $\beta = 0.7$ with periodic boundary conditions at a distance $\ell = 512$. In this regime, the modulus of the envelope $|A|$, the space-time evolution of which is displayed, is enslaved to the local value of the phase gradient $\partial_x \phi$. The phase diffusion coefficient vanishes then changes its sign at the threshold of the Benjamin–Feir instability given by $\alpha\beta = 1$.

where the first term on the r.h.s. is obtained by rescaling the negative diffusion coefficient, the second term accounts for the stabilization of short wavelengths, and the third describes the nonlinearities.

Reviewing the different steps performed, we arrive at a kind of "Russian doll" picture: with instability modes generating cells (Fig. 1), then envelope equations in the generic form (8) describing modulated patterns at a scale much larger than the wavelength of the individual cells, and finally phase equations such as (10) accounting for specific slow envelope modulations close to a phase instability threshold. At each step, we can define an effective aspect ratio that generalizes the "bare" aspect ratio (6) referring to λ_c. To be explicit, at the level of the envelope equations, the coherence length is $1/\sqrt{\delta r}$ times larger than $\xi_0 \sim \lambda_c$, so that the corresponding "dressed" aspect ratio is $\Gamma \times \sqrt{\delta r} \ll \Gamma$. At the next step, close to the phase instability threshold, the length scale of phase fluctuations diverges like $|D|^{-1/2}$, which still reduces the effective aspect ratio. Owing to these successive length rescalings, it becomes clear that close to the secondary phase instability threshold, itself not far from the onset of the primary instability, the effective aspect ratio may be so small that the theory of low dimensional dynamical systems sketched previously can be applied to the phase variables. In this way, one can understand spatio-temporal chaos mostly as temporal chaos affecting a spatial phase. Of course, the picture can be noticeably more complicated when taking into consideration the possibility of a chaotic motion of defects or of some complicated phase/defect dynamics; but we can still keep on with the idea of spatio-temporal chaos viewed as temporal chaos for collective modes, extended or/and localized.[2]

The relevance of low dimensional dynamical systems in the context of spatio-temporal chaos can be slightly generalized. Indeed, the somehow deductive approach followed up until now makes reference to a well-defined regular (i.e., laminar) basic state and works by successive expansions valid close to it. The motion so described takes place on a "true" center manifold and slaving comes to eliminate out-of-purpose modes. Provided that dissipation is in some sense strong enough ("squeezing property"), this can be extended to cases lacking a reference state by the so-called *inertial manifold theory* [20]. The inertial manifold is a finite-dimensional globally attracting manifold on which one can reduce the dynamics of the system. From the point of view developed up to now, it is reassuring that such a reduced dynamics can be shown to exist for some systems (e.g., the KSE (10) and variants of it [20]). Unfortunately, the interesting case of the Navier–Stokes equations in three dimensions is still open.

Let us complete this brief overview of the transition to turbulence in extended systems by remarking that, in addition to assuming that the aspect-ratio Γ is large, the global scenario sketched above takes for granted that every step is essentially *super-critical* in the sense that the regime that sets in remains "close" to the regime that looses stability, in practice because the bifurcated state gently saturates beyond threshold. As a consequence, successive perturbation expansions remain tractable, at least in principle. When an instability is sub-critical, i.e., a bifurcated state of large amplitude can coexist with the basic state below threshold (a situation that is not accessible to linear stability theory), a quite different scenario can develop. Coexistence in phase space as alluded to above is a confined-system concept. In extended systems, this property is translated to a coexistence in physical space, with *domains* of one or the other possible states separated by *fronts*. As first pointed out by Pomeau [21], the global dynamics now involves front propagation that can display stochastic properties akin to a *contamination* process when one of the competing regimes is chaotic. This new route to turbulence called *spatio-temporal intermittency* has many features of *directed percolation*, a stochastic process studied in the framework of critical phenomena in statistical physics [22]. It has been studied at a quantitative level mostly by numerical simulation of models, and its main features have been qualitatively observed in several physical systems [23]. The concept seems particularly suited to the analysis of the transition to turbulence *via* turbulent patches, spots, puffs, slugs,..., as in Couette, Poiseuille, or boundary layer flow.

[2] Though localized, defects keep a "collective" meaning since they affect a pattern whose order results from the space-time coherence of feedbacks: usually, they are singular solutions of the envelope equations.

Climate: why and how?

Systems to which the approach just sketched can be applied are usually called *complex*, both from the "microscopic" point of view, due to the unpredictability of individual trajectories or patterns, and from the "macroscopic" point of view, in connection with the difficulty of relating the statistical properties of the system to its definition. This point was raised very early by Lorenz, who observed that the prediction of the average value of X given by the logistic iteration $X_{n+1} = rX_n(1 - X_n)$ was hard to relate to the value of r [24]. In the same spirit but in a more spatio-temporal context, the prediction of the specific behavior generated by a cellular automaton in a family of apparently similar automata seems difficult from the knowledge of the rule alone [25]. From these standpoints, we must expect difficulties in understanding (and predicting) climatic variations since the system is certainly complex by virtue of its several components: atmosphere, hydrosphere, cryosphere, lithosphere, and biosphere (consult, e.g., [26]). Each is extended in space and coupled to the others through nonlinear feedbacks and mostly at their boundaries. Each is submitted to forcings and involves processes with widely different time scales. In principle, we could think of sets of equations governing these components at the ultimate description level, but we are interested in sufficiently global features, i.e., averaged properties over wide regions and long periods of time. These averages are to be understood as *observables* deriving from a set *climatic variables* having a dynamical content and relating to most global features of general atmospheric and oceanic circulation, glaciers, vegetation, etc.

Ideally, these variables should be extracted from some inertial manifold approach to the complete set of equations alluded to above, so that the remaining fast small-scale fluctuations be considered as a noise deprived from dynamical content. This is clearly not possible since many processes are not even well formalized; but, beyond the empirical answer from observations, the theoretical problem can still be to derive some *effective* dynamics of climate and to further discuss the possible existence of some relevant *climatic attractor*. Instead of an obviously impracticable *deductive* approach, one can try a more *inductive* approach. This is what we shall sketch in the following: first presenting recent developments of dynamical systems theory as applied to the representation and interpretation of observational data, and next giving a concrete example of what can be obtained from the modeling of a specific coupling between variables supposedly relevant to climate oscillations during ice ages, before bringing some concluding remarks.

Dynamical systems and paleo-climate data analysis

The knowledge about (past) climate is gathered mainly in the form of time-series of a finite (and small) number of observables usually integrated over time between two samplings at a rate that may be more or less regular (possibly synchronized with forcing, e.g., daily, seasonal, annual) and at stations belonging to some more or less dense network (in fact, information can be quite fragmentary especially for paleo-data).

The idea borrowed from dynamical systems approach is to *reconstruct* directly from observations the phase space where relevant variables live using the assumption that a drastic reduction of the dimension of the effective system has taken place (n_{eff} small enough), since the original system is *a priori* infinite-dimensional. Taking this for granted, let us explain in a few words the so-called *method of delays* [27, 28] yielding a representation of the system in a pseudo-phase space whose coordinates are derived from the time series of some observable and which, in principle, can be put in correspondence with the true phase space of the system. Proceeding by steps, we assume the existence of a hypothetical n-dimensional discrete-time system (2) studied through some observable $Y = g(\mathbf{X})$ of interest, i.e., through a time series $Y^{(k)}$, $k = 0, 1, \ldots$. Trajectories are reconstructed in a n_{e}-dimensional space whose states are defined by[3]

$$\mathbf{V}^{(k)} = \left\{V_1^{(k)} = Y^{(k)}, V_2^{(k)} = Y^{(k+1)}, \ldots, V_{n_{\text{e}}}^{(k)} = Y^{(k+n_{\text{e}}-1)}\right\} \tag{11}$$

(subscript "e" stands for "embedding"). Of course, the original system is unknown (dimension n, dynamical law $\mathbf{F}$, definition g of the observable) but the reconstruction can be shown to be reliable provided that $n_{\text{e}} \gg n$ [28]. Problems may arise at the stage of choosing the delay, which must be sufficient for a correct account of the fastest significant time scale when taking noise into account (see Fig. 4, left). Another issue is the choice of the width of the sliding window put onto the data: this determines the embedding dimension in (11) which must be just sufficient in order to distinguish between true and false neighbors everywhere (see Fig. 4, right) [29].

Achieving a compromise between reliability (not to say accuracy) and the computational burden is a delicate matter, and the application of the method for a global characterization of the dynamics through a small set of numbers (generalized fractal dimensions) is usually bound to fail, owing to an insufficient amount of data blurred by noise [30]. Less demanding issues can be of interest however, in particular the direct study of pieces of trajectories in the reconstructed phase space, especially after having selectively extracted specific time-scales of motions by *singular spectrum analysis* [31, 32].

This data treatment method is a variant of *principal component analysis* (see [26], Appendix B) adapted to time-series. In the present context, it amounts to a change from the canonical basis in reconstructed phase space to a new basis maximizing the auto-correlation of the vectors $\{\mathbf{V}^{(k)}, k = 1, 2, \ldots\}$ in the least square sense. A whole spectrum of eigen-values is obtained, the values of which relate to the "spectral power" contained in the scales of motion described by the corresponding eigen-vectors. Given a specific spectrum, it is usually easy to distinguish a significant part from the noise and to filter the latter out. As an example, let us consider the last glaciation cycle as seen in Vostok, Antarctica (from Yiou *et al.* [33, 34]). Figure 5 displays a record

[3] The procedure can be worked out for continuous-time systems (then n is the effective dimension n_{eff} after elimination of enslaved variables) and extended to sets of observables (of different nature, or from spatial sampling of the same field).

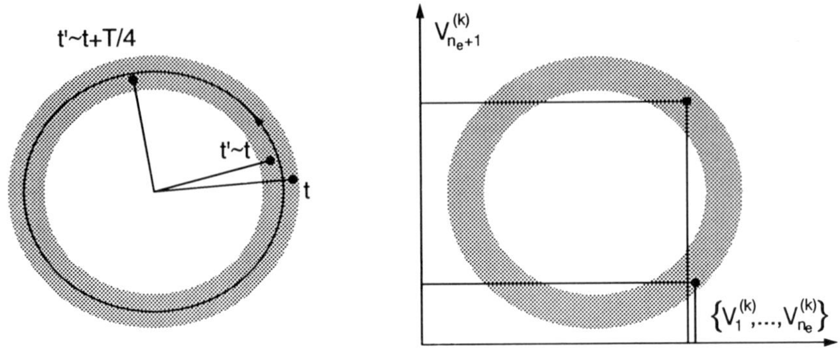

Figure 4: Left: if the delay between two samplings is not large enough, the state space distance between them may be of the same order of magnitude as the noise which affects measurements featured here by the grey band around the hypothetical trajectory corresponding to the fastest relevant oscillation mode; a good rate corresponds to having successive evaluations of the observable at the fastest significant frequency roughly in quadrature so that the amount of new information is maximized. Right: the embedding dimension has to be large enough so that the two neighboring reconstructed points with n_e remain neighbors for $n_e + 1$; if not, n_e is too small.

of the temperature as derived from deuterium measurement of the ice extracted from a drill in the ice cap. Analyzing the data as explained above, one gets the spectrum in Fig. 5 (inset), from which one can isolate a first group of five leading eigen-values corresponding to the trend and low frequency behavior (the light line through the data), then a second group of nine eigen-values associated with higher frequencies, and finally a last group of smaller eigen-values, which it is tempting to associate with stochastic noise. A remarkable fact is the presence of a pair of nearly degenerate eigenvalues (6 and 7) with eigen-vectors roughly in quadrature. Projecting the data on this sub-space yields a slightly modulated periodic component with a period of the order of 6 kyr (dotted line through the data).

For the moment, the analysis has not gone very far beyond this "characterization" stage [35]. It has been developed not only for paleo-data but also, e.g., to study recently recorded "global warming" [32]. The next step would be to raise it to the level of "phenomenological modeling" by building a nonlinear model, i.e., finding maps or differential systems that reproduce the observed trends. Several strategies have been developed to do that, from local/global fitting of flows in the pseudo-phase space [36, 37] to black-box approaches in terms of neural networks, each with its own range of applicability but seeming to work well for truly low dimensional systems [38]. Thus, difficulties are likely to be faced for systems emerging from space-time developing processes but adaptations can be envisaged [39, 40]. After the dynamics has been learned in some way, the third step is of course "prediction" but, as far as climate is concerned, this has been attempted only in very few cases (e.g., the El Niño–Southern

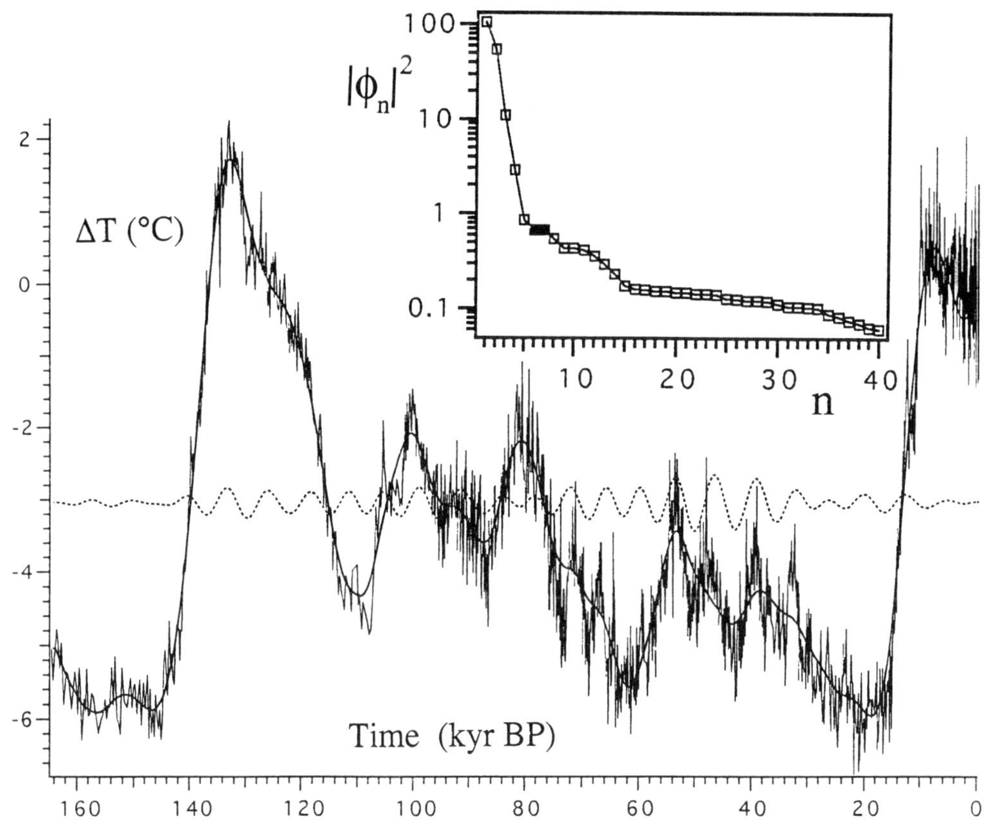

Figure 5: Temperature at Vostok during the last glaciation cycle. Time is running to the right. The last glacial extremum at -18 kyr and the previous inter-glacial stage around -115 kyr are easily identified. Inset: Singular spectrum analysis of the data; embedding dimension is 40; the pair of nearly degenerate eigen-value is marked by solid squares. Courtesy of P. Yiou.

Oscillation index [35]). Were prediction fully reliable, differences with observations over the most recent period could then be reliably attributed to anthropogenic sources.

Dynamical systems and modeling

The modeling of climate dynamics can be developed at several levels of space-time resolution [41]. Leaving for the moment problems associated with the most microscopic approach taken in general circulation models (GCMs), we focus here on the inductive modeling of climate by means of simplified systems involving a small number of relevant variables, since they fit most obviously the framework of dynamical systems theory.

Such modeling always starts with a conceptual description of processes and feedbacks that are supposed to be phenomenologically important in a given context, e.g., the temperature-albedo feedback (increasing temperature lowers the albedo which, by decreasing the amount of energy reflected back to space, further increases the temperature), or thermohaline circulation. General physical (conservation) laws can then be put in the form of coupled balance equations for "macroscopic" variables associated with large domains characterized by some identifying property: a hemisphere, an ocean basin, an ice cap, etc. [42]. Difficulties appear with the physico-mathematical description of sinks and sources within a domain, and transfer terms between neighboring domains.[4] Estimating relevant variation ranges for the parameters entering the specified analytic forms is also an important step, since a complete study of the obtained system, if sophisticated enough, is likely to prove too time-consuming.

As an example of the implementation of a conceptual model, let us consider the ocean response to iceberg surge during glaciations. Recently, a roughly periodic (6-10 kyr) series of iceberg discharges into the North Atlantic ocean has begun to be well documented [43]. These so-called Heinrich events are revealed by layers of detritus in sediments off Western Europe and have been correlated with specific temperature fluctuations over the central Greenland (Dansgaard–Oeschger oscillations). In Figure 6, evidence is given from three indicators related to the temperature of precipitations over the Greenland ice sheet derived from the isotopic fractions of ^{18}O in ice collected by the GRIP drilling project, with the temperature and the salinity of surface waters in the middle of the North-Atlantic basin derived from the composition of foraminifera relics in the sediments.

The conceptual climatic mechanism proposed by Paillard and Labeyrie [43] results from a coupling between thermohaline circulation in the North Atlantic ocean and the Laurentide ice sheet over Canada, both described using simplified models, as sketched in Fig. 7 (top). The ocean model is a three-box system adapted from Birchfield [44] with a high latitude box, and two low-latitude boxes (surface and bottom). They exchange heat and salinity through advection (variable m) and diffusion at their boundaries. They are forced by heat import (export) at low (high) latitude. Fresh water and latent heat are transported by a fictitious atmosphere and part of the fresh water can be stored in the ice sheet. The ice model is a simplified version of that derived by MacAyeal [45]. It has basically two regimes, one of slow accumulation (variable $H(t)$), and one of fast surge, with switching between the two depending on the "frozen/melted" state of the ground (controlled by T_{b}). The fresh water cycle influencing the thermohaline circulation is then enslaved to the accumulation/melting stages of the ice volume variations. When running the coupled model, relaxation oscillations are observed (Fig. 7, bottom). According to Paillard and Labeyrie, this can explain the observed signatures of the events in Fig. 6, especially the temperature increase just after an event.

By varying the parameters, a bifurcation analysis typical of low-dimensional dissipative dynamical systems can then be performed. Of course, the selection of processes

[4]At this stage it should be noticed that the often invoked existence of threshold processes may be hard to reconcile with genericity assumptions yielding smoother mathematical behavior.

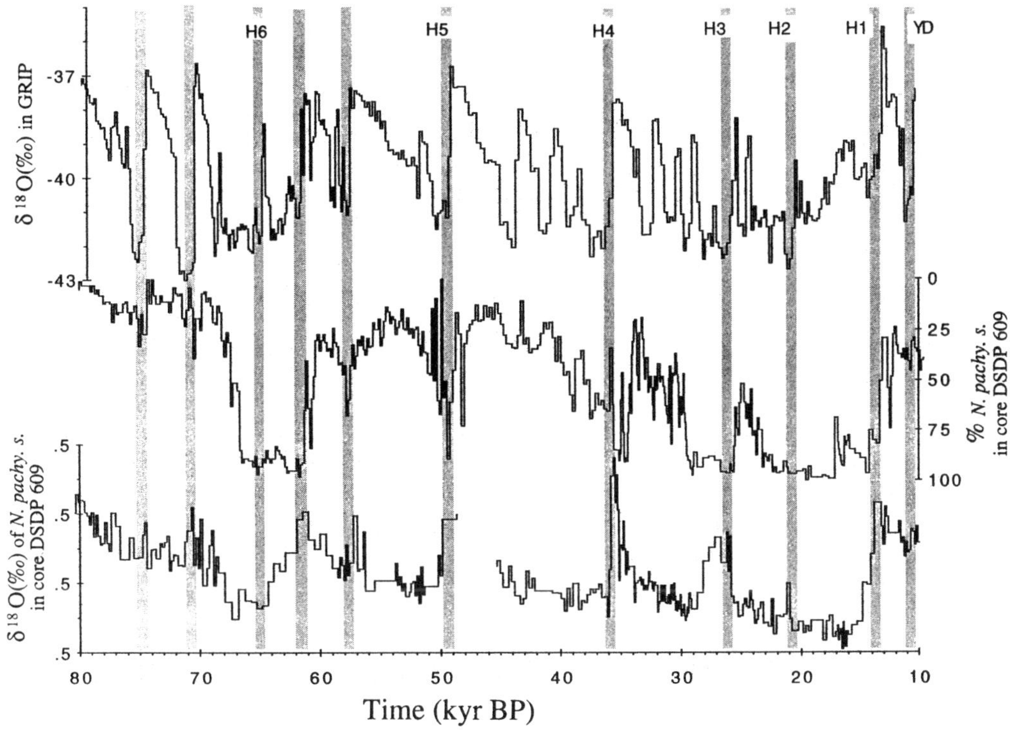

Figure 6: Observation of fast climatic fluctuations during the last glaciation from three series of observations. Top: observable linked to the temperature of precipitation over central Greenland. Middle: observable associated with the surface temperature of the North-Atlantic ocean. Bottom: observable mostly related with the salinity perturbations due to the discharge of fresh water from iceberg melt. Grey bands labeled H_n are the Heinrich events as deduced from the granulometry of sediments: iceberg surge associated with rafting of grosser debris. Courtesy of D. Paillard.

included in the conceptual model may be questioned. The interpretation of data would be more convincing if the variables retained could be shown to derive "spontaneously" *via* slaving from a more resolved model. Furthermore, other couplings also known to produce oscillations with appropriately tuned parameters could be held responsible for the same observed features (e.g., Ghil–Le Treut model coupling the temperature field and an ice sheet *via* albedo feedback with possibly isostatic adjustment of the continent under the ice weight [46]). Some models may be thought more *ad hoc* than others in this respect, but each probably contains a part of the "truth."

Discussion

Dynamical systems theory could be an appealing framework to apply to climate because it deals with the *universal* behavior of *generic* systems. However, in its present

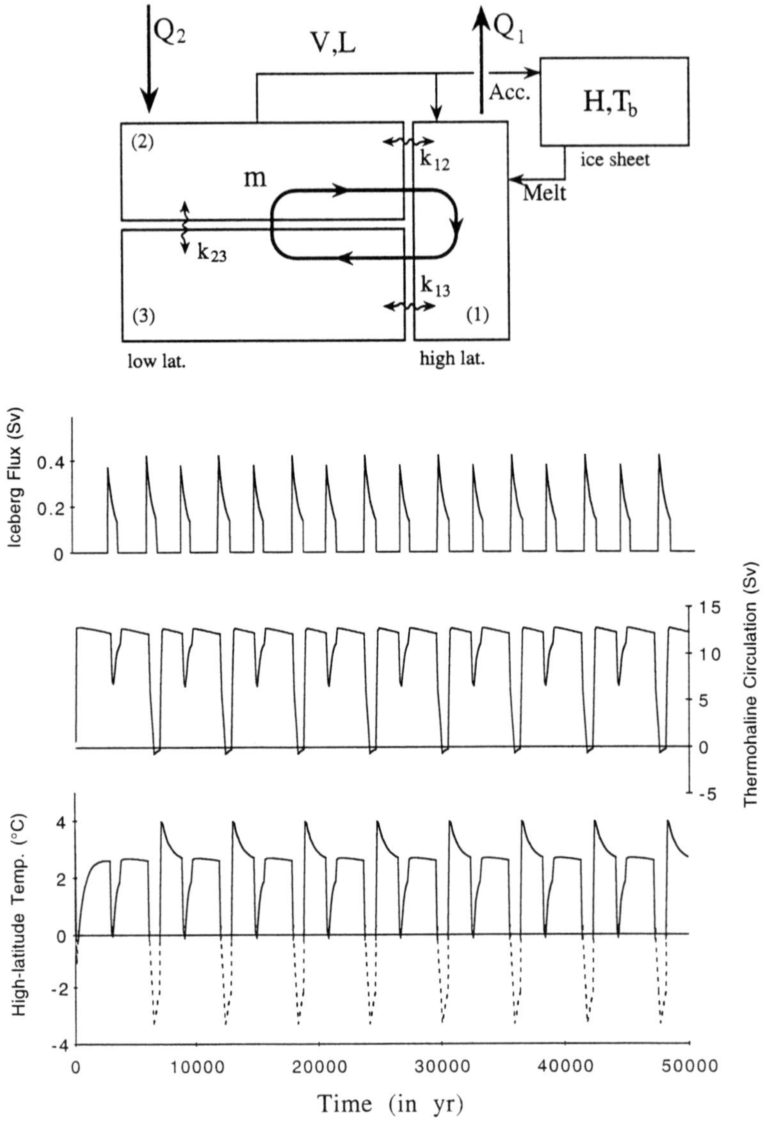

Figure 7: Top: Conceptual model of interaction between the thermohaline circulation in the North Atlantic ocean and the dynamics of the Laurentide ice sheet: three box ocean model with advection m and diffusion of heat and salinity; forcing by heat fluxes ($Q_{1,2}$); the ice sheet is characterized by its height H and the soil temperature beneath T_b. Bottom: Results of the simulation of the model. After a short transient, a regime of regular relaxation oscillations sets in. The iceberg flux mimics Heinrich events. Courtesy of D. Paillard.

state it is straightforwardly applicable only to low-dimensional systems, so that it essentially concerns the reduced models that reflect our understanding of selected feedback processes in the climate system. Another possible application deals with the advanced treatment of observational data which, for the moment, essentially comes in the form of time series of a small number of observables but is potentially extensible to space-time signals. Two illustrations were given, both related to paleo-data on ice-ages. In this restricted context, an obvious aim of the dynamical systems approach could then be to combine these methodologies to get a deeper understanding of what is going on and, in particular, to focus on the issue of universality *vs.* specificity raised by usual model-building practices, i.e., to discuss the question of how *ad hoc* the conceptual feedbacks are when they cannot be derived from first principles in a systematic way.

Of course, in this perspective it is tacitly assumed that the long term evolution involves only a small number of effective variables, whereas an infinity of degrees of freedom should enter the most microscopic description of the climatic system. In practice, one of the dominant orientations followed by present climate research is the study of so-called *general circulation models* (GCMs), i.e., atmospheric and oceanic, with land taken into account through a realistic geography and coupled to the two other components, ice and the biosphere. Such models present themselves as generalized reaction-advection-diffusion systems. The general laws of fluid dynamics are implemented with as few simplifying assumptions as possible so as to describe all scales down to those characterizing the network. The "reaction" part in these models relates to local processes (physical, biological...) and transport by sub-grid eddies, which both require adequate parameterization. At this stage, we should notice that the time-space locality assumption might fail if averaging of the most microscopic processes lead to memory and/or avalanche effects up to and beyond the grid level, and that the deterministic framework might no longer be appropriate if they add an irreducible noise component to the dynamics of scales represented in the model.

Let us however remark that the weight of the parameterization may be less crucial for some applications of GCM simulation. In particular, it has been shown that the reliability of "long term" weather forecasting (i.e., on the scale of 15 days) could be dramatically increased in certain circumstances by implementing the notion of trajectory divergence characteristic of chaotic behavior: By monitoring the divergence of the predicted and natural evolution of the meteorological fields —two trajectories equally compatible with the same set of initial conditions within observational errors— over a short period of time, one is able to correct the input data set and increase the time interval over which the computed orbit "shadows[5]" the observed one [48]. But this

[5]Strictly speaking, in dynamical systems theory "shadowing" is the established term for describing the interrelation between orbits of some well-defined system and orbits of the same system in the presence of noise; usually the observed trajectory is considered as the solution of the dynamical equations perturbed by noise, part of which is due to inadequate modeling, see [47, 36] and references therein. Here the meaning of the word is therefore somewhat deviated to stress the sensitivity to initial conditions and their improvement for better forecasting.

rather special case of a local perspective to nonlinear dynamics in state space contrasts with the needs of climate research which is concerned with global statistical properties and is thus more sensitive to the details of parameterization (cf. Lorenz' remark [24]). For example, inappropriate modeling of processes —such as cloud formation or snow aging, that change the local albedo and hence the energy balance— may produce important drifts in the long term while remaining harmless in the short term.

It might seem that the enslaving principle cherished by low-dimensional dynamical systems practitioners would be difficult to meld into the GCM approach. In fact, it is already implicitly put into practice when simulating coupled GCMs in asynchronous mode: the slow response of oceans and the biosphere with respect to the short atmospheric time scales allows one to run the different components at different rates and couple them only at regular intervals larger than the time steps used in each simulation usually fixed at a small value for numerical stability considerations. In this connection, it could be tempting to think of climatic variables as preferably linked to the naturally slow variables, but it does not seem possible to eliminate the atmospheric degrees of freedom explicitly as was done for obtaining (5). Accordingly, it is more probable that a better approach to the definition of these effective variables follows from semi-empirical state-space reconstruction further improved by singular spectrum analysis [39].

Regarding the usual framework of autonomous dynamical systems we are obliged to observe that the climatic system itself is forced at time scales somewhat larger than its own natural time scales, essentially by ice-sheet dynamics through slow modification of altitudes and sea level, and by astronomy through insolation variations. For example, it seems clear that orbital forcing rooting Milankovitch's theory of ice ages leaves its print in the data, to the point that it can serve to improve datation of sediment records [49], but the magnitude of the climatic response to this forcing is most often interpreted from a linear viewpoint and the role of nonlinear amplification is still a matter of debate since the actual cause of the major feature of ice-ages, namely the alternation of glacial and inter-glacial stages with a period of 10^5 years for about $5\,10^5$ years, remains to be ascertained. In this context, GCMs are only adapted to snapshots highlighting special periods such as the present one, or the last glacial maximum, or the start of an ice age, etc., but their success or failure to reproduce known features of ancient climates is a crucial way to test the reliability of their parameterizations at a qualitative level. By contrast, current "sensitivity experiments" attempt to discuss this reliability from a quantitative evaluation of the average distance between model predictions and some known climatology.

To conclude, dynamical systems theory may help us regarding several issues. As far as modeling and analysis are concerned, it demands essentially that the model definition and the parameterization be robust, and that the dynamical features observed be studied in the proper framework of bifurcation theory, adequately extended to include space-time effects. In the field of data processing, new tools of time series analysis can also help us in discussing the nature of couplings between variables attached to well identified sub-systems (time-scales, forcing) and assessing the role of neglected processes (noise). Finally, dynamical systems theory offers us to interpret the key point of variability at the most naive level as the result of the instability of trajectories characteristic of chaos and understand climate statistics as attractor properties. Definite conclusions about the actual causes of long-term evolution could therefore be searched for in this setting, as arising from (space-time) chaos in a dominantly autonomous *complex* system, or as being a more trivial nonlinearly enhanced response to forcing (which would not fail to have consequences on the debate about the role of mankind on the environment, and especially on global warming). For all these reasons, recent advances in nonlinear dynamics offers opportunities that deserve attention beyond its original community, with a perhaps privileged audience among climatologists.

Acknowledgements: My understanding of temporal and spatio-temporal chaos owes much to Y. Pomeau (LPS–ENS, Paris). Phase turbulence and spatio-temporal intermittency have been studied mostly in collaboration with H. Chaté (SPEC, CE Saclay). Not being a specialist of climate dynamics, I would like to thank the members of the theory group of the Environment and Climate Modeling Laboratory (LMCE) at Saclay for many interesting discussions, and especially P. Yiou and D. Paillard who provided me with original documents; B. Legras, O. Talagrand, R. Vautard from LMD, Paris, and M. Ghil from UCLA, are also deeply acknowledged for interactions. Finally, my thanks also go to the organizers of the Symposium "Nonlinear Phenomena in the Climate System" for their invitation to give this talk.

References

[1] For a introduction, consult: P. Bergé, Y. Pomeau, Ch. Vidal: *Order within Chaos* (Wiley, New York, 1984); and for a more advanced review: H.G. Schuster: *Deterministic chaos* – 2nd edition (Physics Verlag, Weinheim, 1988).

[2] The first part of the present text is a digest of the views developed in P. Manneville, *Dissipative Structures and Weak Turbulence* (Academic Press, 1990).

[3] M. Hénon. Comm. Math. Phys. **50** (1976) 69.

[4] E.N. Lorenz, J. Atmos. Sci. **20** (1963) 130.

[5] O.E. Rössler, Phys. Lett. A bf57 (1976) 397.

[6] D. Ruelle, Ann. N.Y. Acad. Sc. **316** (1979) 408.

[7] D. Ruelle, F. Takens: Comm. Math. Phys. **20**(1971)167, **23**(1971)344.

[8] D. Ruelle, "Where can one hope to profitably apply the ideas of chaos," Physics Today, July 1994.

[9] J. Guckenheimer, P. Holmes: *Nonlinear oscillations, dynamical systems and bifurcation of vector fields* (Springer-Verlag, 1983). See also: G. Iooss, "Reduction of the dynamics of a bifurcation problem using normal forms and symmetries," in E. Tirapegui, D. Villaroel, eds., *Instabilities and nonequilibrium structures* (Reidel, 1987).

[10] H. Haken, *Synergetics* (Springer-Verlag, 1983).

[11] Consult: A.C. Newell, "The dynamics of patterns: a survey," in [12], for an introduction and: A.C. Newell, T. Passot, and J. Lega, Ann. Rev. Fluid Mech. **25** (1993) 399 for more advanced topics.

[12] J.E. Wesfreid *et al.*, Eds.: *Propagation in systems far-from-equilibrium* Springer Series in Synergetics vol. 41 (Springer, Heidelberg, 1988).
[13] A.C. Newell, "Envelope equations," Lectures in Applied Mathematics **15** (1974) 157.
[14] Recent advances are reviewed by: M.C. Cross, P.C. Hohenberg, Rev. Mod. Phys. **65** (1993) 851.
[15] See, e.g., H. Brand, "Phase dynamics — a review and a perspective," in [12].
[16] H. Chaté, P. Manneville, "Phase turbulence," in [18].
[17] P. Coullet, L. Gil, J. Lega, Physica D **37** (1989) 91.
[18] O. Cardoso O., P. Tabeling, Eds.: *Turbulence: A tentative dictionary* (Plenum Press, New York, 1995).
[19] For an introduction with references, consult P. Manneville, "The Kuramoto–Sivashinsky equation: a progress report," in [12].
[20] P. Constantin, C. Foias, B. Nicolaenko, R. Temam, *Integral manifolds and inertial manifolds for dissipative partial differential equations* (Springer-Verlag, 1989).
[21] Y. Pomeau, in: A.R. Bishop, G. Grüner, B. Nicolaenko, Eds., *Spatio-temporal coherence and chaos in physical systems* (North-Holland, 1986).
[22] Directed percolation is reviewed by W. Kinzel, in: G. Deutcher *et al.*, Eds., *Percolation structures and processes*, Annals of the Israel Phys. Soc. vol. 5 (1983).
[23] See H. Chaté, P. Manneville, "Spatio-temporal intermittency," in [18] for a brief theoretical account, and F. Daviaud, "Experiments in 1D turbulence," also in [18] for a description of relevant experiments.
[24] E. N. Lorenz, Tellus **16** (1964) 1.
[25] S. Wolfram, ed., *Theory and application of cellular automata* (World Scientific, 1986); H. Chaté, P. Manneville, Progress of Theoretical Physics **87** (1992) 1.
[26] J. P. Peixoto, A. H. Oort, *Physics of Climate* (American Institute of Physics, 1992).
[27] N.H. Packard, J.P. Crutchfield, J.D. Farmer, R.S. Shaw, Phys. Rev. Lett. **45** (1980) 712.
[28] F. Takens, in *Dynamical Systems and Turbulence* D.A Rand, L.S. Young, eds. (Springer-Verlag, 1981).
[29] N.A. Gershenfeld, "An experimentalist's introduction to the observation of dynamical systems," in: Hao Bai-lin, ed., *Directions in Chaos* Vol. 2 (World Scientific, 1989).
[30] C. Nicolis, G. Nicolis, Nature **311** (1984) 529; P. Grassberger, Nature **323** (1986) 609.
[31] D.S. Broomhead, G.P. King, Physica D **20** (1986) 217.
[32] R. Vautard, P. Yiou, M. Ghil, Physica D **58** (1992) 95.
[33] P. Yiou, M. Ghil, J. Jouzel, D. Paillard, R. Vautard, Climate Dynamics **9** (1994) 371.
[34] P. Yiou, *Approches de la Dynamique du Paléoclimat: des Données et des Modèles*, PhD Thesis, University of Paris (VI), 1994.
[35] M. Ghil, M. Kimoto, J. D. Neelin, *Nonlinear dynamics and predictability in the atmospheric sciences*, Review of Geophysics, Supplement, April 1991 pp. 46–55; U.S. National report to International Union of Geodesy and Geophysics 1987–1990 (American Geophysical Union, 1991).
[36] D. J. Farmer, J. J. Sidorowich, Physica D **47** (1991) 373, and reference herein.
[37] J. P. Crutchfield, B. S. McNamara, Complex Systems **1** (1987) 417.
[38] N. A. Gershenfeld, A.S. Weigend, eds., *The Future of Time Series: Learning and Understanding* (Addison-Wesley, 1993); see in particular the lucid introduction by the editors of the book.

[39] G. Plaut, R. Vautard, J. Atmos. Sci. **51** (1994) 210.
[40] See for example: C. Uhl, R. Friedrich, H. Haken, Z. Phys. B **92** (1993) 211, and Phys. Rev. E, submitted.
[41] A. Henderson-Sellers, K. McGuffie, *A Climate Modelling Primer* (Wiley, 1987).
[42] J.-C. Duplessy, M.-T. Spyridakis, eds., *Long-Term Climatic Variations, Data and Modelling*, NATO ASI Series, Series I: Global Environmental Change, vol. 22 (Springer-Verlag, 1994).
[43] D. Paillard, L. Labeyrie, Nature **372** (1994) 162, and references therein.
[44] G. E. Birchfield, Climate Dynamics **4** (1989) 57.
[45] D.R. MacAyeal, Paleoceanography **8** (1993) 767 & 775.
[46] M. Ghil, H. Le Treut, J. Geophys. Res. **86** (1981) C6-5262.
[47] C. Grebogi, S.M. Hammel, J.A. Yorke, T. Sauer, Phys. Rev. Lett. **65** (1990) 1527.
[48] O. Talagrand, private communication.
[49] R. Peltier, in [42].

Interaction of modes with disparate scales in Rayleigh-Bénard convection

G. MANOGG[1] and P. METZENER[2]

[1] *Department of Computer Science, University of Swansea, Singleton Park, Swansea SA2 8PP, UK*

[2] *Department of Mathematics, EPFL, 1015 Lausanne, Switzerland*

Abstract

A double layer convection model is used to derive a system of equations which describes the interaction between two modes of widely different spatial scales. The mathematical system consists of a Ginzburg-Landau equation coupled with a Kuramoto-Sivashinsky type expression. They are coupled via quadratic terms which give rise to a number of higher codimension singularities. Only some of these have been studied in the past. As a special case the normal form which governs the Takens-Bogdanov singularity in strong spatial 2:1 resonance is derived and its solutions in a special scaling described. The persistence of irregular motion is demonstrated.

1 Introduction

Wave-mode interactions are widely believed to be the underlying cause of complex fluid flow. Bifurcation analyses in such situations are therefore a very popular approach to understand the dynamics that occurs in natural phenomena. Depending on the wave-lengths of the interacting wave-modes the governing normal form equations contain additional nonlinear terms which may give rise to interesting behaviour. In strong resonances (e.g. 2:1, 1:1, 1:2:3) such terms are quadratic, appear in local models and lead generically to complicated solutions. A weakly nonlinear analysis then allows us to describe the fluid flow close to equilibrium in a consistent way and shows the existence of complex dynamics in the physical system.

Fixing the wave-numbers of the fundamental solutions of the system leads to normal form equations which are ordinary differential equations. In the special case of convection this corresponds to fixing the length of the system. In the past the analysis of such expressions has produced interesting results (see e.g. Dangelmayr [3], Armbruster and Dangelmayr [1], Proctor and Jones [22, 23], Julien [10, 11], Manogg and Metzener [16], Manogg [15]). Typically, (as in strong spatial 2:1 resonance) it is shown that time dependent solutions persist close to criticality. If we allow the critical wave-numbers to vary (in the special case of convection this corresponds to assuming the system to be infinitely extended in one direction) multi-scales models give rise to systems of coupled Ginzburg-Landau (GL) equations. In both cases we suppose that both critical wave-modes possess wave-numbers of $O(1)$.

Interaction of two long wave-length (i.e. the critical wave-numbers are asymptotically small) modes (LL) was studied in the context of convection in convective shells by e.g. Metzener and Davis [17]. They found a system which consists of a stationary Ginzburg-Landau equation coupled via third order terms to a Kuramoto-Sivashinsky type expression (KS) with a cubic nonlinearity. Such cubic coupling terms indicate weak coupling and the scope for asymptotically stable time dependent solutions is rather limited.

In the past some attempts have been made to model a short wave / long wave (SL) interaction. Metzener and Proctor [18] suggested a model based on theoretical considerations which consists of a Landau equation coupled quadratically with a GL equation. It models SL interaction of two stationary wave-modes. They also considered a model of directional solidification [19] and derived a system which describes the coupling between a stationary short wavelength and an oscillatory long wavelength mode. This interaction leads to modified coupled GL equations.

The presence of long-wave instabilities leads generically to a fundamental difficulty: The canonical form of the equation which governs the long wave pattern is no longer a Ginzburg-Landau equation but rather a KS expression (see e.g. Knobloch [12] and references therein). Coupling of two wave-modes with widely different wave-numbers seems impossible as the time scales are different. If we assume the wavelength of the long scale wave-mode to be of order $1/\varepsilon$ then the time scale of the GL equation is $1/\varepsilon^2$ and in the KS expression $1/\varepsilon^4$.

This difficulty may be overcome by introducing two different and independent mechanisms which are responsible for the different wave-length instabilities. In the context of Marangoni convection Golovin *et al.* [7] derived a set of equations which describes the interaction of a long scale deformational surface wave-mode with an $O(1)$ convection wave-mode. Their system consists of a GL equation coupled with a Kuramoto-Sivashinsky type expression which contains apart from the coupling term no nonlinearity. In a truncated low order model they show presence of irregular dynamics which may be reinterpreted in terms of the underlying convective system.

Below (see also Manogg [15]) we consider two-dimensional double-layer convection and use widely different substances in the different layers to overcome the time scale difficulty. In one layer the critical wave-mode has an asymptotically small wave-number and in the other the critical pattern possesses an $O(1)$ wave-number (ordinary Rayleigh-Bénard convection). In both layers the instabilities occur simultaneously. We derive asymptotically a coupled GL - KS system that governs the dynamics in this regime.

The GL - KS system possesses quadratic coupling terms and a quadratic nonlinearity in the KS equation. The Landau - GL expression suggested by Metzener and Proctor [18] may be derived consistently from this system in a certain scaling. Their results may therefore be reinterpreted in the present framework of double-layer convection. Furthermore, we show that secondary instabilities on the basic mixed solution lead to a great variety of canonical bifurcation scenarios of higher codimension. The scope for complexity is enormous. One particular case is considered in more detail and we show the persistence of irregular motion under certain circumstances.

This article is structured as follows: We first introduce the formulation of the physical problem and explain the derivation of the coupled GL - KS system. We then derive the Landau - GL expression suggested by Metzener and Proctor [18] consistently from the GL - KS model adopting a special scaling of the coefficients. The GL - KS system possesses a mixed and a pure solution. The competition between these solutions leads to interesting canonical singularities of higher codimension. Of these singularities some have been studied before (e.g. the Takens-Bogdanow singularity (TB) in $O(2)$ symmetry). Here we concentrate on the special case of TB in strong spatial 2:1 resonance in $O(2)$ symmetry and extend results obtained in the literature. We conclude with a brief summary of the results and give an outlook on future work.

2 Formulation

We consider two two-dimensional, horizontal layers of fluid, with the independent space variables x and y. The lower layer is of depth D ($-D \le y \le 0$), the upper of depth d ($0 \le y \le d$). They are separated by a thin, heat-conducting, heated plate where we impose a heat flux F. The lower surface is held at a constant temperature T_B and at the upper horizontal plate we impose a heat flux Q. A similar system has been considered by Proctor and Jones [22, 23] as well as Julien [10]; they modelled strong spatial 2:1 resonance imposing perfectly conducting horizontal plates. The formulation here differs from theirs in the boundary conditions on the temperature and the velocities as well as the scaling.

The motion of the fluids in both layers is governed by the Navier Stokes equations coupled with the heat equation. Quantities referring to the fluid in the bottom layer are indicated with the index 1 and in the upper layer with 2. In both layers we adopt the Boussinesq approximation.

We denote $\nabla = (\partial_x, \partial_y)$ and partial derivatives are indicated by an index. Furthermore we write $r = D/d$ and impose $r = O(1)$. In the streamfunction formulation (we use Φ respectively Ψ for the streamfunctions and T respectively Θ for the temperatures in the lower respectively upper layer) the perturbed, dimensionless equations are given for $-r \le y \le 0$ (lower layer) by

$$\frac{1}{M}\left[-\nabla^2\Phi_t - (\Phi_y \nabla^2\Phi_x - \Phi_x \nabla^2\Phi_y)\right] = -(\nabla^2)^2\Phi + R_1 T_x \tag{1}$$

$$P_1[T_t + (\Phi_y T_x - \Phi_x T_y)] = -r\,\Phi_x + \nabla^2 T \tag{2}$$

and in the upper layer with $0 \le y \le 1$ by

$$-\nabla^2\Psi_t - (\Psi_y \nabla^2\Psi_x - \Psi_x \nabla^2\Psi_y) = -(\nabla^2)^2\Psi + R_2\,\Theta_x \tag{3}$$

$$P_2[\Theta_t + (\Psi_y\,\Theta_x - \Psi_x\,\Theta_y)] = -\Psi_x + \nabla^2\Theta. \tag{4}$$

Hereby we scaled both expressions using the viscous time scale of the upper layer i.e. we wrote for the lengths $(x, y) = d\,(x', y')$, for the time scale $t = \frac{\rho_2^0 d^2}{\mu_2} t'$, for the velocity $(u, v) = \frac{\mu_2}{\rho_2^0 d}(u', v')$ and for the temperatures $T = a_1 D\,T'$ and $\Theta = a_2 d\,\Theta'$. We then dropped the primes. The a_i denote the temperature gradients of the motionless state (linear temperature profiles) across the two layers and may be varied by changing the bottom temperature T_B and the heat fluxes F and Q imposed at the interface and the top of the system respectively. The superscript 0 refers to the quantities of the basic state. The ρ_i denote the densities and the μ_i the viscosities of the two fluids involved.

The dimensionless numbers are the Prandtl numbers

$$P_1 = \frac{\mu_2 \, c_1}{\kappa_1} \frac{\rho_1^0}{\rho_2^0} \;\; ; \;\; P_2 = \frac{\mu_2 \, c_2}{\kappa_2},$$

the Rayleigh numbers

$$R_1 = \frac{\rho_1^{0\,2} \, g \, \alpha_1 \, D \, a_1 \, d^3 \, c_1}{\kappa_1 \, \mu_1} \;\; ; \;\; R_2 = \frac{\rho_2^{0\,2} \, g \, \alpha_2 \, d \, a_2 \, d^3 \, c_2}{\kappa_2 \, \mu_2},$$

and a new parameter M (it is in fact the ratio of the cinematic viscosities of the two fluids) with

$$M = \frac{\mu_1}{\mu_2} \frac{\rho_2^0}{\rho_1^0}.$$

The κ_i stand for the heat conductivities, the c_i for the heat capacities and the α_i for the thermal expansion coefficients.

For the boundary conditions we introduce the abbreviations

$$k_1 = \frac{P_2}{P_1} \frac{a_2 \, d}{a_1 \, D} \;\; ; \;\; k_2 = \frac{P_1}{P_2} \frac{\kappa_1}{\kappa_2} \frac{a_1 \, D}{a_2 \, d}$$

and suppose these quantities to be small so as to have weak coupling between the temperature fields of the two layers. We introduce therefore a small positive number ε, write $k_1 = \varepsilon^2 \, k_1'$, $k_2 = \varepsilon^2 \, k_2'$ and drop the primes. The physical system after the scaling is shown in fig. 1. For the perturbed temperature Θ in the upper layer we yield a Biot condition of the form $\Theta_y + \beta \, \Theta = 0$ at the top horizontal plate i.e. $y = 1$. We require $\beta = \varepsilon^4 \, \beta'$, drop the prime and find

$$\left\{ \begin{array}{lll} y = -r : & \Phi = \Phi_{yy} = 0 \; ; \; T = 0 \\ y = 0 : & \Phi = \Phi_{yy} = 0 \; ; \; T = \varepsilon^2 \, k_1 \, \Theta \\ & \Psi = \Psi_y = 0 \; ; \; \Theta_y = \varepsilon^2 \, k_2 \, T_y \\ y = 1 : & \Psi = \Psi_{yy} = 0 \; ; \; \Theta_y + \varepsilon^4 \, \beta \, \Theta = 0 \, . \end{array} \right. \tag{5}$$

The conditions we impose on the boundaries and the separating horizontal plate represent constraints on the physical properties of the fluids. We comment on these conditions later on. Note that the coupling is purely thermal.

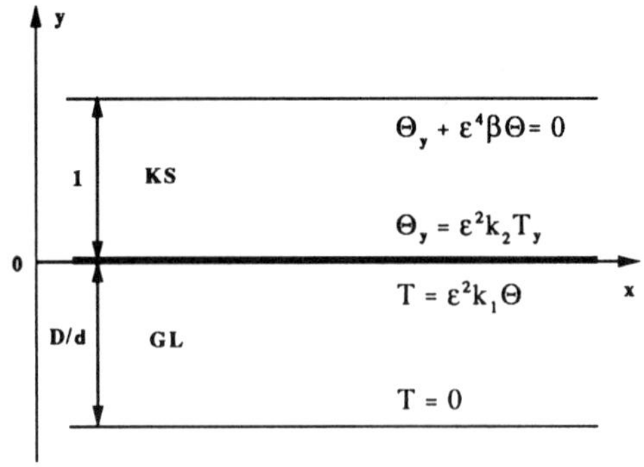

Figure 1: The experimental setup after the scaling. Note the boundary conditions imposed on the horizontal walls; they are a consequence of the physical properties of the two fluids (see below). The amplitude equations describing fluid motion are a Ginzburg-Landau equation (GL) in the lower (adiabatic boundary conditions) and a Kuramoto-Sivashinsky type expression (KS) in the upper layer (poorly conducting plates). The fluids are separated by a thin plate indicated by a bold line.

3 The reduced system

We now derive the reduced system that describes the interaction of the two critical wave-modes in the two layers. The details of the derivation shown below may be found in Manogg [15]. We use the boundary conditions on the temperature to derive the critical wave-mode in each layer assuming that they are uncoupled; in the lower layer the marginal wave-mode has a wavenumber σ_c of $O(1)$ (perfectly conducting plates), in the upper layer we have $\sigma_c << 1$ (almost insulating horizontal boundaries). We would therefore find a Ginzburg-Landau equation for the lower and a Kuramoto-Sivashinsky expression (using a long wave asymptotics) for the upper layer - if the layers were uncoupled i.e. $k_1 = k_2 = 0$. Imposing the coupling to be very weak, as done above, we introduce quadratic coupling terms into the amplitude equations. The small number ε that appeared above measures the critical wavenumber in the upper layer.

In figure 1 we show the physical setup of the system indicating the boundary conditions that we impose to achieve the weak coupling. The conditions on the separating plate are a consequence of the physical properties of the fluids involved.

The following reduction procedure is carried out in an infinitely extended layer in which we impose the solutions to be periodic in x.

3.1 The Ginzburg-Landau equation in the lower layer

Here we use equations (1 - 2) and the boundary conditions for T and Φ at $y = -r$ and $y = 0$ (see equation (5)) to derive a Ginzburg-Landau equation. This equation contains an additional quadratic coupling term which accounts for the weak temperature coupling at the separating plate.

We take R_1 to be the main bifurcation parameter of the problem and carry out a standard Lyapunov-Schmidt reduction using the following expansions

$$\begin{aligned} \Phi &= \varepsilon\Phi_1 + \varepsilon^2\Phi_2 + \dots \\ T &= \varepsilon T_1 + \varepsilon^2 T_2 + \dots \qquad (6) \\ R_1 &= R_{10} + \varepsilon^2 R_{12}\,. \qquad (7) \end{aligned}$$

The small parameter ε is chosen to be the measure of the coupling (see above). Therefore, unlike in the more classical approaches, equation (7) fixes the difference $R_1 - R_{10}$. We then rescale the time variable $s = \varepsilon^2 t$ and define $\xi = \varepsilon x$ (Newell and Whitehead [20], DiPrima *et al.* [5] and more recently Iooss *et al.* [9]). Equating powers of ε we find systems at different orders.

At lowest order we find the linear system. The amplitude of the basic mode A depends on the rescaled variables s and ξ. Φ_1 and T_1 are proportional to $\exp(i\,\sigma_0\,x)$ with $\sigma_0 = \pi/(\sqrt{2}\,r)$.

The coupling between the two layers comes at second order, i.e. ε^2 into play. This is due to the weak temperature coupling that we impose through the boundary conditions (5). Given the heat flux boundary conditions in the upper layer we imose there a long wave asymptotics. The lowest order contribution is therefore of $O(1)$ (Chapman and Proctor [2], Manogg [15]). We denote it with Θ_0. Introducing $F = 1/(2\sqrt{2}\,r)\int_{-\sqrt{2}r}^{\sqrt{2}r}\Theta_0\,dx$ with $F = F(s,\xi)$ (we average over the fundamental wavelength of the lower layer) we find for the quantities Φ_2 and T_2 the following expressions:

$$\begin{aligned} &\Phi_2 = 0 \\ &T_2 = c_1\,|A|^2\,\sin(\frac{2\,\pi}{r}\,y) + k_1\,(c_2\,y + c_3)\,F + c.c. \end{aligned}$$

where the c_i are constants. Clearly we require Φ_2 and T_2 not to lie in the kernel of the linear operator.

At third order we find the equation of bifurcation imposing the solvability condition. We recover a coupling term of the form $A\,F$ and obtain

$$A_s = \lambda\,A + \frac{3}{P_1}\,A_{\xi\xi} - k_1\,\frac{3\,\sigma_0^2}{\sqrt{r}}\,A\,F - P_1\,\frac{9}{4\,r^2}\,27\,\sigma_0^8\,\pi^2\,A|A|^2. \qquad (8)$$

Here λ is an affine function of the variation of the bifurcation parameter R_1. The form of this coupling term is no surprise as it adjusts the temperature gradient across the lower layer depending on the value of F. Its particular form is due to the choice $k_1 = O(\varepsilon^2)$.

3.2 The long wave asymptotics in the upper layer

The boundary conditions on the temperature in the upper layer suggest that the critical wave-mode will have an asymptotically long wavelength (Chapman and Proctor [2],

Knobloch [12], Manogg [15] and references therein). We therefore expect to find a Kuramoto-Sivashinsky type equation that describes the evolution of this critical wave-mode after introducing a long wave asymptotics.

The standard reduction procedure can be applied here in the case of a finite or infinitely long layer but only with some modifications:

- Classically we require a time scale with $\tau = \varepsilon^4 t$ (Chapman and Proctor [2]); this is not possible here because due to the coupling we need the same time scale as in the lower layer i.e. $s = \varepsilon^2 t$. To be able to impose the solvability condition and to find the time derivative at fourth order in ε we choose $P_2 = O(\varepsilon^2)$ for the fluid in the upper layer.

- If we choose $P_2 = O(\varepsilon^2)$ we need to break the up/down symmetry within the upper layer to retrieve nonlinear terms in the reduced expressions. The easiest way to break the up-down symmetry in the upper layer is to impose asymmetric boundary conditions on the velocity on the horizontal boundaries.

- We cannot impose a long wave asymptotics directly here; the heat coupling at the interface requires the quantities to depend on x as well as on $\xi = \varepsilon x$. If we impose $k_2 = O(\varepsilon^2)$ as done above then the dependence of the unknowns on the basic lengthscale x due to the coupling does not appear up to $O(\varepsilon^3)$ (note that $\Phi = O(\varepsilon)$ and $T = O(\varepsilon)$). We therefore introduce the notation $\langle ... \rangle = 1/(2\sqrt{2}\,r) \int_{-\sqrt{2}r}^{\sqrt{2}r} ...dx$ and write anticipating the correct scaling for the long wave asymptotics

$$\Psi = \varepsilon\,\psi(\xi, s, y) + \varepsilon^3\,\Omega(x, \xi, s, y) \tag{9}$$

$$\Theta_2 = \theta(\xi, s, y) + \varepsilon^3\,\Upsilon(x, \xi, s, y) \tag{10}$$

  where $\langle\Omega\rangle = \langle\Upsilon\rangle = 0$. The lowest order contribution of the temperature Θ_2 is $O(1)$ and corresponds to the classical long wave asymptotics scaling. We can now eliminate the lengthscale x up to fourth order from the system by integrating the governing equations over one fundamental wavelength of the lower layer.

- We need to impose $\beta = \varepsilon^4\,\beta'$ (as done above) in the boundary condition at $y = 1$, otherwise we recover additional terms at lower order than $O(\varepsilon^4)$.

Below we rewrite system (3 - 5) using the decompositions (9 - 10). We take $P_2 = \varepsilon^2 P_2'$, drop the prime and use $\nabla^2 = (\partial_x + \varepsilon\partial_\xi)^2 + \partial_y^2$. We then develop only up to the order we require to determine the reduced equation imposing the solvability condition at $O(\varepsilon^4)$ in the heat equation (Chapman and Proctor [2], Manogg [15]). This yields

$$\begin{aligned}
\varepsilon^2\,\{\psi_{yys} + [\psi_y\,\psi_{yy\xi} - \psi_\xi\,\psi_{yyy}]\} &= \psi_{yyyy} - R_2\,\theta_\xi + \varepsilon^2\,[\Phi_{yyyy} + 2\,\psi_{yy\xi\xi} + 2\,\Phi_{yyxx} \\
&\quad - R_2\,\Upsilon_x] + O(\varepsilon^3) \\
\varepsilon^4\,P_2\,\{\theta_s + [\psi_y\,\theta_\xi - \psi_\xi\,\theta_y]\} &= \theta_{yy} + \varepsilon^2\,[-\psi_\xi + \theta_{\xi\xi}] + \varepsilon^3\,[-\Phi_x + \Upsilon_{yy} + \Upsilon_{xx}] \\
&\quad + \varepsilon^4\,[2\,\Upsilon_{x\xi} + \Phi_\xi] + O(\varepsilon^5).
\end{aligned}$$

The boundary conditions read

$$\begin{cases} y = 0 : & \psi = \Phi = \psi_y = \Phi_y = 0\,;\ \theta_y + \varepsilon^3\,\Upsilon_y = \varepsilon^2\,k_2\,T_y \\ y = 1 : & \psi = \Phi = \psi_{yy} = \Phi_{yy} = 0\,;\ \theta_y + \varepsilon^3\,\Upsilon_y + \varepsilon^4\,\beta\,(\psi + \varepsilon^3\,\Upsilon) = 0 \end{cases}$$

Hereby we used $k_2 = \varepsilon^2 k_2'$ and dropped the prime. Before expanding all the terms we integrate the whole equation over one fundamental wavelength of the lower layer and make use of $\langle\Omega\rangle = \langle\Upsilon\rangle = 0$. We find

$$\varepsilon^2\,\{\psi_{yys} + [\psi_y\,\psi_{yy\xi} - \psi_\xi\,\psi_{yyy}]\} = \psi_{yyyy} - R_2\,\theta_\xi + \varepsilon^2\,2\,\psi_{yy\xi\xi} + O(\varepsilon^3) \tag{11}$$

$$\varepsilon^4\,P_2\,\{\theta_s + [\psi_y\,\theta_\xi - \psi_\xi\,\theta_y]\,\} = \theta_{yy} + \varepsilon^2\,[-\psi_\xi + \theta_{\xi\xi}] + O(\varepsilon^5) \tag{12}$$

together with

$$\begin{cases} y = 0 : & \psi = \psi_y = 0\,;\ \theta_y = \varepsilon^4\,k_2\,\langle T_{2\,y}\rangle + O(\varepsilon^5) \\ y = 1 : & \psi = \psi_{yy} = 0\,;\ \theta_y + \varepsilon^4\,\beta\,\theta = 0. \end{cases} \tag{13}$$

where we used equation (6) for T. The boundary conditions at $y = 0$ indicate that the coupling terms only arise when imposing the solvability condition at $O(\varepsilon^4)$ in the heat equation. This is due to the integration over the fundamental wavelength of T_1 (see equation 6) which we carry out on equations (3 - 5), the weak coupling ($k_2 \sim \varepsilon^2$) and the fact that the temperature T in the lower layer is of order ε.

Observe that only even powers in ε appear; we therefore use the following expansions

$$\theta = F_0 + \varepsilon^2\,F_2 + \ldots\ ;\quad \psi = G_0 + \varepsilon^2\,G_2 + \ldots\ ;\quad R_2 = R_{20} + \varepsilon^2\,R_{22} + \ldots \tag{14}$$

and take R_2 to be the bifurcation parameter. Formulating the problems at successive orders of ε^2 we find at lowest order that $F_0 = F(\xi,\,s)$. The evolution equation for F is now found carrying out the standard reduction procedure (Chapman and Proctor [2], Manogg [15]); we find $R_{20} = 320$ and at $O(\varepsilon^4)$ the solvability condition yields the following expression:

$$(P_2\,F - \frac{1}{21}\,F_{\xi\xi})_s = -\beta\,F - \frac{R_{22}}{320}\,F_{\xi\xi} - \frac{58}{693}\,F_{\xi\xi\xi\xi} + K\,|A|^2 - \frac{5}{42}\,(F_{\xi\xi}\,F_\xi)_\xi \tag{15}$$

where $K = k_2\,P_1\,R_{10}\,\sigma_0^2\,3\,\sqrt{2}\,\pi^2/r^2$. The additional term of the form $F_{\xi\xi s}$ arises because the time derivative in the momentum equation comes into play at second order in ε due to the time scale adopted here.

3.3 The coupled system

Before writing down the full coupled system, let us reflect upon the constraints that we imposed on the physical properties of the fluids involved to achieve the weak temperature coupling at the separating plate. We require in the upper layer a fluid with a small Prandtl number. Furthermore the heat conductivities as well as the densities of the two fluids need to be very different. It can be seen (Manogg [15]) that choosing in the lower layer a gas like air and in the upper layer a liquid metal such as Potassium at 340 Kelvin fits the requirements quite well when we fix ε using the Prandtl-number of the fluid in the upper layer.

Combining now equations (8) and (15) and carrying out a suitable scaling we find the following form:

$$A_t = \lambda A + a A_{xx} - A F - A|A|^2 \qquad (16)$$

$$F_t - b F_{xxt} = -\beta F - \mu F_{xx} - F_{xxxx} + \kappa |A|^2 - c(F_{xx} F_x)_x \qquad (17)$$

where

$$\lambda = \tilde{\lambda} \quad ; \quad a = \frac{3}{P_1}\sqrt{\frac{693 P_2}{58}} \quad ; \quad b = \frac{1}{21}\sqrt{\frac{693}{58 P_2}}$$

$$\beta = \frac{\tilde{\beta}}{P_2} \quad ; \quad \mu = \frac{R_{22}}{320}\sqrt{\frac{693}{58 P_2}} \quad ; \quad \kappa = \frac{\sqrt{r}}{\pi^2 P_2} k_1 k_2 4\sqrt{2} \quad ; \quad c = \frac{55\sqrt{r}}{116 k_1 \sigma_0^2} .$$

The coupled system modelling a similar interaction and which was postulated by Metzener and Proctor [18] may now be derived consistently from equations (16 - 17). To achieve this we need to impose additional conditions on the coefficients a and c. These conditions are not in conflict with the derivation so far, but no physical system is as yet known in which these conditions are satisfied. However, it is of interest to note that there exists a consistent link between the two systems. In the following ε is a small number - independent of the one used above.

Assume for now that a is relatively small and define $\varepsilon^2 = a$; we also require c to be relatively large with $c = c'/\varepsilon$ and $\kappa = \varepsilon^2 \kappa'$. We write $L(\mu) = -\beta - \mu\,\partial_{xx} - \partial_{xxxx}$ and fix μ_0 such that $L(\mu = \mu_0) F = 0$ when F is periodic in x. Scaling

$$\partial_t \to \varepsilon^2 \partial_s \; ; \quad \lambda \to \varepsilon^2 \lambda' \; ; \quad A \to \varepsilon A' \; ; \quad \kappa \to \varepsilon^2 \kappa' \; ; \quad c \to c'/\varepsilon \; ; \quad F \to \varepsilon^2 F' ,$$

writing

$$\mu = \mu_0 + \varepsilon \mu_1 + \varepsilon^2 \mu_2 + \ldots \; ; \quad F' = F_0 + \varepsilon F_1 + \varepsilon^2 F_2 + \ldots$$

and dropping primes we find

$$A_s = \lambda A + A_{xx} - A F - A|A|^2$$

for the first and

$$\varepsilon^2 [F - b F_{xx}]_s = L(\mu) F + \varepsilon c ({F_x}^2)_{xx} + \varepsilon^2 \kappa |A|^2$$

for the F equation. For the first equation we immediately find

$$A_s = \lambda A + A_{xx} - A F_0 - A|A|^2.$$

For the second equation we apply a standard reduction technique and use the above expansions to find a Landau equation from the expression in F imposing the solvability condition at $O(\varepsilon^2)$. We assume the motion to be periodic in x. In particular we require A to have the period $2\pi\, N/L$ for some fixed N and L, and we impose with $\sigma = 2\pi/L$

$$F_0 = B\, e^{i\sigma x} + c.c.$$

This corresponds to fixing the length of the domain to L and to imposing periodic boundary conditions on the vertical walls.

At lowest order we find immediately that $\mu_0 = (\sigma^4 + \beta)/\sigma^2$. Imposing the solvability condition at $O(\varepsilon)$ we set $\mu_1 = 0$ and find $F_1 = c_1\, B^2\, e^{2i\sigma x} + c.c.$ where c_1 is a real constant.

At $O(\varepsilon^2)$ we find after projecting the expressions onto $e^{-i\sigma x}$ (i.e. imposing the solvability condition)

$$B_s = \mu\, B + c_2\, \langle |A|^2\, e^{-i\sigma x} \rangle - c_3\, B\, |B|^2$$

with $\langle ... \rangle = \int_{-\pi/\sigma}^{\pi/\sigma} ...dx$ and c_2, c_3 positive constants.

We can now put everything together and write the following coupled system:

$$A_s = \lambda\, A + A_{xx} - A\,(B\, e^{i\sigma x} + \bar{B}\, e^{-i\sigma x}) - A\, |A|^2 \quad (18)$$

$$B_s = \mu\, B + c_2\, \langle |A|^2\, e^{-i\sigma x} \rangle - c_3\, B\, |B|^2 . \quad (19)$$

These expressions are identical to the ones postulated by Metzener and Proctor [18] from theoretical considerations; they investigated analytically some asymptotic cases and proved the existence of stable travelling and standing waves.

4 Stationary solutions and their stability

We return now to equations (16 - 17) and investigate the existence and stability of the fundamental stationary solutions.

At $\lambda = 0$ a mixed solution appears supercritically from the trivial state with $F = |A_0|^2\, \kappa/\beta$ and $|A_0|^2 = \beta\, \lambda/(\kappa + \beta)$. This solution describes convection with $O(1)$ wavelength in the lower and with an asymptotically long wavelength in the upper layer.

There also exists a pure state with $A = 0$ and $F \neq 0$ that bifurcates supercritically off the trivial solution at $\mu = 2\sqrt{\beta}$. This solution is purely governed by the Kuramoto-Sivashinsky expression and we will not investigate its stability further.

Secondary instabilities on the mixed state caused by the competition with the pure solution are stationary as well as oscillatory. This is established carrying out an Eckhaus stability analysis writing (see Proctor [21])

$$A = A_0\, e^{iqx}\, [1 + \alpha_1\, e^{ilx} + \bar{\alpha}_2\, e^{-ilx}] \;\; ; \;\; F = |A_0|^2\, \frac{\kappa}{\beta}\, [1 + \rho\, e^{ilx} + \bar{\rho}\, e^{-ilx}] .$$

The stability with respect to perturbations of a given wavelength is now studied in terms of the parameter μ for λ fixed (note: $|A_0|^2 = (\lambda - aq^2)/(1 + \kappa/\beta)$). In fact, choosing the remaining coefficients appropriately, interactions between stationary and oscillatory instabilities occur (see Manogg [15]). For simplicity we choose in the following $q = 0$ and fix the length of the domain imposing periodic boundary conditions on the vertical walls. This restricts the admissible wavelengths of the perturbations. Judicious choice of the length of the domain and the remaining coefficients leads to the following bifurcation scenarios:

- Takens-Bogdanov (TB) in $Z(2)$ or $O(2)$ symmetry.
- TB in strong spatial 2:1 resonance in $Z(2)$ or $O(2)$ symmetry.
- TB in oscillatory 2:1 resonance in $Z(2)$ or $O(2)$ symmetry.
- Triad in $Z(2)$ or $O(2)$ symmetry.

These interactions are described by ordinary differential equations. Various complicated stable time dependent solutions, in some cases (as we shall see below) irregular motion may be observed. The symmetry group depends on the boundary conditions imposed on the vertical walls: $Z(2)$ symmetry is induced when introducing rigid or stress free boundaries and $O(2)$ symmetry is found with periodic vertical walls.

This list is clearly not exhaustive but shows the wealth of different dynamical behaviour that one expects to see in the framework of this resonance between wave-modes of widely different wavelengths. Furthermore, the complicated dynamical behaviour which occurs in the above listed scenarios may now be reinterpreted to describe fluid motion in the physical context of the two layer convection problem.

Figure 2: The Takens-Bogdanov singularity without spatial resonance (1) and in strong spatial 2:1 resonance (2). We use μ and the length of the box l_c as the main bifurcation parameters keeping λ fixed at the critical value for $\mu = \mu_c$ and $l = l_c$.

5 The Takens-Bogdanov singularity in strong spatial 2:1 resonance

In the remainder of this article we concentrate on the above mentioned interaction that occurs under certain constraints for the secondary instability on the mixed solutions. Recall that Knobloch and Proctor [13] investigated the solutions of the Takens-Bogdanov singularity (TB) in $Z(2)$ symmetry and Dangelmayr and Knobloch [4] in $O(2)$ symmetry.

It is possible, yet a very lengthy and cumbersome process, to derive the governing equations of the Takens-Bogdanov singularity in strong spatial 2:1 resonance directly from system (16 - 17). The idea is to introduce perturbations to the mixed steady state solution (see above) that are proportional to the spatial wave modes e^{ix} and e^{2ix} (below we use the symbols u and w for the perturbations to the mixed solution). To find the normal forms (21 - 22) we then need to determine the correct magnitude (scaling) of these perturbations compared to the time scale and carry out a formal Lyapunov-Schmidt reduction. The procedure is explained for the simpler case of the Takens-Bogdanov singularity (without strong spatial resonance) in $O(2)$ symmetry in [15] (see fig. 2). There the perturbations are of $O(\varepsilon)$. The time scale is of $O(\varepsilon)$ which is in contrast to the classical choice of $O(\varepsilon^2)$ employed in the derivation of Ginzburg-Landau equations. This special scaling gives rise to second order time derivatives in the normal form equations similar to the ones seen in (21 - 22).

Here we derive the normal form equations governing TB in strong spatial 2:1 resonance in $O(2)$ symmetry (see fig. 2) using group theoretical methods. In the case of strong 2:1 resonance Dangelmayr [3] used this approach to determine the normal form equations and in the case of TB in $O(2)$ symmetry Dangelmayr and Knobloch [4] used a similar method.

We combine their method to derive the normal form up to third order. For more details concerning the derivation of the normal form equations the reader is referred to Manogg [15]. We seek a vector field with the general linearisation of the form

$$L_0 = \begin{pmatrix} 0_2 & 1_2 & 0_2 \\ 0_2 & 0_2 & 0_2 \\ 0_2 & 0_2 & 0_2 \end{pmatrix} \tag{20}$$

where

$$1_2 = \begin{pmatrix} 1 & 0 \\ 0 & 1 \end{pmatrix} \quad ; \quad 0_2 = \begin{pmatrix} 0 & 0 \\ 0 & 0 \end{pmatrix}$$

and involving the unknowns u and w with $u, w \in \mathcal{C}$ and $u \sim e^{ix}$, $w \sim e^{2ix}$. Furthermore the expressions should be invariant with respect to the action of $O(2)$ i.e. under rotations $(u, w) \to (ue^{id}, we^{2id})$, d real and under reflections $(u, w) \to (\bar{u}, \bar{w})$.

Expanding the general equivariant functions of the invariant expressions we truncate at third order. Performing then a near identity transformation to remove unnecessary terms we arrive at the following system:

$$\ddot{u} \;=\; \lambda u + \nu \dot{u} + a\,\bar{u}w + b\,w\dot{\bar{u}} + c_1\,u^2\dot{\bar{u}} + c_2\,\dot{u}|w|^2 + d\,u|w|^2 + e\,u|u|^2 \tag{21}$$

$$\dot{w} = \mu w + f u^2 + g w|u|^2 - w|w|^2 . \tag{22}$$

μ, λ and ν represent unfolding parameters. The remaining coefficients are real constants.

Several small amplitude scalings may now be carried out privileging strong stationary 2:1 resonance or TB in $O(2)$ symmetry. Both problems constitute codimension two singularities. An intermediate scaling giving rise to a codimension three system is established when balancing the magnitudes of $\partial_t^2 \sim w$ and $\partial_t w \sim u^2$. After a rescaling this yields

$$\ddot{u} = \bar{u}w - \lambda u + \nu \dot{u} + h.o.t. \tag{23}$$

$$\dot{w} = \chi u^2 + \xi w + h.o.t. \tag{24}$$

where $\chi = sgn(af)$ and $\xi = sgn(\mu)$. We restrict attention in the following to $\xi = -1$.

Equations (23 - 24) are invariant (by construction) under the action of $O(2)$ symmetry. The real subspace is therefore invariant.

Restricting these expressions to the real subspace, we find systems which depend on the sign of χ that have appeared in other physical contexts: For $\chi = -1$ the equations

$$\begin{aligned} \ddot{u} &= uw - \lambda u + \nu \dot{u} \\ \dot{w} &= -u^2 - w \end{aligned}$$

were first put forward by Morioka and Shimizu [28] as an *ad hoc* model for the Lorenz equations. A.L. Shil'nikov [26] studied its solutions in a theoretical context and Rucklidge [25] gave this model in the context of magnetoconvection and analysed its solutions in great detail. For $\chi = +1$ the equations appeared in the work of Proctor and Weiss [24] in the context of thermosolutal convection. Knobloch *et al.* [14] finally analysed this set of equations in the context of double diffusive convection.

The dynamics is quite distinct in the two systems, but they have one feature in common: the occurence of irregular motion due to the presence of heteroclinic cycles in conjunction with the Shil'nikov criterion (see L.P. Shil'nikov [27], Tresser [29]).

In the case $\chi = -1$ a homoclinic figure eight orbit switches to butterfly configuration leading in its vicinity to irregular orbits. There also exist T-points (see Glendinning and Sparrow [6]) consisting of a symmteric pair of heteroclinic orbits. Again, chaotic trajectories appear in their vicinity and may be traced numerically (see A.L. Shil'nikov [26], Rucklidge [25]).

When $\chi = +1$ the driving force of irregular motion is a periodic orbit born in a locally supercritical hopf bifurcation on the trivial solution that undergoes symmetrybreaking transitions on its approach to a symmetric heteroclinic cycle where numerically observable irregular motion arises (see Proctor and Weiss [24], Knobloch *et al.* [14]).

To treat the complex system we need to investigate the equations more in detail. The stationary solutions are the trivial solution (T) and the mixed state (M) with

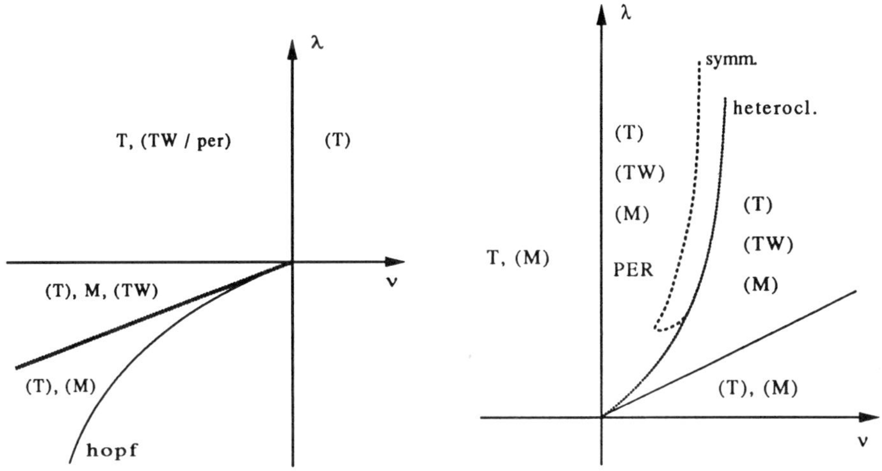

Figure 3: The simplified bifurcation diagram for the case $\chi = -1$ (left) and $\chi = +1$ (right). With $\chi = -1$ (left) and $\lambda < \nu/2$ (broad line) there do not exist asymptotically stable bounded solutions. The complicated dynamics in the real subspace arises from the periodic solution born at the "hopf" line and is unstable with respect to perturbations in the complex subspace. With $\chi = +1$ (right) the travelling waves do not influence the dynamics and the standing waves undergo the same bifurcation cascade as in the real system. Along the line "symm" the periodic solution undergoes a symmetrybreaking transition. The line "heterocl" indicates the locus of the heteroclinic cycle which is approached by the standing waves. The remaining abbreviations are explained in the text.

$u \neq 0$, $w \neq 0$. The basic asymptotic time dependent solutions are the standing waves SW and the travelling waves TW. The latter only exist in the complex system. Both waves appear in a double hopf bifurcation on T along the line $\lambda = 0$, $\nu > 0$. The travelling waves break the $Z(2)$ and the standing waves the $SO(2)$ symmetry (see Golubitsky *et al.* [8]). To understand what happens as we change the unfolding parameters we need to consider the cases $\chi = -1$ and $\chi = +1$ separately. For the missing details in the account the reader is referred to Manogg [15].

When $\chi = -1$ the double hopf bifurcation on T is subcritical and gives rise to unstable TW and SW. The travelling waves connect onto M along the line $\lambda = \nu/2$ in a subcritical stationary bifurcation. The complicated dynamics in the invariant real subspace is caused by a supercritical hopf bifurcation on the mixed solution along the line $\lambda = \nu(1-\nu)/2$. It is the subcritical connection of TW to M that destabilises the real subspace and therefore the complicated dynamics described above (see fig. 5). To quadratic order the solutions diverge and higher order terms are needed to keep the dynamics finite. The analysis of a system which contains saturating third order terms is left to future work.

When $\chi = +1$ the bifurcation scenario is very different. Here we observe a super-

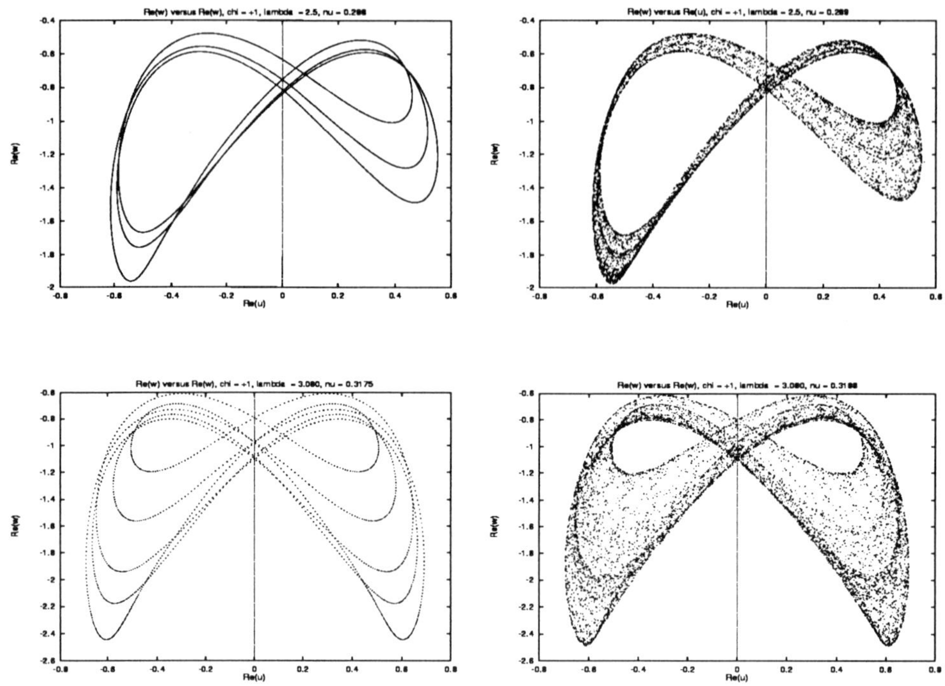

Figure 4: The dynamics in the complex system with $\chi = +1$ is similar to the dynamics in the real subspace. Here we show the projections of the trajectories onto the $Re(u)$ vs $Re(w)$ space. In the top pictures $(\lambda, \nu) = (2.5, 0.288)$ (left) and an asymmetric period three orbit appears; with $(\lambda, \nu) = (2.5, 0.289)$ (right) apparently irregular behaviour ensues. In the bottom pictures $(\lambda, \nu) = (3.08, 0.3175)$ (left) and a symmetric period five orbit appears which breaks up when $(\lambda, \nu) = (3.08, 0.3188)$ (right). Similar transitions are observed in the projections onto the $Im(w)$ vs $Im(u)$ space.

critical double hopf bifurcation on T which yields unstable TW but asymptotically stable standing waves SW. Interestingly the travelling waves do not influence the dynamics at all. It can be shown that locally the solutions are asymptotic to a behaviour which is analogous to the dynamics confined to the real subspace. In fact numerical calculations reveal that even in nonlocal regions the bifurcation structure is identical to the cascade observed in the real subspace and indicated in fig. 5. The standing waves undergo a symmetrybreaking transition and approach a heteroclinic cycle which connects the two unstable mixed solutions. On their approach they suffer period doubling cascades. Symmetric period five and asymmetric period three orbits with subsequent apparently irregular trajectories indicate observable highly irregular behaviour (see fig. 5). In this case we may reinterpret such solutions to describe the fluid flow in the physical double-layer convection system.

6 Conclusion

Two-dimensional double-layer convection is a physical system that contains a large variety of interesting wave-mode interactions. In the past Proctor and Jones [22, 23] as well as Julien [10] derived models for strong spatial resonances between $O(1)$ wavelength convective motion. Here we show that imposing appropriate boundary conditions on the temperature leads to the strong interaction between an asymptotically long wave and ordinary convective motion. A system consisting of a Kuramoto-Sivashinski type equation coupled quadratically with a Ginzburg-Landau expression is derived and describes the dynamics close to the simultaneous onset of convection in both layers. We show that competition between two fundamental types of convective motion leads to a large variety of canonical bifurcation scenarios of higher codimension and therefore to a large variety of asymptotically stable solutions. We study the particular case of the Takens-Bogdanow singularity in strong spatial 2:1 resonance in $O(2)$ symmetry and derive the normal form equations using group theoretical arguments. The analysis of the equations truncated at lowest order reveals the persistence of irregular behaviour under certain circumstances. This dynamics may be reinterpreted to occur also in the underlying physical double-layer convection system.

Weakly nonlinear reduction techniques allow in the present situation the detailed description of the fluid flow close to the onset of motion. It is possible to link many canonical higher codimension singularities to the fluid motion in the double-layer convective system. As such we therefore show that strong interaction between critical wave-modes of disparate scales occurs in physical systems (see also in the context of Marangoni convection Golovin *et al.* [7]) and leads to a large variety of possible dynamics which is amenable to analysis by means of dynamical systems theory.

The immediate extension of this work would be to shed more light onto the competition between the mixed and pure solutions of the coupled system. This may be done by analytical as well as numerical means and includes the description of the unfoldings of several higher codimension singularities. Furthermore, the extension of the present two-dimensional system to three dimensions clearly represents a challenge and would clarify also the competition between three- and two-dimensional convection patterns.

References

[1] D. Armbruster and G. Dangelmayr. Steady state mode interactions in the presence of O(2) symmetry and non flux boundary value problems. *Cont. Math.*, 56:53 – 68, 1986.

[2] C.J. Chapman and M.R.E. Proctor. Nonlinear Rayleigh-Bénard convection between poorly conducting boundaries. *J. Fluid Mech.*, 101(4):759 – 782, 1980.

[3] G. Dangelmayr. Steady state mode interactions in the presence of O(2) symmetry. *Dyn. and Stab. of Systems*, 1:159 – 185, 1986.

[4] G. Dangelmayr and E. Knobloch. The Takens-Bogdanov bifurcation with O(2)-symmetry. *Phil. Trans. R. Soc. Lond.*, A 322:243 – 279, 1987.

[5] R.C. DiPrima, W. Eckhaus, and L.A. Segel. Nonlinear wave-number interaction in near critical two-dimensional flows. *J. Fluid Mech.*, 49:705 – 744, 1971.

[6] P. Glendinning and C. Sparrow. T-Points: A Codimension Two Heteroclinic Bifurcation. *J. Stat. Phys.*, 43(3 / 4):479 – 488, 1986.

[7] A.A. Golovin, A.A. Nepomnyashchy, and L.M. Pismen. Interaction between short-scale Marangoni convection and long-scale deformational instability. *Phys. Fluids*, 6(1), 1994.

[8] M. Golubitsky, I. Stewart, and D.G. Schaeffer. *Singularities and Groups in Bifurcation Theory*, volume II. Springer Verlag, Berlin, 1988.

[9] G. Iooss, A. Mielke, and Y. Demay. Theory of steady Ginzburg-Landau equation, in hydrodynamic stability problems. *Eur. J. Mech. B/Fluids*, 8(3):229 – 268, 1989.

[10] K.A. Julien. *Strong spatial resonances in convection.* PhD thesis, University of Cambridge, 1991.

[11] K.A. Julien. Strong spatial interactions with 1:1 resonance: a three layer convection problem. *Nonlinearity*, 7(6):1655 – 1693, 1994.

[12] E. Knobloch. Pattern selection in long-wavelength convection. *Physica D*, 41:450 – 479, 1990.

[13] E. Knobloch and M.R.E. Proctor. Nonlinear periodic convection in double diffusive systems. *J. Fluid Mech.*, 108:291 – 316, 1981.

[14] E. Knobloch, M.R.E. Proctor, and N.O. Weiss. Heteroclinic bifurcations in a simple model of double-diffusive convection. *J. Fluid Mech.*, 239:273 – 292, 1992.

[15] G. Manogg. *Strong spatial resonance in Rayleigh-Bénard convection.* PhD thesis, Ecole Polytechnique Fédérale de Lausanne, Switzerland, 1994.

[16] G. Manogg and P. Metzener. Strong resonance in two dimensional non Boussinesq convection. *Phys. Fluids*, 6(9):2944 – 2955, 1994.

[17] P. Metzener and S.H. Davis. Secondary bifurcation in convective shells. *Phys. Rev. A*, 33(6):4326 – 4333, 1986.

[18] P. Metzener and M.R.E. Proctor. Interaction of patterns with disparate scales. *Eur. J. Mech., B / Fluids*, 11(6):759 – 778, 1992.

[19] P. Metzener and M.R.E. Proctor. Non-local effects on pattern selection in directional solidification. *Preprint*, 1994.

[20] A. Newell and J. Whitehead. Finite bandwidth, finite amplitude convection. *J. Fluid Mech.*, 38:279 – 303, 1969.

[21] M.R.E. Proctor. Instability of roll-like patterns for degenerate marginal curves. *Phys. Fluids A*, 3(2):299 – 302, 1991.

[22] M.R.E. Proctor and C.A. Jones. Strong spatial resonance and travelling waves in Bénard convection. *Phys. Lett. A*, 121:224 – 228, 1987.

[23] M.R.E. Proctor and C.A. Jones. The interaction of two spatially resonant patterns in thermal convection. Part 1. Exact 1:2 resonance. *J. Fluid Mech.*, 188:301 – 335, 1988.

[24] M.R.E. Proctor and N.O. Weiss. Normal forms and chaos in thermosolutal convection. *Nonlinearity*, 3:619 – 637, 1990.

[25] A. Rucklidge. Chaos in a low-order model of magnetoconvection. *Physica D*, 62:323 – 337, 1993.

[26] A.L. Shil'nikov. On bifurcations of the Lorenz attractor in the Shimizu-morioka model. *Physica D*, 62:338 – 346, 1993.

[27] L.P. Shil'nikov. A case of the existence of a countable number of periodic motions. *Soviet Math. Dokl.*, 6:163 – 167, 1965.

[28] T. Shimizu and N. Morioka. On the bifurcation of a symmetric limit cycle to an asymmetric one in a simple model. *Phys. Lett. A*, 76:201 – 204, 1980.

[29] C. Tresser. About some theorems by L.P. Shil'nikov. *Ann. Inst. H. Poincaré*, 40, Phys. théor.(4):441 – 461, 1984.

A New Approach to Sideband–Instabilities Using the Principle of Reduced Instability

A. MIELKE

Institut für Angewandte Mathematik, Universität Hannover, Welfengarten 1, 30167 Hannover, Germany

Abstract

First we develop the theory of reduced instability in order to analyze stability of bifurcating solutions via the Liapunov–Schmidt reduction. Next, this theory is generalized to cover sideband instabilities using a Floquet ansatz. The method is applied to roll patterns in the Swift–Hohenberg equation which serves as a model for pattern formation. We analyze the stability problem of small rolls analytically by calculating expansions of the critical eigenvalues in terms of the sideband vector. Thus we recover the Eckhaus and the zigzag instability. Finally, we compare the results with the predictions of the Newell–Whitehead equation and obtain good agreement.

1 Introduction

The mathematical theory of pattern formation has attracted much attention over the last four decades. The beginnings were laid in the late 50s and early 60s when the bifurcation of the Rayleigh–Bénard convection rolls and the Taylor vortices was put on a mathematical sound basis. These two phenomena remain until today the paradigm of spontaneous pattern formation in physical systems. We refer to [NPL93] a recent physical review of the subject. An interesting study of modulations of hexagonal patterns is done in [ST94].

The newly developed theory, however, predicted that the spatially periodic patterns exists in a larger wavenumber regime than could be observed in experiments. This strange phenomenon was only resolved by Wiktor Eckhaus through his outstanding work in stability theory in his book in 1965. The main point is that the patterns are unstable with respect to perturbations which have a spatial periodic which is slightly larger or smaller than that of the stationary pattern. This mechanism is nowadays called modulational or sideband instability, and in the particular context of reflectionally symmetric dissipative systems the term *Eckhaus instability* is used. Even more striking, Eckhaus was able to show that close to the threshold of instability the regime of stable periodic patterns is exactly $(k_0 - \delta/\sqrt{3}, k_0 + \delta/\sqrt{3})$ when k_0 is the critical wavenumber and $(k_0 - \delta, k_0 + \delta)$ the existence interval of periodic patterns.

It took another five years until the so-called modulation equations where found which are even able to predict the slow spatial and temporal modulations of nearly periodic patterns [NW69, DES71]. However, the sideband instability theory as well as the method of modulation equations was not really accepted in the mathematics community for almost twenty years, since there seemed to be an unbridgeable gap between the original problem (e.g. the Navier–Stokes equation) and the modulation equation. Only in the last four years, it was shown that the formally derived modulation equations really make correct predictions over a sufficiently long time interval, see e.g. [vH91, KSM92, Sch94a].

However, the mathematical theory of sideband instability is still not very well developed. By now there is only a theory for Schrödinger operators with periodic potentials [RS78, CE90], however these results are very special and depend heavily on the self–adjointness, which is not present in most pattern forming systems. A recent approach was developed in [BM94a, BM94b], which is based in the theory of spatial center manifolds. In the second work a rigorous proof of the Benjamin–Feir instability for water waves was given; this sideband instability is the Hamiltonian counterpart of the Eckhaus instability. The main idea of this method is to write the problem as a differential equation with respect to the unbounded spatial direction and to use a spatial center manidold reduction in order to study the bifurcation of periodic patterns. Next, the spectral problem is also written in spatial dynamics form and then reduced to a (linear) spatial center manifold. This yields a reduced spectral problem which coincides in lowest order with the spectral problem obtained from the modulation theory approach. Thus, classical formal instability results can now be viewed in a rigorous mathematical setting.

Here we develop a method which is very similar to the reduction onto spatial center manifolds, however, it is solely based on the classical Liapunov–Schmidt reduction. Thus, the present method is easier and more flexible in applications, since the construction of center manifolds is always much more difficult than the application of the Liapunov–Schmidt reduction. In fact, the present method uses the Floquet theory employed by Busse in his important work of instabilities of periodic patterns in hydrodynamics, see e.g. [Bu71, BB84, BS89]. Thus, the aim here is to develop a general theory which includes Busse's approach and, moreover, to study the mathematical aspects of the theory.

In this paper, we will explain the main philosophy of the method and work out a first example, namely the sideband instability for the roll patterns in the Swift–Hohenberg equation (SHE):

$$\partial_t u \;=\; -(1+\partial_x^2+\partial_y^2)^2 u + \varepsilon^2 u - u^3, \quad t \geq 0,\ (x,y) \in I\!R^2. \tag{1.1}$$

There are roll solutions $\widetilde{u}_{\varepsilon,k}$ which are (t,y)–independent and periodic in x with periodic $2\pi/k$. These solutions exists for all $k \in (\sqrt{1-\varepsilon}\,, \sqrt{1+\varepsilon})$, but some of these roll patterns are unstable, namely the rolls with $k < 1 + \mathcal{O}(\varepsilon^2)$ are *zigzag unstable* (see [Bu71] for a first study) and those with $k \in (1+\varepsilon/(2\sqrt{3})+\mathcal{O}(\varepsilon^2), \sqrt{1+\varepsilon})$ are *Eckhaus unstable* (see [Eck65]). For more details we refer to Section 4.

The principle of reduced instability is based on the idea that the spectral problem for a steady state $\widetilde{u}$ can be reduced in the very same way as the nonlinear problem was reduced to study the bifurcation of steady states. To be more specific, let X be some function space of 2π–periodic functions, then the bifurcations of rolls can be studied by finding zeros of the mapping

$$N(\varepsilon, k, \widetilde{u}) = -(1+k^2\partial_\xi^2)^2\widetilde{u} + \varepsilon^2\widetilde{u} - \widetilde{u}^3, \quad \widetilde{u} \in X.$$

The Liapunov–Schmidt reduction relies on the splitting $\widetilde{u} = \widetilde{u}_0 + \widetilde{u}_1$ based on a projection P, which leads to the coupled system $PN(\varepsilon, k, \widetilde{u}_0 + \widetilde{u}_1) = 0$ and $(I-P)N(\varepsilon, k, \widetilde{u}_0 + \widetilde{u}_1) = 0$. The second equation can be solved for $\widetilde{u}_1 = \mathcal{U}(\varepsilon, k, \widetilde{u}_0)$ and insertion into the first equation yields the reduced problem, called bifurcation equation

$$0 = n(\varepsilon, k, \widetilde{u}_0) = PN(\varepsilon, k, \widetilde{u}_0 + \mathcal{U}(\varepsilon, k, \widetilde{u}_0)).$$

To show spectral instability of a solution $\widetilde{u}$ we show that the spectral problem $\lambda v = L(\partial_x)v + 3\widetilde{u}^2(x)v$ has a solution (λ, v) with $\operatorname{Re}\lambda > 0$ and $v \neq 0$. The main difference to classical approaches is that we allow v to lie in $L^\infty(I\!R^2)$ rather than restricting it to the space X containing the patterns with the same periodicity as $\widetilde{u}$. In fact, we assume that v takes the form $v(x,y) = e^{i(\sigma x + \tau y)}V(\xi)$ with $\xi = kx$ and $V \in X$. Thus, we arrive at

$$0 = B(\varepsilon, k, \sigma, \tau, \lambda)V := -(1 + (k\partial_\xi + i\sigma)^2 - \tau^2)^2 V + \varepsilon^2 V - 3\widetilde{u}_{\varepsilon,k}^2 V - \lambda V.$$

Such Floquet type methods are in use for computational methods for several years (cf. [BS89]); however, a systematic mathematical approach has not been developed. For steady bifurcations related methods are constructed in [Mi92].

To this problem we again apply the Liapunov–Schmidt reduction with the same splitting and projection as above, and we find a reduced spectral problem

$$0 = B_{00}(\varepsilon, k, \sigma, \tau, \lambda)V_0 := PB(\varepsilon, k, \sigma, \tau, \lambda)[V + \mathcal{V}(\varepsilon, k, \sigma, \tau, \lambda)V], \tag{1.2}$$

where $\mathcal{V}$ is the associated reduction function. In our case $X_0 = PX$ is two–dimensional and thus B_{00} corresponds to a 2×2-matrix. The *principle of reduced instability* means that $\widetilde{u}$ is a (spectrally) unstable steady state of (1.1) if the reduced spectral problem (1.2) has a solution $(\sigma, \tau, \lambda, V_0)$ with $\operatorname{Re}\lambda > 0$ and $V_0 \neq 0$.

We establish this principle in Section 2 (for the case $(\sigma, \tau) = 0$) and compare it with the classical principle of reduced stability which was developed in the mid 80s, see [KL83, Va84, Rec89, Dam91]. Moreover, we derive an expansion for the function B_{00} which shows that the leading order terms are easy to calculate. In Section 3 we generalize the approach to the sideband instability question. In Section 4 we then treat, in all detail, the instability of the rolls in the Swift–Hohenberg problem using the previous abstract results. In the last section we make a comparison with the stability predictions of the so-called Newell–Whitehead equation [NW69] and find a perfect agreement for the lowest order terms.

The method developed in this work is sufficiently simple in order to be applicable to more general situations. First of all it can be used to study the full three–dimensional Navier–Stokes equations or other systems in continuum mechanics. Moreover, more complex patterns such as hexagons or rectangles can be treated. This will be the topic of future research.

2 The principle of reduced instability

We consider general evolution problems in the form

$$M\partial_t u = L(\partial_x)u + F(\mu, u), \tag{2.1}$$

where u lies in some Banach space X, $M : X \to Y$ is a bounded linear operator; $L : D(L) \subset X \to Y$ is a closed unbounded operator, $\mu \in \mathbb{R}^n$ are the parameters, and F is a smooth mapping from $\mathbb{R}^n \times D(L)$ into Y with $\|F(\mu, u)\|_Y = \mathcal{O}(\|u\|_{D(L)}(|\mu|+\|u\|_{D(L)}))$.

We are interested in the stability of bifurcating steady solutions. The bifurcating equilibria can be found by the Liapunov–Schmidt reduction which we sketch here only and refer to [CH88] for the details. We assume that L is a Fredholm operator, that is the kernel of L, $X_0 = \ker L$, is finite–dimensional and the range of L, $Y_1 = \operatorname{range} L$, has finite codimension. Then, there exist continuous projections $P : X \to X$ and $Q : Y \to Y$ with $PX = X_0$, $(I - Q)Y = Y_1$. Moreover, defining $X_1 = (I - P)X$, the linear operator L is a bijection between $X_1 \cap D(L)$ and Y_1. (In cases where $X \supset Y$ are Hilbert spaces and L is self–adjoint we take $Q = P$.)

With these linear operators the equilibria, which satisfy $0 = Lu + F(\mu, u)$, can be studied by splitting $u = u_0 + u_1$, where $u_0 = Pu \in X_0$ and $u_1 = (I - P)u \in X_1$. We find

$$\begin{aligned} 0 &= QL(u_0 + u_1) + QF(\mu, u_0 + u_1) = QF(\mu, u_0 + u_1), \\ 0 &= (I-Q)L(u_0 + u_1) + (I-Q)F(\mu, u_0 + u_1) = Lu_1 + (I-Q)F(\mu, u_0 + u_1). \end{aligned} \tag{2.2}$$

We used here $QX_1 = \{0\}$ and $LX_0 = \{0\}$. The second equation in (2.2) can now be solved locally for $u_1 = \mathcal{U}(\mu, u_0) \in X_1 \cap D(L)$ by the implicit function theorem due to the boundedness of $L^{-1} : Y_1 \to X_1 \cap D(L)$. Hence, all small equilibria can be characterized by the *bifurcation equation* which is obtained by inserting $u_1 = \mathcal{U}$ into the first equation of (2.2):

$$0 = f(\mu, u_0) = QF(\mu, u_0 + \mathcal{U}(\mu, u_0)) \in Y_0. \tag{2.3}$$

Thus, we are left with a finite–dimensional algebraic problem in $\mathbb{R}^n \times X_0$.

Now assume we have found a solution $(\mu, \tilde{u}_0)$ of (2.3), then the question arises whether the stability of the associated solution $(\mu, \tilde{u}) = (\mu, \tilde{u}_0 + \mathcal{U}(\mu, \tilde{u}_0))$ of (2.1) can be decided by means of the function f in the reduced problem, at least in those cases where $X = Y$ and $M = I$(=identity on X). This question attracted some attention in the early 1980s and lead to the

Principle of Reduced Stability. *Assume that Y is densely contained in X, that $M = I$, and that the spectrum of L is contained in $\{0\} \cup \{ z \in \mathbb{C} : \operatorname{Re} z < -\alpha \}$ for some $\alpha > 0$. Moreover, assume that the eigenvalues of $D_{u_0} f(\mu, \tilde{u}_0)$ have all negative real part, then the spectrum of $L + D_u F(\mu, \tilde{u})$ is contained in $\{ z \in \mathbb{C} : \operatorname{Re} z < -\beta \}$ for some $\beta > 0$.*

See [KL83] for a discussion and a first example which shows that the principle does not hold in this general setting. In subsequent work additional conditions on the spectrum of L were derived which make the conclusion correct, cf. [KL83, Va84, Rec89, Dam91]. However, these conditions seem much more involved as the method proposed below.

Here we study the exact spectral problem of the linearization in $(\mu, \tilde{u})$ by constructing a reduced spectral problem which allows us to controll all eigenvalues near 0; this exact reduction will be necessary when we study the sideband instability below. To this end we consider the spectral problem for the full problem, namely

$$\lambda M v = Lv + D_u F(\mu, \tilde{u}_0 + \mathcal{U}(\mu, \tilde{u}_0))v. \tag{2.4}$$

We call $\tilde{u} = \tilde{u}_0 + \mathcal{U}(\mu, \tilde{u}_0)$ *spectrally unstable* if (2.4) possesses a solution (λ, v) with $\operatorname{Re}\lambda > 0$ and $v \neq 0$.

Considering λ as a small parameter we are able to do a Liapunov–Schmidt reduction for (2.4), since the dominating operator is still L. With $v = v_0 + v_1$, where $v_0 = Pv$, we obtain

$$\begin{aligned} \lambda QM(v_0 + v_1) &= QD_u F(\mu, \tilde{u})(v_0 + v_1), \\ \lambda(I - Q)M(v_0 + v_1) &= Lv_1 + (I - Q)D_u F(\mu, \tilde{u})(v_0 + v_1). \end{aligned} \tag{2.5}$$

The second equation can be solved by the implicit function theorem (linear case!) for v_1 in the form $v_1 = \mathcal{V}(\lambda, \mu, \tilde{u})v_0$, where $\mathcal{V}(\lambda, \mu, \tilde{u}) \in \mathcal{L}(X_0, X_1 \cap D(L))$ depends smoothly on $(\lambda, \mu, \tilde{u})$ small. Thus, we obtain a reduced spectral problem

$$\lambda QMv_0 = QF(\mu, \tilde{u})v_0 + Q\{F(\mu, \tilde{u}) - M\}\mathcal{V}(\lambda, \mu, \tilde{u})v_0, \tag{2.6}$$

which is a finite-dimensional spectral problem on X_0 with the spectral parameter λ appearing nonlinearly. However, these nonlinear terms are relatively small, as we will see below. It is clear now, that the justification of the principle of reduced stability is concerned with the question under which conditions the correction term involving $\mathcal{V}$ can be dropped. This leads to the main observation:

> **Principle of Reduced Instability.** *If* (2.6) *has a solution* (λ, v_0) *with* $v_0 \neq 0$ *and* $\operatorname{Re}\lambda > 0$, *then the basic state* $\tilde{u} = \tilde{u}_0 + \mathcal{U}(\mu, \tilde{u}_0)$ *is (spectrally) unstable as an equilibrium solution of* (2.1).

We note here, that this principle is formulated as an instability result; and in this setting it is a rigorous statement, since from the nontrivial unstable v_0 in (2.6) we easily obtain the unstable eigenvector v for (2.4), namely $v = v_0 + \mathcal{V}(\lambda, \mu, \tilde{u})v_0$. We do not make any stability conclusions here for the simple reason that our theory only controls the small eigenvalues λ of the problem. Under general conditions we can not guarantee that eigenvalues pass the imaginary axis far away from the origin.

We now derive some expansions for the reduced spectral problem showing that in most cases the exact spectral properties can be obtained with only minor additional efforts compared to the construction of the nonlinear reduced problem $0 = f(\mu, u_0)$. To formulate the result we write $M : X \to Y$ in block form $\begin{pmatrix} M_{00} & M_{01} \\ M_{10} & M_{11} \end{pmatrix}$ where $M_{0j} = QM|_{X_j}$ and $M_{1j} = (I-Q)M|_{X_j}$ for $j = 0, 1$. In our application below we have $X \supset Y$, $M = I$, and $P = Q$, which implies $M_{jj} = I_{X_j}$ and $M_{01} = M_{10} = 0$.

Theorem 2.1
Assume that the nonlinearity satisfies $\|F(\mu,u)\|_Y = \mathcal{O}(\|u\|_{D(L)}(\|\mu\|+\|u\|^n_{D(L)}))$. *Then, the linear reduction function* $\mathcal{V}$ *has the expansion*

$$\mathcal{V}(\lambda,\mu,\tilde{u}) \;=\; D_u\mathcal{U}(\mu,\tilde{u}) + \lambda\mathcal{V}_1(\mu,\tilde{u}) + \mathcal{O}(|\lambda|^2\|\mathcal{V}_1(\mu,\tilde{u})\|_{X_1}),$$

where $\mathcal{V}_1(\mu,\tilde{u}) = L^{-1}M_{10} + \mathcal{O}(|\mu| + \|\tilde{u}\|^n)$, *and the reduced spectral problem reads*

$$\begin{aligned} \lambda M_{00}v_0 &= \Big[D_{u_0}f(\mu,\tilde{u}_0) + \lambda f_1(\mu,\tilde{u}) + \lambda^2 r(\lambda,\mu,\tilde{u})\Big]v_0, \\ \textit{where } f_1(\mu,\tilde{u}) &= -M_{01}D_{u_0}\mathcal{U}(\mu,\tilde{u}_0) + QD_uF(\mu,\tilde{u})\mathcal{V}_1(\mu,\tilde{u}) \\ &= \mathcal{O}((\|M_{01}\| + \|M_{10}\| + |\mu| + \|\tilde{u}\|^n)(|\mu| + \|\tilde{u}\|^n)), \\ \textit{and } r(\lambda,\mu,\tilde{u}) &= \mathcal{O}\Big([\|M_{01}\| + |\lambda| + \|QD_uF(\mu,\tilde{u})|_{X_1}\|]\|\mathcal{V}_1(\mu,\tilde{u})\|\Big). \end{aligned} \tag{2.7}$$

Proof: We recall the definition of the reduction function $\mathcal{U}$ of the nonlinear problem in (2.2) and take the derivative with respect to u_0 in the direction v_0, which results in

$$0 \;=\; LD_{u_0}\mathcal{U}(\mu,\tilde{u})v_0 + (I-Q)D_uF(\mu,\tilde{u}_0+\mathcal{U}(\mu,\tilde{u}))[v_0 + D_{u_0}\mathcal{U}(\mu,\tilde{u}_0)v_0].$$

Now, we insert the expansion $v_1 = [\mathcal{V}_0 + \lambda\mathcal{V}_1 + \lambda R(\lambda,\mu,\tilde{u})]v_0$ into the equation defining $\mathcal{V}$. Comparing equal powers of λ we obtain

$$\begin{aligned} \lambda^0 :\;\; 0 &= L\mathcal{V}_0v_0 + (I-Q)D_uF(\mu,\tilde{u})[v_0 + \mathcal{V}_0v_0], \\ \lambda^1 :\;\; M_{10}v_0 + M_{11}\mathcal{V}_0v_0 &= [L + (I-Q)D_uF(\mu,\tilde{u})]\mathcal{V}_1v_0, \\ \lambda^2 :\;\; M_{11}(\mathcal{V}_1 + \lambda R)v_0 &= [L + (I-Q)D_uF(\mu,\tilde{u})]Rv_0. \end{aligned}$$

The uniqueness of the coefficients yields $\mathcal{V}_0 = D_{u_0}\mathcal{U}$, and the smallness of the nonlinearity implies $\mathcal{U}(\mu,u_0) = \mathcal{O}(\|u_0\|(|\mu| + \|u_0\|^n))$ as well as $D_uF(\mu,u_0+\mathcal{U}(\mu,u_0)) = \mathcal{O}(|\mu| + \|u_0\|^n)$ and $D_{u_0}\mathcal{U}(\mu,u_0) = \mathcal{O}(|\mu| + \|u_0\|^n)$. Therefore, the desired estimates for $\mathcal{V}_1$ and R follow. Inserting the results for $\mathcal{V}$ into the first equation of (2.5) the expansion of the reduced spectral problem (2.7) is established. □

Through this theorem we have exact control on the maximal size of the correction terms in the reduced spectral problem. In the important case $M_{01} = M_{10} = 0$ we have

$$\lambda f_1(\mu,u_0) + \lambda^2 r(\lambda,\mu,u_0) = \mathcal{O}(|\lambda|(|\lambda|^2 + |\mu|^2 + \|u_0\|^{2n})).$$

Hence, in most cases the calculation of these correction terms will not be necessary, and we can replace the principle of reduced stability through a rigorous reduced spectral theory which does not afford much more computational work than the nonlinear Liapunov–Schmidt reduction for the constructon of the solutions $\tilde{u}$ itself: the additional work reduces to the trivial task to calculate M_{00}.

The spatial center manifold approach to the principle of reduced instability The principle of reduced instability was developed first for steady solutions of elliptic systems on infinitely long cylindrical domains. There, all small bifurcating solutions are constructed by reducing the problem onto a spatial center manifold. This method originates in [Ki82] and was further developed in [Mi86, Mi88]. For the principle of reduced instability in this context, we refer to [BM94a, BM94b] for a general theory and nontrivial applications. Here we only sketch the main ideas to show the analogy to the corresponding theory based on the Liapunov–Schmidt reduction.

Assume that (2.1) is posed on an infinitely long cylinder $\Omega = I\!R \times \Sigma$, where $x \in I\!R$ is the axial variable and Σ is a bounded cross–section. The associated spatial dynamical system reads

$$\frac{d}{dx} w = Aw + E\partial_t w + \mathcal{F}(\mu, w), \quad w \in Z, \tag{2.8}$$

where no x–derivatives appear in the right–hand side of (2.8), cf. [Mi88, IM91]. Now all small bounded, t–independent solutions $w : I\!R \to Z$ can be found by reducing the system to a spatial center manifold as follows: A defines a splitting of the phase space Z into the finite–dimensional center space Z_c and the hyperbolic part Z_h such that (2.8) linearly decouples into

$$\begin{aligned} \tfrac{d}{dx} w_c &= A_c w_c + P_c \mathcal{F}(\mu, w_c + w_h), \\ \tfrac{d}{dx} w_h &= A_h w_h + (I - P_c)\mathcal{F}(\mu, w_c + w_h), \end{aligned} \tag{2.9}$$

where $P_c : Z \to Z$ is the projection onto Z_c. The main point is that Z_c is finite dimensional and that all eigenvalues of A_c lie on the imaginary axis while the spectrum of A_h is bounded away from the imaginary axis. Under suitable technical assumptions (cf. [Mi88]) all small solutions of (2.9) lie on a spatial center manifold which is the graph of a reduction function; namely $w_h = \mathcal{H}(\mu, w_c)$. Thus, it remains to analyze the reduced spatial dynamical system

$$\frac{d}{dx} w_c = A_c w_c + P_c \mathcal{F}(\mu, w_c + \mathcal{H}(\mu, w_c)) \tag{2.10}$$

in order to find all bifurcating solutions.

Now assume we have found such a solution $\tilde{w} = \tilde{w}_c + \mathcal{H}(\mu, \tilde{w}_c)$, and we are interested in the spectral stability of $\tilde{w}$ (and of $\tilde{u}$ solving (2.1)). As above $\tilde{w}$ is called spectrally unstable, if there exists (λ, W) with $\operatorname{Re}\lambda > 0$ and $W \neq 0$ satisfying the spectral problem

$$\frac{d}{dx} W = AW + \lambda EW + D_w \mathcal{F}(\mu, \tilde{w}(x))W. \tag{2.11}$$

No assumptions are made on W except for boundedness over $x \in I\!R$. To this linear equation, where $(\lambda, \mu, \tilde{w}(\cdot))$ is small, we again apply the spatial center manifold reduction with $W_c = P_c W$. As the problem is x–dependent we have to use a non–autonomous version of the spatial center manifold (cf. [Mi86, Mi88]) and find the linear reduction function $W_h = \mathcal{W}_{\tilde{w}(\cdot)}(\lambda, \mu, x)W_c$ and the reduced spatial spectral problem

$$\frac{d}{dx} W_c = A_c W_c + P_c[\lambda E + D_w \mathcal{F}(\mu, \tilde{w}(x))](W_c + \mathcal{W}_{\tilde{w}}(\lambda, \mu, x)W_c). \tag{2.12}$$

We now have the following

> **Principle of Reduced Instability.** *Let $(\mu, \widetilde{w}_c)$ be a given solution of* (2.10) *and assume that $(\lambda, W_c) \in \mathbb{C} \times L^\infty(\mathbb{R}, Z_c)$ is a solution of the reduced spatial spectral problem* (2.12) *with* $\operatorname{Re}\lambda > 0$ *and nontrivial W_c, then the solution $\widetilde{w} = \widetilde{w}_c + \mathcal{H}(\mu, \widetilde{w}_c)$ of* (2.8) *is spectrally unstable.*

We close this section with some remarks concerning the difference between the Liapunov–Schmidt reduction and the spatial center manifold approach which both can be used to study periodic patterns in extended systems. First we mention that the Liapunov–Schmidt reduction is restricted to periodic solutions, while the spatial center manifold allows all kind of spatial behavior, such as quasiperiodic, asymptotically periodic, and heteroclinic and homoclinic solutions. The latter ones correspond to pulses and travelling fronts. As a consequence the principle of reduced instability can be applied to non–periodic structures without principle problems; yet for such patterns the reduced spatial spectral problem (2.12) may be very hard to analyze. Moreover, the spatial dynamics approach does not only deliver the existence of certain periodic or non–periodic structures, it also give a the reduced spatial dynamical systems which can be interpreted as the steady part of a modulation equation, see [IMD89, IM91]. The main restriction of the spatial dynamics approach is the one–dimensionality of x, which means that only problems on cylindrical domains can be treated.

The Liapunov–Schmidt reduction has two major advantages: it is much easier to calculate than the spatial center manifold, and it can be applied to the study for periodic pattern on domains with 2 or 3 unbounded directions like the hexagons in Bénard convection.

3 The general framework of sideband instability

Sideband instablities occur in systems which are posed over unbounded domains of the form $\Omega = \mathbb{R}^d \times \Sigma$, where $x \in \mathbb{R}^d$ denotes the unbounded directions while $y \in \Sigma$ is the cross–sectional variable in the bounded open set $\Sigma \subset \mathbb{R}^s$. The problem may be written in the general form

$$M\partial_t u \;=\; L(\partial_x)u + F(\mu, u), \quad t > 0,\ (x, y) \in \mathbb{R}^d \times \Sigma, \tag{3.1}$$

where we display the unbounded variable $x \in \mathbb{R}^d$ but not the cross–sectional variable $y \in \Sigma$.

We are interested in the stability of a periodic steady solution $\widetilde{u}$ where the periodicity is specified by the lattice group $\mathcal{G} \subset \mathbb{R}^d$ such that $\widetilde{u}(x + g, \cdot) = \widetilde{u}(x, \cdot)$ for all $g \in \mathcal{G}$. The dual group Γ is defined via

$$\Gamma = \{\, \gamma \in \mathbb{R}^d \;:\; \gamma \cdot g \in 2\pi\mathbb{Z} \text{ for all } g \in G \,\}.$$

If Γ is discrete, the periodic pattern $\widetilde{u}$ can be written as a generalized Fourier series $\widetilde{u}(x, \cdot) = \sum_{\gamma\in\Gamma} e^{i\gamma\cdot x} b_\gamma(\cdot)$, where $b_\gamma(y) = \frac{1}{|\mathcal{P}|}\int_{\mathcal{P}} e^{-i\gamma\cdot x}\widetilde{u}(x, y))\, dx$, where $\mathcal{P} \subset \mathbb{R}^d$ is a suitable periodicity domain. The points $\gamma \in \Gamma$ play the role of wave vectors for the periodic pattern $\tilde{u}$.

The stability of these solutions in the space of functions with the same periodicity can be studied by well-known theories since the spectrum of the linearization is discrete. For periodic patterns which bifurcate from a trivial solution the above-mentioned Liapunov–Schmidt reduction and the subsequent application of the principle of reduced instability is possible. The standard application of the (temporal) center manifold theory is another alternative.

However, in many physical situations the solutions appear to be stable with respect to perturbations in the same periodicity class while they are unstable with respect to perturbations which have slightly different wave vectors, $k = \gamma + \sigma$ with $\sigma \in \mathbb{R}^d$ small. It is exactly this phenomenon of sideband instability which we want to study in this paper. We introduce a new simple method which allows us to attack very general problems like instabilities in hydrodynamics and is a slight generalization of the principle of reduced instability. Assume that the solution $(\mu, \tilde{u})$ of (3.1) was obtained by a Liapunov–Schmidt reduction on the space of $\mathcal{G}$–periodic functions by solving the reduced problem $f(\mu, u_0) = 0$ and letting $\tilde{u} = \tilde{u}_0 + \mathcal{U}(\mu, \tilde{u}_0)$. The associated spectral problem is

$$\lambda M v = L(\partial_x)v + D_u F(\mu, \tilde{u}(x))v, \tag{3.2}$$

and $\tilde{u}$ is called (spectrally) unstable if there exists a solution (λ, v) of (3.2) with $\operatorname{Re}\lambda > 0$ and $0 \not\equiv v \in L^\infty(\mathbb{R}^d, Z)$.

The idea is to look for solutions v in the form $v(x) = e^{i\sigma\cdot x}V(x)$ where V is again $\mathcal{G}$–periodic. This ansatz is also called a Floquet ansatz and $\sigma \in \mathbb{R}^d$ is the sideband vector. In certain cases it can be shown that all bounded solutions of (3.2) have to have this form; yet, for general elliptic systems where Ω is more than one–dimensional there are examples which allow for other types of bounded solutions, see [Kuc93, Mi92]. One advantage of our instability principle is, that it is sufficient to find one solution v in Floquet form and that it is not necessary to exclude other bounded solutions.

For $v(x) = e^{i\sigma\cdot x}V(x)$ the spectral problem (3.2) can be rewritten in the form

$$\begin{aligned} &\lambda M V = L(\partial_x + i\sigma)V + D_u F(\mu, \tilde{u}(x))V(x) \\ &\text{with } L(\partial_x + i\sigma)V = e^{-i\sigma\cdot x}L(\partial_x)[e^{i\sigma\cdot x}V(x)]. \end{aligned} \tag{3.3}$$

This problem is again posed completely on the set of $\mathcal{G}$–periodic functions where the sideband vector $\sigma \in \mathbb{R}^d$ appears simply as an additional vector of parameters. Hence, we are able to study the question of the existence of nontrivial solutions (λ, V) by applying the Liapunov–Schmidt reduction to (3.3). Using the same notation as in Section 2.1 we find the reduced stability problem

$$\lambda M_{00} V_0 \; = \; P_0\big[L(\partial_x + i\sigma) + D_u F(\mu, \tilde{u}(x))\big](V_0 + \mathcal{V}(\mu, \sigma, \lambda)V_0) \tag{3.4}$$

4 The Swift–Hohenberg problem

We work out the details of the method for a simple model problem showing the same theoretical behavior as many pattern forming systems, such as for example the Bénard problem. The main advantage is that the whole theory can be explained with a minimal amount of mathematical technicalities.

The two–dimensional Swift–Hohenberg equation (SHE) is given by

$$u_t = -(1+\Delta)^2 u + \varepsilon^2 u - u^3, \qquad \text{for } t>0,\ (x,y)\in \mathbb{R}^2, \tag{4.1}$$

where $\Delta = \partial_x^2 + \partial_y^2$ is the Laplace operator. The linearization at zero admits the solutions $v(t,x,y) = e^{\lambda(k,l)t + i(kx+ly)}$, where $\lambda(k,l) = -(1-k^2-l^2)^2 + \varepsilon^2$. Hence, $u \equiv 0$ is weakly unstable with unstable modes having wavevectors with $k^2 + l^2 \approx 1$.

We will now show that there exist bifurcating stationary solutions which are spatially periodic. We will restrict our analysis to the case of rolls, i.e., solutions which do not depend on one of the variables. For the other bifurcating patterns, such as rectangular or hexagonal patterns, a similar analysis can be work out, however, most solutions will be unstable even with respect to perturbations lying in the same space of periodic functions as the basic periodic pattern.

Without loss of generality we assume that the rolls are periodic in x and independent of y. Hence, we have $u_{\text{rolls}}(t,x,y) = \widetilde{u}(\varepsilon,k,\xi)$ where $\xi = kx$ and $\widetilde{u}$ is 2π–periodic in ξ. The symmetry group is $\mathcal{G} = \frac{2\pi}{k}\mathbb{Z}\times\mathbb{R}$ with the discrete dual group $\Gamma = k\mathbb{Z}\times\{0\}$.

In order to obtain the existence of rolls via the Liapunov–Schmidt reduction, we define the basic spaces $X^j = H^j_{\text{per}}([0,2\pi],\mathbb{R})$ for $j\in\mathbb{N}_0$, where X^4 plays the role of X and X^0 the role of Y as given in Section 2. The problem for $\widetilde{u}$ reads

$$N(\varepsilon,k,\widetilde{u}) := -(1+k^2\partial_\xi^2)^2\widetilde{u} + \varepsilon^2\widetilde{u} - \widetilde{u}^3, \tag{4.2}$$

where $N : \mathbb{R}^2\times X^4 \to X^0$ is an analytical mapping. The linearization $L_{\text{per}}U := \partial_{\widetilde{u}}N(0,1,0)[U] = -(1+\partial_\xi^2)^2 U$ is self–adjoint, has the discrete spectrum $\Sigma(L_{\text{per}}) = \{-(1-m^2)^2 : m\in\mathbb{Z}\}$, and the kernel is spanned by $U_1(\xi) = \cos\xi$ and $U_2(\xi) = \sin\xi$. We define the orthogonal projection $P_0 u = \frac{1}{2\pi}(\langle U_1,u\rangle U_1 + \langle U_2,u\rangle U_2)$, where $\langle u,v\rangle = \int_0^{2\pi} uv\,d\xi$. We often identify P_0X^j with $\mathbb{R}^2$ and then use the mapping $\mathcal{Q} : X^j \to \mathbb{R}^2; u\mapsto \frac{1}{2\pi}(\langle u,U_1\rangle,\langle u,U_2\rangle)^T$.

Decomposing $\widetilde{u}\in X^4$ into $\alpha_1U_1 + \alpha_2U_2 + u_1$, where $Pu_1 = 0$, we see that $L_{\text{per}} : (I-P)X^4 \to (I-P)X^0$ is a bijection. Moreover, we rewrite (4.2) in the form

$$\begin{aligned} &\mathcal{Q}N(\varepsilon,k,\alpha_1U_1+\alpha_2U_2+u_1) = 0,\\ &(I-P)N(\varepsilon,k,\alpha_1U_1+\alpha_2U_2+u_1) = 0. \end{aligned} \tag{4.3}$$

The second equation can be solved for $u_1\in(I-P)X^4$ as a function of (ε,k,α) by the help of the implicit function theorem. From $(I-P)L_{\text{per}}U_j = 0$ we obtain $u_1 = \mathcal{U}(\varepsilon,k,\alpha) = \mathcal{O}(|\alpha|^3)$ for $\alpha\to 0$. Hence, all small solutions of (4.2) satisfy the bifurcation equation

$$0 = f(\varepsilon,k,\alpha) = \mathcal{Q}N(\varepsilon,k,\alpha_1U_1+\alpha_2U_2+\mathcal{U}(\varepsilon,k,\alpha)) \in \mathbb{R}^2.$$

A simple calculation leads to

$$f(\varepsilon,k,\alpha) = [-(1-k^2)^2 + \varepsilon^2 + g(\varepsilon,k,|\alpha|^2)]\alpha, \quad \text{where } g(\varepsilon,k,|\alpha|^2) = -\frac{3}{8}|\alpha|^2 + \mathcal{O}(|\alpha|^4),$$

for $\alpha \to 0$, uniformly for $(\varepsilon, k-1)$ small.

The reduced function f is a scalar function times the vector $\alpha \in \mathbb{R}^2$. This is due to the fact that the original problem is translation invariant and reflection symmetric. Shifting ξ in (4.2) to $\xi + \eta$ leads in PX^4 to the mapping $\alpha \to R(\eta)\alpha$, where $R(\eta)$ is the rotation in the α-plane by the angle η. The reflection $\xi \to -\xi$ corresponds to $\alpha \to S\alpha = (\alpha_1, -\alpha_2)$. As the original problem was equivariant under these operations, the same is true for the reduced problem. Hence, $f(\varepsilon, k, R(\eta)\alpha) = R(\eta)f(\varepsilon, k, \alpha)$ and $f(\varepsilon, k, S\alpha) = Sf(\varepsilon, k, \alpha)$. These symmetries imply the special form of f.

To find all small periodic solutions it now suffices to set the factor in front of α equal to zero. This leads to the following existence result.

Theorem 4.1

(Existence) There is an $\varepsilon_0 > 0$ such that for all $\varepsilon \in (0, \varepsilon_0]$ all $k \in (\sqrt{1-\varepsilon}, \sqrt{1+\varepsilon})$ there is a unique solution $u = \widetilde{u}_{\varepsilon,k}$ of (4.2) which is even in x and positive at $x > 0$. All other small bounded solutions are obtained from translates of this family. We have the expansion

$$\widetilde{u}_{\varepsilon,k}(\xi) = \widetilde{a}\cos\xi + \mathcal{O}(\widetilde{a}^2) \text{ for } (\varepsilon, k-1) \to 0,$$

where $\widetilde{a} = \widetilde{a}(\varepsilon, k) = \frac{1}{2\pi}\langle \widetilde{u}_{\varepsilon,k}, U_1\rangle = \sqrt{\frac{8}{3}(\varepsilon^2 - (1-k^2)^2)} + \mathcal{O}(\varepsilon^2 - (1-k^2)^2)$.

Without loss of generality we restrict our further analysis for given ε and k in the existence domain to the uniquely defined roll solution $\widetilde{u}_{\varepsilon,k} = \widetilde{a}U_1(\xi) + \mathcal{U}(\varepsilon, k, (\widetilde{a}, 0))(\xi)$ which is even in x and satisfies $\widetilde{u}_{\varepsilon,k}(0) > 0$.

To study the instability with respect to sideband perturbations we go back to the spectral problem (2.4) which reads in our case

$$\lambda v = -(1+\partial_x^2)^2 v + \varepsilon^2 v - 3\widetilde{u}_{\varepsilon,k}^2(x)v.$$

We look for solutions of the form $v(x,y) = e^{i(\sigma x + \tau y)}V(\xi)$ where $\xi = kx$ and $V \in X^4$. This leads to

$$B(\varepsilon, k, \sigma, \tau, \lambda)V := -(1 + (k\partial_\xi + i\sigma)^2 - \tau^2)^2 V + \varepsilon^2 V - 3\widetilde{u}_{\varepsilon,k}^2 V - \lambda V = 0,$$

where $(\varepsilon, k, \sigma, \tau, \lambda) \in \mathbb{R}^4 \times \mathbb{C}$ and $B(\ldots) \in \mathcal{L}(X^4, X^0)$. It is important to note that B is strongly related to N, namely $B(\varepsilon, k, 0, 0, 0) = \partial_{\widetilde{u}}N(\varepsilon, k, \widetilde{u}_{\varepsilon,k})$. In particular, $B(0, 1, 0, 0, 0)$ coincides with the the linear part L_{per} of N. Hence, we solve $B(\ldots)V = 0$ again by using the Liapunov–Schmidt reduction with the same splitting which was used for the nonlinear problem. Let $V = \beta_1 U_1 + \beta_2 U_2 + V_1$ with $PV_1 = 0$, then, for sufficiently small $(\varepsilon, k-1, \sigma, \tau, \lambda)$, the equation $(I-P)B(\ldots)V = 0$ can be solved uniquely for $V_1 = \mathcal{V}(\varepsilon, k, \sigma, \tau, \lambda)\beta = \mathcal{O}(\widetilde{a}^2|\beta|)$. Obviously the relation between β and V_1 is linear, and the estimate follows easily from the fact that the coupling only occurs through the term $-3\widetilde{u}_{\varepsilon,k}^2 V$ and the associated estimate in Theorem 4.1.

Inserting this expansion function into the projection $PB(\ldots)V = 0$ leads to the reduced eigenvalue problem which corresponds to (3.4). It is given by a 2×2–matrix, which depends nonlinearly on λ:

$$m(\varepsilon, k, \sigma, \tau, \lambda)\beta = \mathcal{P}B(\varepsilon, k, \sigma, \tau, \lambda)[\beta_1 U_1 + \beta_2 U_2 + \mathcal{V}(\varepsilon, k, \sigma, \tau, \lambda)\beta].$$

By the construction and Theorem 2.1 we know

$$m(\varepsilon,k,0,0,0)=\begin{pmatrix} c(\varepsilon,k) & 0\\ 0 & 0\end{pmatrix} \text{ where } c(\varepsilon,k)=\partial_3 g(\varepsilon,k,\tilde{a}^2)2\tilde{a}^2=-\frac{3\tilde{a}^2}{4}+\mathcal{O}(\tilde{a}^4).$$

Hence, for $\sigma=\tau=0$ there is always an eigenvalue $\lambda_1=0$ which is associated to the translational invariance. The other eigenvalue $\lambda_2=c(\varepsilon,k)$ is strictly negative in our case. These eigenvalues show that the periodic solutions are (neutrally) stable with respect to perturbations of the same period $2\pi/k$. To study sideband instability we have to investigate how the critical eigenvalue $\lambda_1\approx 0$ behaves for small σ and τ.

Lemma 4.2
The critical eigenfunction λ_1 has the expansion

$$\begin{aligned}\lambda_1 &= c_1(\varepsilon,k)\sigma^2+c_2(\varepsilon,k)\tau^2+\mathcal{O}(\sigma^4+\tau^4), \quad \textit{with}\\ c_1 &= \tfrac{2}{c(\varepsilon,k)}\Big[8k^2(1-k^2)^2-(1+k^2)c(\varepsilon,k)\Big]+\mathcal{O}(\tilde{a}^2) \textit{ and } c_2 = \tfrac{2(1-k^2)}{c(\varepsilon,k)}+\mathcal{O}(\tilde{a}^2).\end{aligned}$$

Hence, instability holds whenever c_1 or c_2 is positive.

Proof: In order to expand the matrix m we note that $\mathcal{V}$ satisfies the expansion

$$\mathcal{V}(\varepsilon,k,\sigma,\tau,\lambda)=D_\alpha\mathcal{U}(\varepsilon,k,(\tilde{a},0))+i\sigma Z(\varepsilon,k,\tilde{a})+\tilde{a}^2\mathcal{O}(\sigma^2+\tau^2+|\lambda|),$$

where $D_\alpha\mathcal{U}(\ldots)=\mathcal{O}(\tilde{a}^2)$ and $Z(\ldots)=\mathcal{O}(\tilde{a}^2)$. These estimates follow easily by considering the equation defining V_1 where the coupling of β and V_1 is only active through $-3\tilde{u}_{\varepsilon,k}^2$ which is of order $\tilde{a}^2$.

Inserting $\mathcal{V}$ into the equation $\mathcal{Q}B(\ldots)(\beta_1U_1+\beta_2U_2+\mathcal{V})=0$ multiplies the terms again by a factor of order $\tilde{a}^2$. We arrive at the expansion

$$m(\varepsilon,k,\sigma,\tau,\lambda)=\begin{pmatrix}\rho+c(\varepsilon,k)-\lambda & i\delta\\ -i\delta & \rho-\lambda\end{pmatrix}+\tilde{a}^4\begin{pmatrix}\mathcal{O}(\sigma^2+\tau^2+|\lambda|) & \mathcal{O}(|\sigma|+\tau^2+|\lambda|)\\ \mathcal{O}(|\sigma|+\tau^2+|\lambda|) & \mathcal{O}(\sigma^2+\tau^2+|\lambda|)\end{pmatrix},$$

where $\rho=-4\sigma^2+2(1-k^2)(\sigma^2+\tau^2)-(\sigma^2+\tau^2)^2$ and $\delta=4\sigma k(1-k^2-\sigma^2-\tau^2)$. The only nontrivial part in this expansion is that the diagonal terms of m do not contain terms proportial to σ, which follows again from the reflection symmetry. The diagonal terms are real–valued and even in σ, whereas the off–diagonal terms are purely imaginary and odd in σ, such that m is Hermitian.

The determinant of m has the expansion $\det m(\varepsilon,k,\sigma,\tau,\lambda)=a_0+a_1\lambda+\mathcal{O}(|\lambda^2|)$, where the coefficients are given by

$$\begin{aligned}a_0(\varepsilon,k,\sigma,\tau)&=\rho^2+c(\varepsilon,k)\rho-\delta^2+\mathcal{O}(\tilde{a}^4(\sigma^2+\tau^2)(|1-k|+\sigma^2+\tau^2+\tilde{a}^2)),\\ a_1(\varepsilon,k,\sigma,\tau)&=-c(\varepsilon,k)-2\rho+\mathcal{O}(\tilde{a}^4(\sigma^2+\tau^2+\tilde{a}^2)).\end{aligned}$$

For small (σ,τ) the coefficient a_0 is much smaller than a_1, and the critical eigenvalue λ_1 has the expansion $\lambda_1=-a_0/a_1+\mathcal{O}(a_0^2)$. Since $a_1>0$ we have instability whenever $a_0<0$. Expanding in terms of σ and τ we obtain the desired result. □

The above expansion of the reduced stability problem yields in fact much more structure, and we can ask the question what is the most unstable perturbation. To answer this question we have to find the maximum of λ_1 with respect to σ and τ. In the Swift–Hohenberg problem we see that only two cases occur.

The first kind of instability is the so–called Eckhaus instability (see [Eck65]) which is purely one–dimensional, that is the rolls as well as the most unstable mode do not depend on y. In our linear analysis this means that $\lambda_1(\varepsilon, k, \sigma, \tau)$ assumes its maximum at a point $(\sigma, \tau) = (\sigma_E, 0)$. The second kind of instability is the so–called "zigzag" instability (see [Bu71] for a first investigation), here the most unstable perturbation is periodic in y but independent of x, which corresponds to the most unstable sideband vector $(\sigma, \tau) = (0, \tau_Z)$. The physical relevance of these instabilities is discussed in [NPL93].

We attack this question by a suitable scaling for the sideband parameters σ and τ:

$$\sigma = \varepsilon\widehat{\sigma}, \; \tau = \sqrt{\varepsilon}\,\widehat{\tau}, \; k^2 - 1 = 2\varepsilon\theta, \; \lambda = \varepsilon^2\widehat{\lambda},$$

where the parameter $\theta \in [-1/2, 1/2]$ measures the relative size of ε and $k-1$. Then, the 2×2–matrix m of the reduced stability problem has the expansion

$$m(\varepsilon, \sqrt{1+2\varepsilon\theta}\,, \varepsilon\widehat{\sigma}, \sqrt{\varepsilon}\,\widehat{\tau}, \varepsilon^2\widehat{\lambda}) = \varepsilon^2\widehat{m}(\theta, \widehat{\sigma}, \widehat{\tau}, \widehat{\lambda}) + \mathcal{O}(\varepsilon^3), \tag{4.4}$$

where the scaled limit $\widehat{m}$ is given by

$$\widehat{m}(\theta, \widehat{\sigma}, \widehat{\tau}) = \begin{pmatrix} -4\widehat{\sigma}^2 - 4\theta\widehat{\tau}^2 - \widehat{\tau}^4 - 2 + 8\theta^2 - \widehat{\lambda} & -4i\widehat{\sigma}(2\theta + \widehat{\tau}^2) \\ 4i\widehat{\sigma}(2\theta + \widehat{\tau}^2) & -4\widehat{\sigma}^2 - 4\theta\widehat{\tau}^2 - \widehat{\tau}^4 - \widehat{\lambda} \end{pmatrix}.$$

Thus, in the limit $\varepsilon = 0$, the critical eigenvalue $\widehat{\lambda}_1$ is given by

$$\widehat{\lambda}_1(\theta, \widehat{\sigma}, \widehat{\tau}) = -4\widehat{\sigma}^2 - 4\theta\widehat{\tau}^2 - \widehat{\tau}^4 - 1 + 4\theta^2 + \left((1-4\theta^2)^2 + 16\widehat{\sigma}^2(2\theta + \widehat{\tau}^2)^2\right)^{1/2}, \tag{4.5}$$

and $\widehat{\lambda}_2 < 0$ for all $(\theta, \widehat{\sigma}, \widehat{\tau})$. Some elementary calculations show that the maxima of λ_1 with respect to $(\widehat{\sigma}, \widehat{\tau})$ always lie on the axis $\widehat{\sigma} = 0$ or $\widehat{\tau} = 0$. From

$$\begin{aligned} \widehat{\lambda}_1(\theta, \widehat{\sigma}, 0) &= -4\widehat{\sigma}^2 - 1 + 4\theta^4 + ((1-4\theta^2)^2 + 64\theta^2\widehat{\sigma}^2)^{1/2} \\ &\leq \begin{cases} 0 & \text{for } \theta^2 \leq \frac{1}{12}, \\ \frac{1-24\theta^2+160\theta^4}{16\theta^2} & \text{for } \theta^2 \in [\frac{1}{12}, \frac{1}{4}]; \end{cases} \\ \widehat{\lambda}_1(\theta, 0, \widehat{\tau}) &= -4\theta\widehat{\tau}^2 - \widehat{\tau}^4 \leq \begin{cases} 0 & \text{for } \theta \in [0, \frac{1}{2}], \\ 4\theta^2 & \text{for } \theta \in [-\frac{1}{2}, 0], \end{cases} \end{aligned}$$

we conclude that, for $\theta > 1/(2\sqrt{3})$, the Eckhaus instability occurs. For $\theta < 0$ we always have the zigzag instability which is stronger than the Eckhaus instability [which also occurs for $\theta \in (-1/2, -1/(2\sqrt{3}))$]. These results are stable under small perturbations with $\varepsilon > 0$, and we obtain the following result:

Corollary 4.3
There are three curves $k = K_j(\varepsilon)$ which have the expansions

$$K_0(\varepsilon) = 1 + \mathcal{O}(\varepsilon^4) \quad \text{and } K_{\pm 1}(\varepsilon) = 1 \pm \frac{\varepsilon}{2\sqrt{3}} + \mathcal{O}(\varepsilon^2),$$

such that the rolls $\widetilde{u}_{\varepsilon,k}$ are Eckhaus unstable for $k \notin [K_{-1}(\varepsilon), K_1(\varepsilon)]$ and zigzag unstable for $k < K_0(\varepsilon)$.

Our present analysis allow us to to calculate the width of the unstable sideband. This is important in cases where we are interested in instability in finite but very large boxes. Assume we have rolls $u_{\varepsilon,k}$ of period $2\pi/k$ in a rectangular box $(x, y) \in [0, L_x] \times [0, L_y]$ with side lengths $L_x = N2\pi/k$, $N \in \mathbb{N}$, and L_y, and we study the problem restricted to the spaces $X^j(L_x, L_y)$ of functions which are periodic with these periods. Let

$$S(\varepsilon, k) = \{ (\sigma, \tau) \in \mathbb{R}^2 \ : \ \lambda_1(\varepsilon, k, \sigma, \tau) > 0 \}$$

be the set of unstable sideband wave numbers. Then, instability of $\widetilde{u}_{\varepsilon,k}$ in $X^j(L_x, L_y)$ only holds if

$$S(\varepsilon, k) \cap \{ (2\pi m_1/L_x, 2\pi m_2/L_y) \in \mathbb{R}^2 \ : \ m_1, m_2 \in \mathbb{Z} \} \neq \emptyset.$$

Using the above scalings we define analogously $\widehat{S}(\theta) = \{ (\widehat{\sigma}, \widehat{\tau}) \in \mathbb{R}^2 \ : \ \widehat{\lambda}(\theta, \widehat{\sigma}, \widehat{\tau}) \geq 0 \}$. It can be shown, that $\widehat{S}(\theta)$ is bounded only for $\theta^2 < 1/8$, for all other θ there exists thin unbounded branches extending to infinity along the parabolae $\widehat{\sigma} = \pm \widehat{\tau}^2$.

Define for $\theta^2 < 1/8$ the length measures $\widehat{L}_x^\theta$ and $\widehat{L}_y^\theta$ such that $\widehat{S}(\theta)$ is contained in the rectangle $[-2\pi/\widehat{L}_x^\theta, 2\pi/\widehat{L}_x^\theta] \times [-2\pi/\widehat{L}_y^\theta, 2\pi/\widehat{L}_y^\theta]$, then we are lead to the following conclusion.

Corollary 4.4
For each $\theta \in (-1/\sqrt{8}, 1/\sqrt{8})$ and for arbitrarily small $\delta > 0$ there exists an $\varepsilon_0 > 0$ such that for every $\varepsilon \in (0, \varepsilon_0)$ the roll pattern $\widetilde{u}_{\varepsilon,1+2\varepsilon\theta}$ is stable with respect to perturbations which are (L_x, L_y)-periodic in (x, y) where the periods have to satisfy the relations

$$\frac{\widehat{L}_x^\theta}{(1-\delta)\varepsilon} \geq L_x \in \frac{2\pi}{1 + 2\varepsilon\theta}\mathbb{Z} \quad \text{and} \quad \frac{\widehat{L}_y^\theta}{(1-\delta)\sqrt{\varepsilon}} \geq L_y.$$

Note that this result contains parts of the zigzag unstable domain $\theta \in (-1/\sqrt{8}, 0)$ as well as parts of the Eckhaus unstable domain $|\theta| \in (1/\sqrt{12}, 1/\sqrt{8})$. In the remaining parts of the θ–interval we can not make conclusions about the box sizes L_x and L_y, since $\widehat{\lambda}_1$ does not generate a bounded set $\widehat{S}(\theta)$. We conjecture that in this regime both length scales, L_x and L_y, have to be smaller than C independent of ε.

Another interesting feature appears for the stability on large rectangular boxes. For the infinite domain the most unstable mode is either parallel or perpendicular to the rolls, that is, the sideband vector is (σ, τ) with $\sigma\tau = 0$. For finite boxes, these maxima of $\lambda(\varepsilon, k, \sigma, \tau)$ are not necessarily relevant and we find cases where the most unstable sideband vector (σ, τ) satisfies $\sigma\tau \neq 0$.

5 The Newell–Whitehead equation

Finally we want to compare our results with another method which was designed exactly to study the dynamics of rolls. According to the modulation theory, the dynamics of roll–like patterns in the SHE (cf. (4.1)) can be described by the so–called Newell–Whitehead equation (NWE), which is based on the assumption that the relevant solutions can be written in the form

$$u(t,x,y) = \varepsilon(A(\varepsilon^2 t, \varepsilon x, \sqrt{\varepsilon} y)e^{ix} + \text{c.c.}) + \mathcal{O}(\varepsilon^2). \tag{5.1}$$

Here $A = A(\widehat{t}, \widehat{x}, \widehat{y})$ is the complex amplitude function for the basic roll pattern e^{ix} and it depends on the slow time $\widehat{t} = \varepsilon^2 t$ and the large spatial scales $\widehat{x} = \varepsilon x$ and $\widehat{y} = \sqrt{\varepsilon} y$.

Inserting the ansatz (5.1) into the SHE and equating the terms multiplying $\varepsilon^3 e^{ix}$ to zero we obtain the partial differential equation

$$\partial_{\widehat{t}} A = -(2i\partial_{\widehat{x}} + \partial_{\widehat{y}}^2)^2 A + A - 3|A|^2 A, \tag{5.2}$$

which is called the Newell–Whitehead equation. See [NW69] for the derivation of this equation in the context of the Bénard problem.

The NWE is only derived formally and it is not clear in how far its dynamics represents any of the dynamics in the SHE. For the one–dimensional Ginzburg–Landau equation, which is obtained by restricting everything to y–independent solutions, a series of paper ([vH91, KSM92, Sch94a]) shows that the modulation produces good approximate solutions for the original problem. Yet, for the NWE it was shown in [Sch94d] that predictions from the NWE may describe the dynamics of the SHE incorrectly. This means the following: let A be a solutions of the NWE and form a function $\widetilde{u}(t,x,y)$ according to (5.1), then $\widetilde{u}$ is not necessarily close to any true solution u of the SHE on a time interval $[0, \widehat{t}_0/\varepsilon^2]$ for some $\widehat{t}_0 > 0$. The problem arises through the fact that solutions u of the SHE may start near the roll pattern, but may develop such that it loses the roll–like structure, see [NPL93, Sch94d].

In this paper we are only concerned with the stability question of rolls, and we show now, that the NWE gives in lowest order the correct stability information. The rolls in the NWE are the $\widehat{y}$–independent periodic solutions given in the form

$$\widetilde{A}_\theta(\widehat{x}) = \sqrt{(1-4\theta^2)/3}\, e^{i\theta\widehat{x}}, \quad |\theta| < 1/2,$$

where we again normalized $\widetilde{A}_\theta$ such that the real part is even and positive at $\widehat{x} = 0$. From (5.1) we see that $\widehat{A}_\theta$ corresponds to the roll pattern $\widetilde{u}(x) = 2\varepsilon\sqrt{(1-4\theta^2)/3} \cdot \cos((1+\varepsilon\theta)x) + \mathcal{O}(\varepsilon^2)$ in the SHE. Thus, $1+\varepsilon\theta$ corresponds to k and $2\varepsilon\sqrt{(1-4\theta^2)/3}$ to $\widetilde{a}(\varepsilon, k)$. We find, to lowest order a perfect agreement of the predictions of the NWE with the results for the original SHE as far as local existence of rolls is concerned.

In order to study the linearized stability of $\widetilde{A}_\theta$ we use the ansatz $A = \widetilde{A}_\theta + e^{i\theta\widehat{x}}(b_r + ib_i)$ in the NWE and linearize in $b_r + ib_i$. Then we obtain

$$\partial_{\widehat{t}} \begin{pmatrix} b_r \\ b_i \end{pmatrix} = \begin{pmatrix} 4\partial_{\widehat{x}}^2 + 4\theta\partial_{\widehat{y}}^2 - \partial_{\widehat{y}}^4 - 2 + 8\theta^2 & -8\theta\partial_{\widehat{x}} + 4\partial_{\widehat{x}}\partial_{\widehat{y}}^2 \\ 8\theta\partial_{\widehat{x}} - 4\partial_{\widehat{x}}\partial_{\widehat{y}}^2 & 4\partial_{\widehat{x}}^2 + 4\theta\partial_{\widehat{y}}^2 - \partial_{\widehat{y}}^4 \end{pmatrix} \begin{pmatrix} b_r \\ b_i \end{pmatrix}. \tag{5.3}$$

This $(\widehat{x},\widehat{y})$–independent problem can be treated by spatial Fourier transformation, and the stability can be analyzed completely by studying the growth rate $\widehat{\lambda} = \widehat{\lambda}(\theta,\widehat{\sigma},\widehat{\tau})$ for solutions of the form $b_r + ib_i = e^{\widehat{\lambda t}+i(\widehat{\sigma x}+\widehat{\tau y})}(B_r + iB_i)$ with constants $B_r, B_i \in \mathbb{C}$. The exponential growth rates $\widehat{\lambda}$ are found by simply replacing the spatial derivative $(\partial_{\widehat{x}},\partial_{\widehat{y}})$ by $(i\widehat{\sigma},i\widehat{\tau})$, which leads to the linear problem $\widehat{m}(\theta,\widehat{\sigma},\widehat{\tau},\widehat{\lambda})\binom{B_r}{B_i} = 0$, where $\widehat{m}$ is exactly the scaled limit in (4.4). Hence, we have made precise that the NWE leads to the same stability analysis as the principle of reduced instability, when the appropriate scaling limit is considered.

References

[BB84] **F.H. Busse, E.W. Bolton.** Instabilites of convection rolls with stress–free boundaries near threshold. *J. Fluid Mech.*, **146** (1984) 115–125.

[BM94a] **T.J. Bridges, A. Mielke.** Instability of spatially–periodic states for a family of semilinear PDE's on an infinite strip. Manuscript Universität Hannover, July 1994. *Math. Nachrichten.* To appear.

[BM94b] **T.J. Bridges, A. Mielke.** A proof of the Benjamin–Feir instability. Manuscript Universität Hannover, August 1994. *Archive Rational Mech. Analysis.* To appear.

[Bu71] **F.H. Busse.** Stability regions of cellular fluid flow. In "*Instability of Continuous Systems*, H. Leipholz (edr), Proc. of the IUTAM Symposium in Bad Herrenalb 1969, Springer–Verlag 1971." pages 41–47.

[BS89] **F.H. Busse, M. Schnaubelt.** On the stability of two–dimensional convection rolls in an infinite Prandtl number fluid with stress–free boundaries. *Zeits. angew. Math. Physik* **40** (1989) 153–162.

[CE90] **P. Collet, J.-P. Eckmann.** *Instabilities and Fronts in Extended Systems.* Princeton University Press, Princeton, 1990.

[CH88] **S.–N. Chow, J.K. Hale.** *Methods of Bifurcation Theory.* Springer–Verlag, 1985.

[Dam91] **K. Damerow.** A sufficient condition for the validity of the principle of reduced stability. *Math. Nachrichten* **154** (1991) 243–252.

[DES71] **R.C. DiPrima, W. Eckhaus, L.A. Segel.** Non–linear wave–number interaction in near–critical two–dimensional flows. *J. Fluid Mech.*, **49** (1971) 705–744.

[Eck65] **W. Eckhaus.** *Studies in Non–Linear Stability Theory.* Springer–Verlag. Springer Tracts in Nat. Phil. Vol. **6**, 1965.

[vH91] **A. van Harten.** On the validity of Ginzburg–Landau's equation. *J. Nonlinear Science*, **1** (1991) 397–422.

[IM91] **G. Iooss, A. Mielke.** Bifurcating time–periodic solutions of Navier–Stokes equations in infinite cylinders. *J. Nonlinear Science,* **1** (1991) 107–146.

[IMD89] **G. Iooss, A. Mielke, Y. Demay.** Theory of steady Ginzburg–Landau equation, in hydrodynamic stability problems. *Europ. J. Mech., B/Fluids*, **8** (1989) 229–268.

[KL83] **H. Kielhöfer, R. Lauterbach.** On the principle of reduced stability. *J. Functional Analysis* **53** (1983) 99–111.

[Ki82] **K. Kirchgässner.** Wave solutions of reversible systems and applications. *J. Diff. Eqns.*, **45** (1982) 113–127.

[KSM92] **P. Kirrmann, G. Schneider, A. Mielke.** The validity of modulation equations for extended systems with cubic nonlinearities. *Proc. Royal Society Edinburgh*, **122A** (1992) 85–91.

[Kuc93] **P.A. Kuchment.** *Floquet theory for partial differential equations.* Birkhäuser Verlag, 1993. See also *Russ. Math. Surveys*, **37**(4) (1982) 1–60.

[Mi86] **A. Mielke.** A reduction principle for nonautonomous systems in infinite-dimensional spaces. *J. Diff. Eqns.*, **65** (1986) 68–88.

[Mi88] **A. Mielke.** Reduction of quasilinear elliptic equations in cylindrical domains with applications. *Math. Meth. Appl. Sci.*, **10** (1988) 51–66.

[Mi91] **A. Mielke.** *Hamiltonian and Lagrangian flows on center manifolds with applications to elliptic variational problems.* Lecture Notes in Mathematics Vol. **1489**, Springer–Verlag. 1991.

[Mi92] **A. Mielke.** A center manifold approach to steady bifurcation from spatially periodic patterns. Manuscript Univ. Hannover, Oct. 1992.

[NPL93] **A.C. Newell, T. Passot, J. Lega.** Order parameter equations for patterns. *Annual Reviews Fluid Mech.,* **25** (1993) 399–453.

[NW69] **A. Newell, J. Whitehead.** Finite bandwidth, finite amplitude convection. *J. Fluid Mech.,* **38** (1969) 279–303.

[Rec89] **L. Recke** On linear stability of bifurcating equilibria. *Math. Nachrichten* **140** (1989) 59–68.

[RS78] **M. Reed, B. Simon.** *Methods of Modern Mathematical Physics IV*, Academic Press, 1978.

[Sch94a] **G. Schneider.** A new estimate for the Ginzburg–Landau approximation on the real axis. *J. Nonlinear Science*, **4** (1994) 23–34.

[Sch94d] **G. Schneider.** Validity and limitation of the Newell–Whitehead equation. IfAM–Preprint Universität Hannover 1994. Submitted to *Math. Nachrichten.*

[ST94] **M.M. Sushchik, L.S. Tsimring.** The Eckhaus instability in hexagonal patterns. *Physica D* **74** (1994) 90–106.

[Va84] **A. Vanderbauwhede.** Stability of bifurcating equilibria and the principle of reduced stability. In "*Bifurcation Theory and Applications*, Proceedings of the CIME meeting in Montecatini 1983, edr. L. Salvadori, Springer Lecture Notes Math. **1057** (1984)", pages 209–223.

Pattern formation in activator-inhibitor systems

M. OR-GUIL, E. AMMELT, F.-J. NIEDERNOSTHEIDE, and H.-G. PURWINS
Institute of Applied Physics, University of Münster, Correnstr. 2/4, D-48149 Münster, Germany

Abstract

We study a two-component reaction-diffusion system of activator-inhibitor type with global inhibition that has been developed to understand qualitatively self-organized pattern formation in electrical two-layer systems. It is shown numerically that the global inhibition can change the expected bifurcation behaviour essentially; in particular, in a two-dimensional medium solitary structures may arise instead of hexagonal patterns and target patterns may arise instead of homogeneous oscillations. Additional solutions of the reaction-diffusion system are presented and compared with experimental results obtained from a semiconductor and a gas discharge system.

1 Introduction

There is a huge variety of biological, chemical and physical systems for which reaction-diffusion equations of activator-inhibitor type have been derived in order to describe spatio-temporal patterns arising spontaneously when an external parameter is varied. Well-known biological examples are, e. g., animal coat and shell patterns ([Meinhardt, 1982]; [Murray, 1989]; [Rensing, 1993]; [Mosekilde and Mouritsen, 1995]). In chemical systems interesting temporal and spatio-temporal behaviour has been found, e. g., in the Belousov-Zhabotinsky reaction or during oxidation on reactive surfaces. The observed structures extend from regular patterns as autowaves or spiral waves to turbulent behaviour (see, for example, [Kuramoto, 1984]; [Field and Burger, 1985]; [Winfree, 1987]; [Mikhailov, 1990]). In physics, various semiconductor devices and gas discharge systems can be described in terms of reaction-diffusion equations and good agreement between numerical simulations and experimental results has been obtained in particular for one-dimensional systems (e. g.,[Kerner and Osipov, 1994]; [Purwins and Radehaus, 1988]; [Niedernostheide et al, 1994a]; [Niedernostheide, 1995]).

In this work we focus on a two-component set of reaction-diffusion equations that has been derived for electrical systems consisting of an active and a passive layer. In the active layer charge carrier multiplication leads to an autocatalytic increase of the current density whereas the passive layer counteracts such current density fluctuations. The phenomenological model was derived for electrical systems driven by a dc voltage source ([Radehaus et al, 1987]). However, the basic mechanism for self-organized pattern formation in such systems, the competition between an activator and an inhibitor, is also valid for ac driven systems. These have been chosen for experimental studies because they have the advantage of better reproducibility.

In what follows, we discuss the basic set of equations and present numerical results selected with regard to experimental results which have been obtained by investigating two different systems, each consisting of an active and a passive layer. A point of special interest concerns the influence of an additional global inhibition that appears as an integral term in the reaction function due to the external circuit and, therefore, is present in many electrical systems. The remainder of the paper is devoted to experimental results. We have studied a semiconductor system consisting of a zinc sulphide layer doped with manganese and sandwiched between two insulating layers of yttriumoxide and a gas discharge system which is constructed in a similar way: an inert gas operating as active medium is enclosed between two insulating glasses. The systems are driven by an ac voltage source via a load resistor and in both devices luminescence patterns give evidence for complex spatio-temporal current-density distributions which can be discussed in terms of activator-inhibitor mechanisms.

2 The Activator-Inhibitor System

Basic Equations. The sytem we consider consists of a set of two-component reaction-diffusion equations with global inhibition, with the help of which a variety of self-organization phenomena can be understood:

$$\begin{aligned} \dot{v} &= D_v \Delta v + \lambda v - v^3 - w + T, \\ \delta \dot{w} &= D_w \Delta w + v - w, \end{aligned} \tag{1}$$

where

$$T = \kappa_1 - \kappa_2 \int_\Omega v d\boldsymbol{r},$$

with

$$\boldsymbol{r} \in \Re^2, \quad \Omega = [0,1] \times [0,1] \quad \text{and} \quad \lambda, \delta, \kappa_2 > 0$$

and the no-flux boundary conditions

$$\left.\frac{\partial v}{\partial \boldsymbol{r}}\right|_{\partial\Omega} = \left.\frac{\partial w}{\partial \boldsymbol{r}}\right|_{\partial\Omega} = 0.$$

The assumption of a rather simple nonlinearity v^3 suffices to achieve our aim, which is to have a model where v acts, within certain parameter limits, as an activator, and where w and the integral coupling term T act as an inhibitor to the system. The spatially uncoupled system may show mono- and bistability, destabilization of stable homogeneous solutions by means of Turing- and Hopf-bifurcation can be found, and parameters can be set as to obtain an excitable media where pulses may spread.

The analogies between the model and two-layer experimental systems have been shown by Radehaus et al (1987). There the activator v represents the current density and the inhibitor w the potential drop across one of the device layers. An increase of

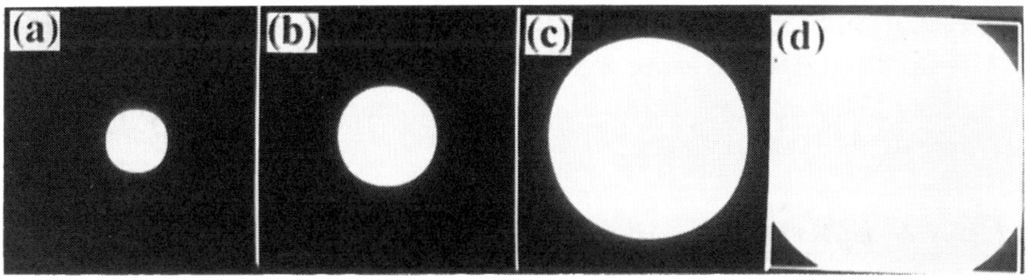

Figure 1: Numerical calculations of the temporal evolution of system (1), showing $v(\boldsymbol{r}, t)$ as a moving circular front in the area Ω. The time between consecutive pictures is 0.03, 0.07 and 0.09. Parameters: $\delta = 40$, $D_v = 10^{-4}$, $D_W = 4 \cdot 10^{-3}$, $\lambda = 2$, $\kappa_1 = 0$, $\kappa_2 = 0$ and no-flux boundary conditions. White regions correspond to regions with the highest value of v.

κ_1 implies an incrase in the total voltage, while κ_2 represents a load resistor and T the voltage drop across the device.

Numerical Results. The numerical results were obtained using a finite difference implicit method. In all computations the domain was set to be a square with the sidelength set to unity and with no-flux boundary conditions.

Front propagation. Fronts which mediate between two homogeneous systems can often be encountered in bistable systems if these are sufficiently large. Without global inhibition, moving front solutions are typical. This is the case in Fig. 1, where the parameter set defines a bistable system and a front is shown which delimits a circular "high current density" domain and moves out of the system leaving a homogeneous "high current density" state behind it. On the other hand, a global inhibition modifies the front velocity because the parameter T depends on the front position, i. e., T becomes time dependent. The front velocity is slowed down, and finally, the front becomes stationary. Experimental examples for both cases will be shown in Secs. 3 and 4.

Spiraling travelling pulses in excitable media. The phenomenon of travelling two-dimensional pulses which in distributed excitable media evolve to a spiral is well known (Mikhailov, 1990; Meron, 1992). Perturbations placed on such a system will lead to slightly curled travelling pulses, which are the precursors of double spirals. These travelling pulses can by collision annihilate each other partially, at the same time creating new travelling pulses moving in different directions. These can in turn produce more pulses by collision, and so on. The pulses will grow to spirals until they are hindered to do so. So it can be assumed that the pulse density is restricted by the refractory phase, leading to a maximum pulse density. If the mean free path allows for the pulses to curl, the pulse generations will not cease. In Fig. 2 a snapshot of such a dynamical behaviour is shown. In the same parameter set also a single periodically rotating spiral would exist.

A similar phenomenon was observed in the slime mold Dictyostelium discoideum. There the formation of spirals is caused by pacemaker regions, fixing the spiral origins to particular spots and so creating a less complex system. In thin ZnS-films, though, a correspondent phenomenon to the above described has been found and is shown in Fig. 7.

Filament cluster patterns favoured by global inhibition. Ignoring the global coupling in equations (1) means setting $\kappa_2 = 0$. Numerical calculations show that in this case destabilization of the stable homogeneous state by means of Turing-bifurcation in the two-dimensional system (1) leads to the formation of stable hexagonal structures. This result agrees with theoretical (Malomed and Tribel'skii, 1987) and numerical results for similar systems (Dufiet and Boissonade, 1992). In system (1) a Turing-bifurcation can be carried out, e. g., by increasing the "voltage" κ_1 until a critical value κ_{1_c}, respectively T_c, is reached. In contrast to the case without global inhibition, Turing-bifurcation in the presence of a global coupling leads to the formation of well localized solitary filament patterns at the critical parameter T_c. This solution is favoured by the system with global inhibitor, since a growing inhomogeneous structure will diminish $T = \kappa_1 - \kappa_2 \int_\Omega v d\boldsymbol{r}$ to a subcritical value. That implies that the homogeneous domain is not unstable any more, and a situation where solitary filaments coexist with homogeneous regions is stabilized. Further increasing of κ_1 will so lead to a repeated reaching of critical values of T, in which filaments arise, leading to a successive appearence of filaments.

Fig. 3 shows three of many different possible stationary solutions with different numbers of filaments at different κ_1, created by means of Turing-bifurcation. Note that the filaments arrange in clusters and form locally hexagonal structure, whereas the homogeneous region is stable. Characteristic for the chosen case is the spatial oscillatory behaviour of the values of $v(\boldsymbol{r})$ and $w(\boldsymbol{r})$ around the homogeneous state in the vicinity of the filaments. This is a consequence of the fact that the homogeneous state is not too far from the Turing-bifurcation: The perturbation of the stable stationary homogeneous structure causes an oscillatory spatial transition from the perturbation to the stationary homogeneous solution, as can be shown by linear stability analysis. Analytical investigations show that the spatial distribution in the transition region defines the interaction between the filaments (Bode and Purwins, 1995). When the transition oscillates spatially, the interaction between two filaments is also oscillating, alternating in sign. Thus, the filaments are fixed to stationary positions, which are stable, corresponding to discrete distances between them.

A similar phenomenon was observed in thin ZnS-films (Fig. 8) and in gas discharge systems (Fig. 12), although there no clustering of the filaments can be seen. However, numerical investigations lead to a similar behaviour if noise is added, especially if the parameters are chosen so that spatial oscillations decay more rapidly.

Fig. 4 shows smaller stable stationary clusters consisting of two, three and four filaments. The oscillatory regions are clearly distinctible. A small shift in the filament position will decay, bringing the cluster to its original formation. At the same time,

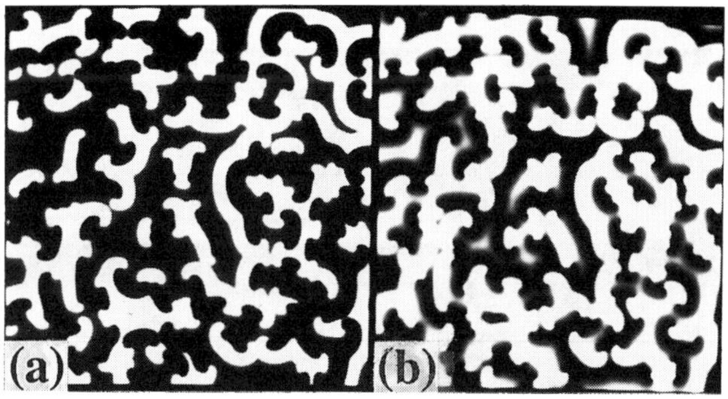

Figure 2: Snapshot of spiraling travelling pulses as numerical solutions of system (1). The figures a) and b) show respectively $v(\boldsymbol{r},t)$ and $w(\boldsymbol{r},t)$. Parameters: $\delta = 30$, $D_v = 10^{-6}$, $D_W = 10^{-10}$, $\lambda = 0.9$, $\kappa_1 = -0.2357$, $\kappa_2 = 0$ and no-flux boundary conditions. White regions correspond to regions with the highest value of v respectively w.

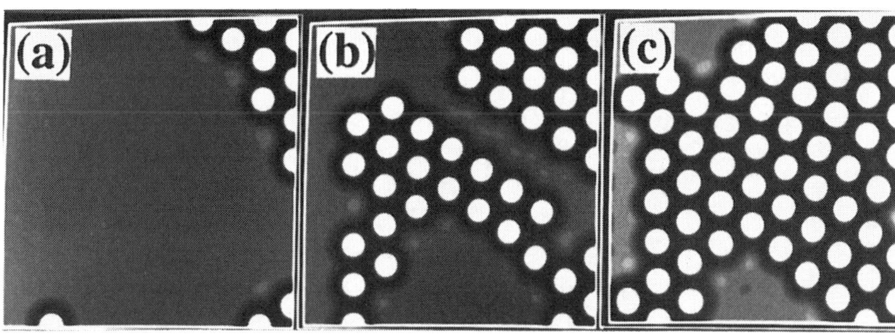

Figure 3: Three different stationary stable numerical solutions $v(\boldsymbol{r})$ of system (1) where homogeneous regions coexist with clusters of filaments. Parameters: $\delta = 1$, $D_v = 6 \cdot 10^{-5}$, $D_W = 6 \cdot 10^{-4}$, $\lambda = 0.8$, $\kappa_2 = 5$, κ_1 = -2.09 (a), -2 (b), -1.8 (c) and no-flux boundary conditions.

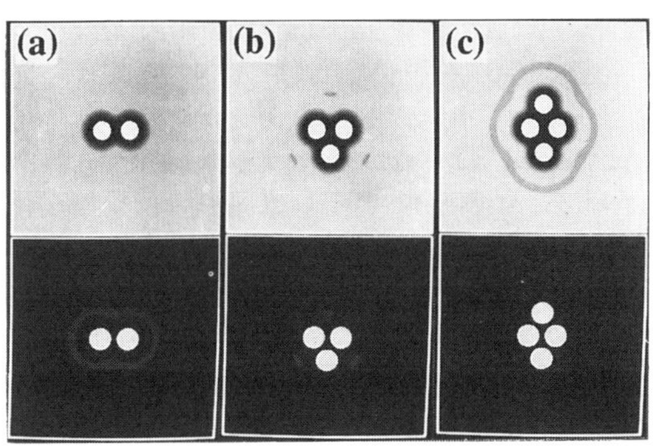

Figure 4: Stable stationary filament clusters as numerical solutions of system (1). The upper pictures show the distribution of $v(\boldsymbol{r})$ while the lower show $w(\boldsymbol{r})$. Parameters: $\delta = 1$, $D_v = 6 \cdot 10^{-5}$, $D_w = 6 \cdot 10^{-4}$, $\lambda = 0.8$, $\kappa_1 = -2.12$, $\kappa_2 = 5$ and no-flux boundary conditions.

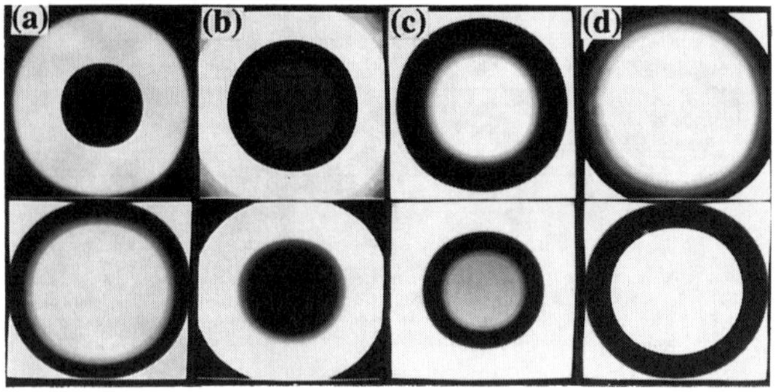

Figure 5: Numerically calculated temporal evolution of moving target patterns in system (1) as an active media with global inhibition. The upper pictures show the distribution of $v(\boldsymbol{r})$ while the lower show $w(\boldsymbol{r})$, both in the area Ω. The time between consecutive pictures is 25, 25 and 15. Parameters: $\delta = 30$, $D_v = 10^{-6}$, $D_w = 10^{-10}$, $\lambda = 0.9$, $\kappa_1 = 0$, $\kappa_2 = 4$ and no-flux boundary conditions. White regions correspond to regions with the highest value of v respectively w.

multistability can be observed in Fig. 4, as the clusters exist at the same parameter set. A striking experimental example of the self-organized appearance of current density filament clusters is the gas discharge system, in which several forms of filament clusters have been observed (Fig. 11).

Target patterns favoured by global inhibition. Without a global inhibitor a Hopf-bifurcation of the stable stationary homogeneous state leads to homogeneous oscillations. Global inhibition on the other hand suppresses oscillations of the homogeneous mode: The trace of the linearization matrix for the homogeneous mode is diminished, leading, therefore, to destabilization by means of an inhomogeneous mode. Therefore one expects an inhomogeneous oscillation in the case of a Hopf-bifurcation in system (1) with $\kappa_2 \neq 0$. Fig. 5 shows inhomogeneous oscillations caused by the repeated formation of a target pattern. The parameter set coincides with that of the pulse structures in Fig. 2, refrained from the "voltage" κ_1 and the "load resistor" κ_2. The homogeneous, Hopf-unstable stationary solution was given a perturbation consisting of concentric circles as initial condition. From this situation a repeatedly appearing target pattern bearing the largest possible wavelenght in the given system is created. The circular pulse generated in the centre of the system suffers there almost no influence from the boundary and thus preserves its symmetry. The pulse spreads leaving behind an almost homogeneous refractory phase, which will again form a pulse as the homogeneous state is unstable. to find expected, adjustable et al, 1995). with It should be stressed that the assumption of system inhomogeneities has not been made, in contrary to many explanations of target patterns found in Belouzov-Zhabotinsky-reactions (Field and Burger, 1985; Winfree, 1987; Meron, 1992).

Subsequently, we shall turn to experimental results gained at two similar constructed systems in which the autocatalytic increase of charge carriers plays an important role.

3 Patterns in Thin ZnS-Films

Thin film electroluminescence devices based on manganese-doped zinc sulphide have been investigated intensively in the seventies and eighties, in particular, in order to apply them for flat display panels. First detailed studies concerning self-organized pattern formation in these devices have been done by Marello et al ([1977]), Rühle et al ([1979a]; [1979b]) and Rüfer et al (1980). Recently, this system has been rediscovered as an exemplary system for self-organization in nonlinear dissipative systems ([Beletskii and Vlasenko, 1993a]; [1993b]; [Niedernostheide et al, 1994b]; [Goßen et al, 1995]).

Experimental Set-Up. The device consists of a glass substrate covered with a transparent electrode of indium tin oxide (ITO). On this electrode an yttrium oxide layer with a thickness of the order of 300–600 nm has been fabricated by electron beam evaporation. The active layer has been produced by thermal coevaporation of zinc sulphide and manganese. The thickness of this layer is of the same order as that of the dielectric layer. The concentration of manganese is between 1 and 2 mol%. The sandwich structure is completed by evaporating a second yttrium oxide layer and, finally, an aluminium layer as second electrode. More details concerning the preparation procedure are described elsewhere ([Goßen et al, 1995]).

A sinusoidal driving voltage or bipolar pulse trains are applied to the device via a load resistor R_v. Typical voltage amplitudes V_0 are of the order of 150–200 V, pulse durations τ_p are between 10 and 100 μs and the repetition rate f is chosen between 1 and 100 kHz. The spatio-temporal distribution of the electroluminescence radiation is observed through the ITO-coated glass substrate by using a microscope and an intensifying CCD camera.

Experimental Results. Before discussing the spatio-temporal behaviour of the electroluminescence radiation, we briefly explain the origin of the electroluminescence and the relevant charge transport processes in the ZnS:Mn device. In thermodynamic equilibrium, electrons are bound to localized states in the interface between the insulator and the ZnS-layer. For sufficiently high amplitudes of the driving voltage, the electric field at the interface near the momentary cathode becomes so large that electrons from the interface states can tunnel into the ZnS-layer. Drifting towards the momentary anode, the electrons are accelerated and may excite manganese centres when their energy exceeds 2.2 eV. The de-excitation of these excited states is connected with the emission of photons causing a yellow luminescence. When the electron energy reaches 3.6 eV, i. e., the gap-energy of ZnS, band-band impact ionization leads to an autocatalytic carrier multiplication inducing also an increase of the luminescence and, therefore, indicating regions of high electron density hereafter re-

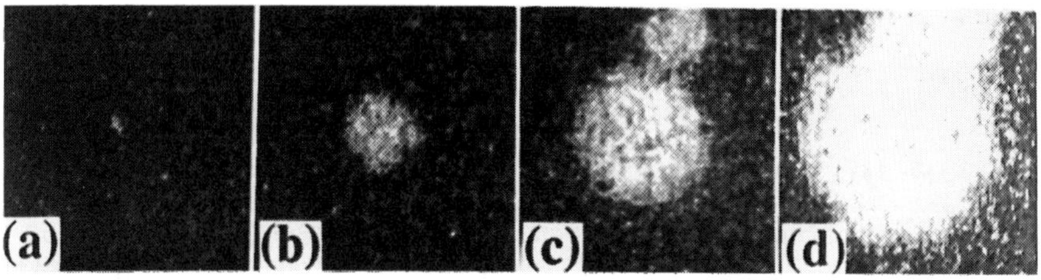

Figure 6: Temporal evolution of a circular luminescence domain in ZnS-films. Bright regions correspond to regions of high radiation density. The time between consecutive pictures is 0.4 s. Parameters: sine-wave excitation, V_0=117.6 V, f=1 kHz, R_v=5 kΩ, imaged device area = 30 × 30 μm^2.

ferred to as 'on'-states. The generated holes may drift to the momentary cathode or captured instantaneously by hole-traps, build-up a space charge and hence modify the electric field distribution (cf. [Howard et al, 1982]). The electrons reaching the momentary anode are trapped again in interface states. As the charge accumulation near the anode leads to a decrease of the electric field near the cathode, carrier-injection is reduced and, consequently, the total charge transfer during a half-cycle is limited. The accumulated negative charge at the momentary anode supports the electron injection at the beginning of the next half-cyle when the polarity of the applied voltage changes.

Frontpropagation. When the device is driven at rather low frequencies of about 1 kHz, we observe that a spatially uniform luminescence distribution develops when the applied voltage exceeds a certain threshold value. As shown in Fig. 6, the system switches to the on-state at positions where small material inhomogeneities reduce the critical electric field slightly. From these preferential sites the luminescence spreads out in form of circlular area. When the radius of the circle exceeds a certain value, the front propagates with a constant velocity of the order of 10 μm/s which increases with increasing driving amplitude. If two such expanding circles arise independently, their luminescence density is the same. In the case of a collision of two circles, the luminescence density in the overlapping region is the same as in the isolated domains. The physical process determining the spreading of the on-state can be attributed to the charge in the semiconductor-insulator interface ([Beletskii and Vlasenko, 1993a]), the spreading of which along the interface causes a field-enhancement in the environment of the front and, consequently, leads to a switching of the neighbouring off-state to the on-state due to the field-dependent impact-ionization. Obviously, for low driving frequencies this activating process determines the evolution of the luminescence pattern while possible inhibiting processes are of minor importance.

In our experiments, the once generated high-luminescence regions spread out over the whole device area. However, Beletskii and Vlasenko (1993b) report on domain expansion in similar devices, but there the domains reach a critical size at which they

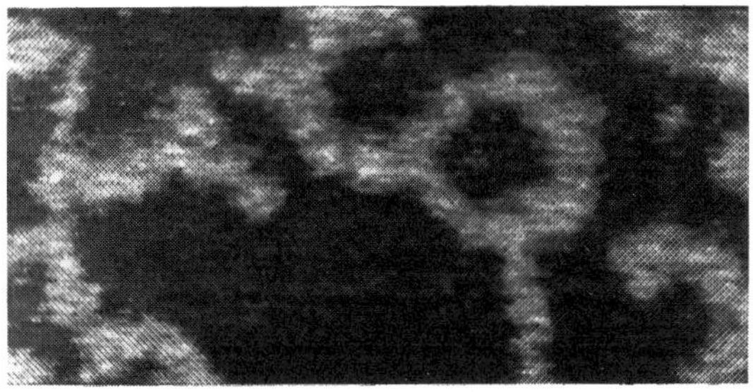

Figure 7: Luminescence radiation density of randomly moving autowaves in ZnS-films. Parameters: pulsed excitation, τ_p=13.5 μs, f=1 kHz, V_0=262 V, R_V=5 kΩ, imaged device area = 30 $\times$ 60 μm^2.

become static. They attribute this behaviour to the influence of the distributed resistance from the transparent electrode which prevents a free domain expansion because it limits the total current as has been demonstrated in the numerical simulations of Sec. 2, too (Fig. 1). Furthermore, Beletskki and Vlasenko have shown, that the velocity of the expanding circle increases with frequency. From this result we can conclude that the effective relaxation time of the activating process decreases with frequency.

Travelling pulses. For larger driving frequencies autowaves appear. Their width is of the order of several micrometers. When the amplitude of the driving voltage is increased the width, the velocity and the density of the autowaves increases. Fig. 7 shows a typical momentary luminescence pattern. For sufficiently large amplitudes the motion becomes irregular. Obviously, for higher frequencies the system behaves as an excitable medium; consequently, inhibiting processes counteracting the autocatalytic increase of the transferred charge are of importance in this regime. Essential inhibiting processes are connected with an increase of the temperature shifting the breakdown voltage to larger values and decreasing the width of the hysteresis loop of the brightness-voltage characteristic. Such a temperature increase is caused by increasing the frequency. On the other hand, the activating process becomes faster with increasing frequency favouring the appearance of autowaves in the simple activator-inhibitor frame. We suppose that the irregular dynamic behaviour of the luminescence distribution (Fig. 7) results from the interaction of many autowaves that move in various directions, collide and annihilate each other by their refractory zones similar to the behaviour shown in (Fig. 2).

We remark that target patterns (Fig. 5) have also been observed in thin film devices ([Rüfer, 1980]; [Beletskii and Vlasenko, 1993a]).

Filament cascades. The appearance of microfilaments in ZnS-films has been demonstrated for the first time by Marello et al (1977). Similar behaviour has been found in our samples. In Fig. 8 three luminescence distributions for different amplitudes of the

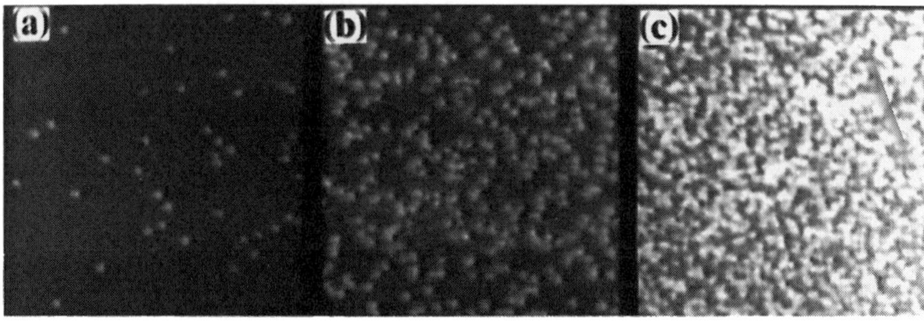

Figure 8: Luminescence radiation density distributions in ZnS-films for three different values of the driving voltage V_0=194 V (a), 195 V (b), 196 V (c). Other parameters: pulsed excitation, τ_p=13.5 μs, f=5.3 kHz, R_V=5 kΩ, imaged device area = 40 × 40 μm^2.

driving voltage are shown. These results have been obtained with a slightly modified device consisting of a single dielectric layer, the thickness of which is approximately twice that of the previously reported samples. Furthermore, the thickness of the active layer was reduced by approximately 90 nm to 310 nm. When the amplitude of the applied voltage exceeds the value at which impact-ionization in the active layer sets in, small filaments appear. The density of filaments is small near the threshold value (Fig. 8a) and increases with increasing voltage amplitude (Figs. 8b,c). For sufficiently large amplitudes the luminescence pattern becomes uniform again. When the voltage amplitude is decreased, the scenario is vice versa, i. e., filaments arise and the filament density decreases when decreasing the voltage. However, comparing the filament densities observed for the same value of the voltage amplitude in a certain voltage range, we find larger filament densities when starting from the uniform high-state and decreasing the voltage. This indicates multi-stable behaviour. In the global brightness-voltage characteristic this behaviour is reflected in a hysteresis curve (for further details, see [Goßen et al, 1995]). It is worth to note that the filaments reported here are static for low excitation frequencies. When the frequency is increased, the filaments become more mobile, but the described scenario remains qualitatively unchanged.

The appearance of filaments with less mobility is connected with a reduction of the autocatalytic process for the following reasons. The width of the active layer was reduced. By this, less charge can be accumulated near the semiconductor/insulator interface. In one half-cycle no charge accumulation is possible due to the missing second dielectric layer. Thus, in this case charge spreading and support of the activation in the subsequent half-cycle cannot occur.

4 Pattern Formation in a Gas Discharge System

Among pattern formation phenomena in the field of electric gas discharges one may distinguish between the longitudinal structurization along the main current and the lateral structurization perpendicular to the main current. We focus on lateral structures observed in a high frequency discharge system with planar electrode configuration. Boyers and Tiller (1973) were the first who investigated a similar system with respect to pattern formation. Their experiment was an arrangement of a planar discharge layer embedded in two dielectric layers being contacted on their external surfaces. Driving this system with a high frequency voltage they observed the appearance of different nonhomogeneous light density distributions as stripes, spots, and domains of intense light radiation. Further investigations revealed some more complex structures with respect to spatio-temporal behaviour ([Ammelt et al, 1993]).

Experimental Set-Up. The experimental set-up of the system we focus on is similar to the above discussed ZnS device. Again the system is driven by a real ac voltage source. The active medium consists of a planar discharge layer with a thickness a of about 0.5 mm being filled with helium at a gas pressure p in the range between 150 and 450 hPa. This gas layer is enclosed between two glass plates of a total thickness b of about 1 mm serving as dielectric layers. Each plate carries an electrical contact on the outward surface. One consists of aluminium with circular or rectangular shape and a diameter d varying from 12 to 50 mm. The other contact consists of a transparent alloy of indiumoxide and tinoxide covering the whole plate. Through this contact the light density distribution can be detected with a photosensitive equipment. It can be shown to a good approximation that the radiation density is proportional to the current density. Due to its smaller size it is the aluminium contact that defines the active region of the system and, therefore, the symmetry of the boundary is determined by the shape of this contact. The system is supplied with a high frequency voltage at a frequency f of a few hundred kHz and a voltage amplitude V_0 of about 400 to 500 V. This high voltage is provided by an electrical oscillator being fed by a variable signal generator.

Experimental Results. All the experiments shown in the following have been carried out at constant values of all parameters except for the driver amplitude which has been varied during each run of experiment. All the photographs shown below have been recorded using a standard b/w CCD-camera and also can be seen with the naked eye.

Domains. With the first experiment the bistable character of the underlying system is demonstrated. Under suitable conditions the light density distribution consists of homogeneous regions with high current densities and regions with low current densities. In a large range of parameters the high current domain is a circular area where the size of the domain can be controlled by the amplitude V_s of the feeding signal generator. Fig. 9. shows the light density distribution related to four different amplitudes V_s. Although V_s changes during this experiment the voltage amplitude V_0 across the discharge system was nearly constant at V_0=472 V.
Stripes of filaments. In a wide range of parameters the current density distribution consists of an arrangement of solitary like spots that we shall call filaments. In this

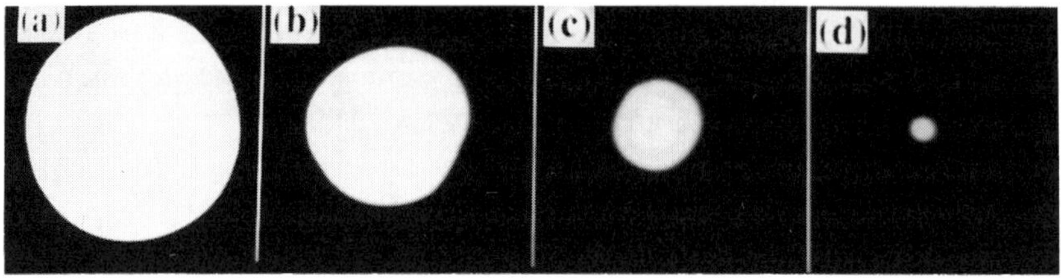

Figure 9: Photograph of a high current domain observed in HF gas discharge systems at different driver amplitudes V_s. The size of the domain decreases with the lowering of V_s=47.2 V (a), 42.4 V (b), 40.0 V (c), 38.5 V (d). Parameters: V_0=472 V, f=370 kHz, p=155 hPa, a=0.5 mm, b=0.85 mm, d=20 mm (circular).

case the spatio-temporal behaviour of the system can be described by the interaction between the filaments. For most sets of parameters their interaction is dominated by repulsion, therefore, filaments hold a large distance between each other. In the following some phenomena will be discussed where, apart from the repulsive interaction, also an attraction between filaments becomes evident. Usually the considered temporal dynamic is slow compared to the period of excitement but on a larger timescale filaments may move as the result of interfilamentary forces. Fig. 10 shows several filaments forming stripes with a characteristic distance between each of the filaments but also between the stripes. These stripes move slowly interacting with neighbouring stripes and separated filaments.

Clusters. Another phenomenon where the attraction between filaments becomes important is the creation of filament clusters. When increasing the driver amplitude V_0 gradually beyond a certain threshold value a large number of filaments appears abruptly. Decreasingthe amplitude filaments vanish one after another while the residual filaments form clusters of filaments being attached together at a characteristic distance. Fig. 11 shows typical examples for clusters at eight different driver amplitudes in the range between 438 and 495 V. Apart from the number of filaments also the symmetry formed by them depends on the driver amplitude. For example, the linear order of the three filaments can only exist at a larger value of V_s in comparison to V_s for the triangular arrangement. The clusters shown in Fig. 11 are rather stable against perturbations and are small compared to the diameter of the active region. Obviously, these clusters are very similar to those shown in Fig. 4. Spatial modulations in the tails of the single filaments are not resolvable; this might be a result of the weak charge carrier density outside the filaments leading to extremely weak or vanishing radiation densities. Usually, clusters drift along the active region and may even rotate. This

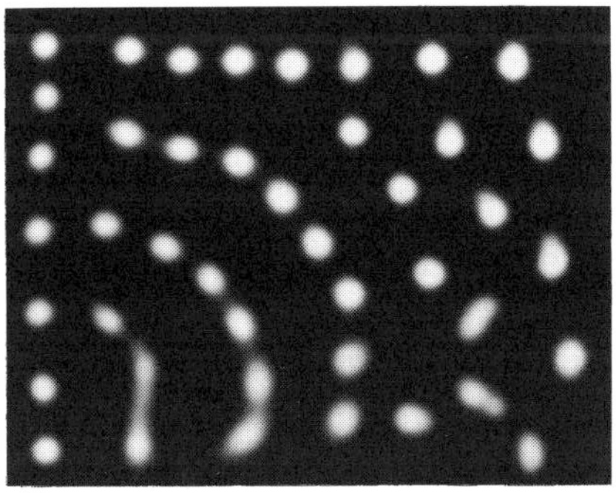

Figure 10: Photograph of stripes formed by solitary filaments in HF gas discharge systems. Parameters: V_0=415 V, f=200 kHz, p=313 hPa, a=0.45 mm, b=1.0 mm, device area = 12 × 14 mm^2 (rectangular).

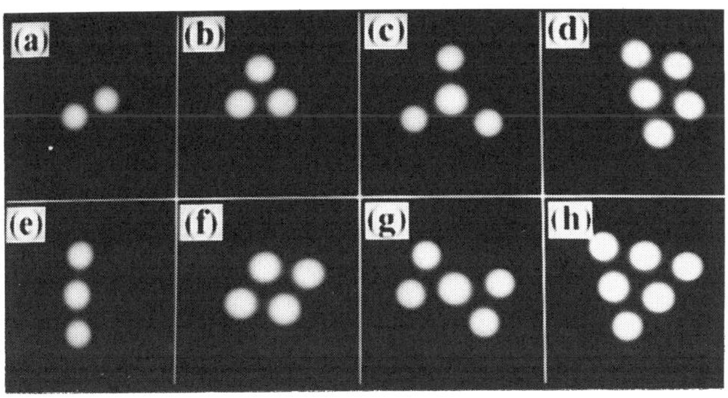

Figure 11: Series of stationary clusters consisting of radiation density filaments observed in the HF gas discharge system. These photographs have been taken at eight different values of V_0 in the range between 438 and 495 V. Further parameters: f=200 kHz, p=419 hPa, a=0.55 mm, b=1.0 mm, d=13 mm.

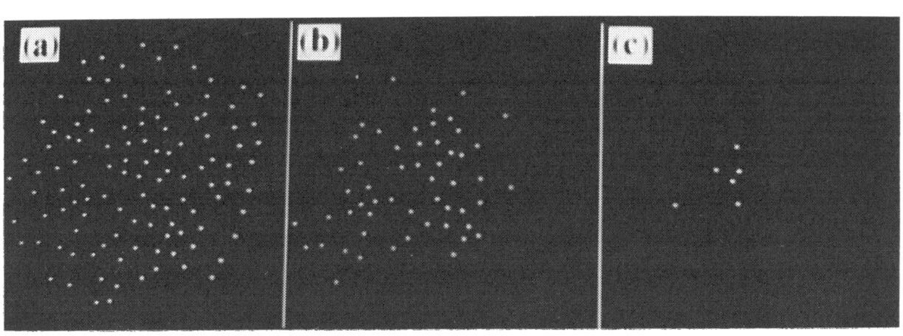

Figure 12: Cascade of vanishing filaments when decreasing the driver amplitude. The corresponding amplitudes are V_0=514.8 V, 506.1 V, and 505.6 V from left to right. Further parameters: f=370 kHz, p=250 hPa, a=0.5 mm, b=0.85 mm, d=50 mm.

movement is probably due to the existence of slight inhomogenities and proves that clusterization is an intrinsic effect of the filaments and their interaction rather than a consequence of inhomogenities.

Cascade of filaments. As a fourth example of pattern formation in the described high frequency gas discharge system the occurrence of a cascade of filaments is presented. Starting from a situation with a large driver amplitude V_0 and a large number of existing filaments the driver amplitude is decreased gradually. Fig. 12 (left) shows the initial state at a voltage of 514.8 V. Decreasing the supply voltage makes the filaments vanish successively. Further stages of this process are shown in Fig. 12 at amplitudes of 506.1 and 505.6 V.

5 Conclusions

We have studied a two-component set of reaction-diffusion equations that has been derived for two-layer systems containing an active medium and an inhibiting component. The numerical results presented in the first part of the article show that in such systems various spatio-temporal patterns can arise. Important parameters are in particular the ratio of the time constants δ and the diffusion lengths D_v/D_w of the two components and the values of the global excitation κ_1 and global inhibition κ_2. Although the specific physical processes in the electroluminescence and the gas discharge systems investigated experimentally are of different nature, various patterns observed experimentally can be described qualitatively in terms of the equations (1) when choosing appropriate parameters. Beyond that, the numerical simulations have demonstrated the importance of the consideration of the global inhibition, which may favour the appearance of single isolated filaments or target patterns in parameter ranges where vanishing global inhibition leads to hexagonal patterns or uniform oscillations, respectively.

Acknowledgements. We thank the *Zentralinstitut für Angewandte Mathematik des Forschungszentrums, Jülich* for the support and for providing CPU time on the Paragon, and the *Deutsche Forschungsgemeinschaft* for financial support.

References

[Ammelt et al, 1993] Ammelt, E, Schweng, D, and Purwins, H-G (1993) Phys. Lett. A **179**, 348–354

[Beletskii and Vlasenko, 1993a] Beletskii, A I and Vlasenko, N A (1993a) Tech. Phys. Lett. **19**, 13–14

[1993b] Beletskii, A I and Vlasenko, N A (1993b) Tech. Phys. Lett. **19**, 798–799

[Bode and Purwins, 1995] Bode, M and Purwins, H-G (1995) *Pattern Formation in Reaction-Diffusion Systems - Dissipative Solitons in Physical Systems*, Physica D, in print

[Boyers and Tiller, 1973] Boyers, D G and Tiller, W A (1973) J. Appl. Phys. **44**, 3102

[Dufiet and Boissonade, 1992] Dufiet, V and Boissonade, J (1992) J. Chem. Phys. **96** (1), 664–673

[Field and Burger, 1985] Field, R J and Burger, M (ed.) (1985) *Oscillations and Traveling Waves in Chemical Systems*, John Wiley & Sons, New York

[Goßen et al, 1995] Goßen, C, Niedernostheide, F-J, and Purwins, H-G (1995) *Pattern Formation of the Electroluminescence in AC ZnS:Mn Devices* in ([Niedernostheide, 1995]), 112–132

[Howard et al, 1982] Howard, W E, Sahni, O, and Alt, P M (1982) J. Appl. Phys. **53**, 639–647

[Kerner and Osipov, 1994] Kerner, B S and Osipov, V V (1994) *Autosolitons*, Fundamental Theories in Physics, Vol. 61, Kluwer, Dordrecht

[Kuramoto, 1984] Kuramoto, Y (1984) *Chemical Oscillations, Waves, and Turbulence*, Springer Series in Synergetics, Vol. 19, Springer, Berlin

[Malomed and Tribel'skii, 1987] Malomed, B A and Tribel'skii, M I (1987) JETP **65**, 305

[1977] Marello, V, Rühle, W, and Onton, A (1977) Appl. Phys. Lett. **31**, 452–454

[Meinhardt, 1982] Meinhardt, H (1982) *Models of Biological Pattern Formation*, Academic Press, London

[Meron, 1992] Meron, E (1992) Phys. Rep. **218**, 1–66

[Mikhailov, 1990] Mikhailov, A S (1990) *Foundations of Synergetics I*, Springer Series in Synergetics, Vol. 51, Springer, Berlin

[Mosekilde and Mouritsen, 1995] Mosekilde, E and Mouritsen, O G (eds.) (1995) *Modelling the Dynamics of Biological Systems*, Springer Series in Synergetics, Vol. 65, Springer, Berlin

[Murray, 1989] Murray, J D (1989) *Mathematical Biology*, Biomathematics, Vol. 19, Springer, Berlin

[Niedernostheide et al, 1994a] Niedernostheide, F-J, Ardes, M, Or-Guil, M, and Purwins, H-G (1994a) Phys. Rev. B **49**, 7370–7384

[Niedernostheide et al, 1994b] Niedernostheide, F-J, Goßen, C, and Purwins, H-G (1994b) *Dynamics of current filaments studied in two examplary systems: dc Si p-n-p-n devices and ac ZnS:Mn films*, Advances in Synergetics, Vol. 2, ed. by V Kuvshinov and C Krylov, Belarusian State University Press, Minsk

[Niedernostheide, 1995] Niedernostheide, F-J (ed.) (1995) *Nonlinear Dynamics and Pattern Formation in Semiconductors and Devices*, Springer Proceedings in Physics, Vol. 79, Springer, Berlin

[Purwins and Radehaus, 1988] Purwins, H-G and Radehaus, C (1988) *Pattern Formation on Analogue Parallel Networks*, Neural and Synergetic Computers, ed. by H Haken, Springer, Berlin

[Radehaus et al, 1987] Radehaus, C, Kardell, K, Baumann, H, Jäger, D, and Purwins, H-G (1987) Z. Phys. B **65**, 515–525

[Rensing, 1993] Rensing, L (ed.) (1993) *Oscillations and Morphogenesis*, Cellular Clocks, Vol. 5, Marcel Dekker, New York

[Rüfer, 1980] Rüfer, H, Marello, V, and Onton, A (1980) J. Appl. Phys. **51**, 1163–1169

[1979a] Rühle, W, Marello, V, and Onton, A (1979a) J. Electronic Materials **8**, 839–853

[1979b] Rühle, W, Marello, V, and Onton, A (1979b) J. Luminescence **18/19**, 729–738

[Winfree, 1987] Winfree, A T (1987) *When Time Breaks Down*, Princeton University Press, Princeton

Instabilities in Two-Layer Channel Flows

Y.Y. RENARDY and M. RENARDY
Department of Mathematics, Virginia Polytechnic Institute and State University, Blacksburg, VA 24061-0123, U.S.A.

Abstract

A Couette-Poiseuille flow of two immiscible liquids in layers between two horizontal parallel plates is considered. The flow is driven by a combination of the motion of the upper plate and a pressure gradient. The two liquids have different viscosities, densities, and surface tension. The relative depths of the liquids are prescribed. There is a steady solution consisting of a flat interface and a quadratic velocity profile in each fluid. The onset of instability is investigated where the steady solution loses stability to a traveling interfacial wave. In the case of a supercritical bifurcation, we proceed to examine the stability of the primary traveling wave solution to sideband perturbations. In this analysis, the evolution of the long-wave mode and its interaction with the traveling wave solution are described. Finally, the criteria for sideband stability of the primary solution are developed.

1 Introduction

Two-fluid flows exhibit phenomena which form a subject rich in interdisciplinary science. Industrial applications include the formation of bicomponent fibers such as nylons, the production of photographic films and the pipeline transport of crude oil. Theoretically, there are often many possibilities for the shape of the boundary separating the two fluids. On the other hand, experiments show that some shapes are preferred over others. The question thus arises as to which interface shapes can be observed, and what physical processes govern the selection of interface shapes. A first step towards the answer concerns the stability of the interface, and the appearance of new arrangements from unstable ones.

The equations governing flows composed of two immiscible fluids with different viscosities and densities are the Navier-Stokes equations and incompressibility; the volumes of the fluids are prescribed, and at the unknown interface position, we have the kinematic free surface condition, the continuity of velocity and shear stress, and the jump in the normal stress is balanced by surface tension. Appropriate boundary conditions are imposed to complete this formulation, e.g., the no-slip condition at solid walls, periodicity in the unbounded direction. There are at least five dimensionless parameters: a Reynolds number, a surface tension parameter, volume ratio, viscosity ratio and density ratio.

In this paper, we focus on the stability of two-layer plane channel flow, which has been instructive in understanding phenomena that occur in more complicated flows. Computational and asymptotic results on linear stability will be discussed. We proceed from the linear stability analysis to the study of nonlinear bifurcations. Bifurcation theory proceeds basically in two steps. First, one must show that in the neighborhood of a point where an instability arises, the problem of solving the differential equation can be reduced to a finite dimensional problem. The center manifold theorem is used in this respect. Secondly, the relevant coefficients in the finite dimensional equations must be calculated and the resulting equations must be solved. We have investigated the stability of the bifurcating travelling wave solutions for plane Couette-Poiseuille flow. Periodicity is prescribed in the horizontal direction. Numerical results on whether the bifurcation is supercritical or subcritical have been obtained for several flow profiles.

The stability of the primary traveling wave solution to sideband perturbations has been examined in various contexts, such as water waves. The usual idea is to derive an amplitude evolution equation known as the Ginzburg-Landau equation. This equation involves an amplitude factor for the critical mode which is allowed to vary slowly as a function of rescaled space and time variables. The situation with two fluids presents a complication to the usual framework of the Ginzburg-Landau equation. The reason is that in addition to the critical mode leading to traveling waves, there is another neutral mode at zero wavenumber, which corresponds to a shift of the interface. We incorporate the evolution of this long-wave (or mean-flow) mode, which has been neglected in past work on two-layer flows. If slow modulations are allowed, we obtain, instead of the Ginzburg-Landau equation, a coupled set of three partial differential equations for the amplitudes (of the traveling wave mode, the long-wave mode, and the pressure mode). Recent experiments on pressure-driven gas-liquid systems [15],[19] show the existence of the sideband instability of a traveling wave, with the possibility of subsequent feeding of energy into a mean flow mode. Past analyses do not account for the growth of the mean flow mode, and this motivated us to undertake this study. The sideband instability also provides a mechanism for three-dimensional disturbances to grow.

2 Governing Equations

Two fluids of densities ρ_i (i=1,2), and viscosities μ_i lie in layers between infinite parallel walls located at $z^* = 0, l^*$. Asterisks are used for dimensional variables. The location of the interface is denoted by $z^* = h^*(x^*, y^*, t^*)$. The velocity is denoted by $\mathbf{v}^* = (u^*, v^*, w^*)$, and the pressure by p^*.

The Navier-Stokes equations in fluid i are:

$$\rho_i(\dot{\mathbf{v}}^* + (\mathbf{v}^* \cdot \nabla)\mathbf{v}^*) = \mu_i \Delta \mathbf{v}^* - \nabla p^* - \rho_i g \mathbf{e}_z, \tag{2.1}$$

$$\text{div } \mathbf{v}^* = 0.$$

At the interface, we must have continuity of velocity and balance of tractions. To formulate these conditions, we let $\mathbf{t}_1$, $\mathbf{t}_2$ denote two unit vectors parallel to the interface, and let $\mathbf{n}$ be a unit normal to the interface:

$$\mathbf{t}_1 = (1, 0, \frac{\partial h^*}{\partial x^*})/\sqrt{1 + (\frac{\partial h^*}{\partial x^*})^2},$$

$$\mathbf{t}_2 = (0, 1, \frac{\partial h^*}{\partial y^*})/\sqrt{1 + (\frac{\partial h^*}{\partial y^*})^2},$$

$$\mathbf{n} = (-\frac{\partial h^*}{\partial x^*}, -\frac{\partial h^*}{\partial y^*}, 1)/\sqrt{1 + (\frac{\partial h^*}{\partial x^*})^2 + (\frac{\partial h^*}{\partial y^*})^2}. \tag{2.2}$$

By $[\![\cdot]\!]$ we denote the jump of a quantity across the interface, i.e. its value in fluid 1 minus its value in fluid 2. Finally, $\mathbf{T}^*$ denotes the stress tensor $\mu[\nabla\mathbf{v}^* + (\nabla\mathbf{v}^*)^T] - p^*\mathbf{1}$.

The conditions to be satisfied across the interface are the continuity of velocity

$$[\![\mathbf{v}^*]\!] = \mathbf{0},$$

continuity of shear stress

$$[\![\mathbf{t}_i \cdot \mathbf{T}^* \cdot \mathbf{n}]\!] = 0, \; i = 1, 2,$$

the jump in the normal stress is balanced by surface tension and curvature

$$[\![\mathbf{n} \cdot \mathbf{T}^* \cdot \mathbf{n}]\!] = \frac{S^*\left\{\frac{\partial^2 h^*}{\partial x^{*2}}(1 + (\frac{\partial h^*}{\partial y^*})^2) + \frac{\partial^2 h^*}{\partial y^{*2}}(1 + (\frac{\partial h^*}{\partial x^*})^2) - 2\frac{\partial^2 h^*}{\partial x^*\partial y^*}\frac{\partial h^*}{\partial x^*}\frac{\partial h^*}{\partial y^*}\right\}}{(1 + (\nabla h^*)^2)^{3/2}},$$

where S^* is the dimensional surface tension coefficient, and the kinematic free surface condition

$$\dot{h}^* + u^*\frac{\partial h^*}{\partial x^*} + v^*\frac{\partial h^*}{\partial y^*} = w^*. \tag{2.3}$$

At the walls, we have the no-slip boundary condition. The upper plate moves with velocity $(U_p^*, 0, 0)$ and the bottom plate is at rest. For the basic flow, we have a steady solution with a flat interface; fluid 1 occupies $0 \leq z^* \leq l_1^*$ and fluid 2 occupies $l_1^* \leq z^* \leq l^*$. The velocity of the interface in the basic flow is denoted $(U^*(l_1^*), 0, 0)$ and for brevity, we denote $U^*(l_1^*)$ by U_i. Dimensionless variables (without asterisks) are as follows:

$$(x, y, z) = (x^*, y^*, z^*)/l^*, \tag{2.4}$$

$$t = t^*U_i/l^*,$$

$$\mathbf{v} = \mathbf{v}^*/U_i,$$

$$p = p^*/(\rho_1 U_i^2).$$

In the basic flow, we allow for a pressure gradient $-G^*$ in the x-direction. Reynolds numbers in each fluid are denoted by

$$R_1 = U_i l^* \rho_1/\mu_1, \quad R_2 = U_i l^* \rho_2/\mu_2. \tag{2.5}$$

There are seven dimensionless parameters: a Reynolds number, say R_1, the undisturbed depth l_1 of fluid 1, a surface tension parameter

$$T = (\text{surface tension coefficient})/(\mu_2 U_i), \tag{2.6}$$

a Froude number F given by

$$F^2 = U_i^2/gl^*, \tag{2.7}$$

where g is the gravitational acceleration constant, a dimensionless pressure gradient

$$G = G^* l^*/(\rho_1 U_i^2), \tag{2.8}$$

the viscosity ratio

$$m = \mu_1/\mu_2, \tag{2.9}$$

and a density ratio

$$r = \rho_1/\rho_2. \tag{2.10}$$

We set

$$h(x,y,t) = h^*(x^*,y^*,t^*)/l^* - l_1. \tag{2.11}$$

The interface is then at $z = l_1 + h(x,y,t)$. Throughout, it is assumed that h is small, and the interface conditions are approximated by Taylor expansions about $z = l_1$. This allows us to rewrite these conditions, which are imposed at the unknown position $z = l_1 + h$, as a condition to be imposed at the known position $z = l_1$.

The velocity $(U(z), 0, 0)$ for the basic flow is given by

$$U(z) = \begin{cases} -GR_1 z^2/2 + c_1 z, & 0 \le z \le l_1, \\ -rGR_2(z-1)^2/2 + c_2(z-1) + U_p, & l_1 \le z \le 1, \end{cases} \tag{2.12}$$

where

$$c_1 = (1 + GR_1 l_1^2/2)/l_1, \quad l_2 = 1 - l_1,$$
$$c_2 = m(-GR_1 + c_1), \tag{2.13}$$

and the upper plate speed U_p is

$$U_p = 1 + \frac{ml_2}{l_1} - \frac{1}{2} m l_2 G R_1. \tag{2.14}$$

The basic pressure field P satisfies $dP/dx = -G$ and

$$\frac{dP}{dz} = \begin{cases} -1/F^2, & 0 \le z \le l_1, \\ -1/(rF^2), & l_1 \le z \le 1. \end{cases} \tag{2.15}$$

Solutions that are perturbations of the above basic flow are sought. The perturbations to the velocity, pressure and interface position are denoted by (u, v, w), p and h, respectively. We let $\mathbf{u}$ denote the set of unknowns (u, v, w, p, h). We can represent the governing equations in the schematic form

$$M(R)\mathbf{u}_t = L(R, D_x, D_y)\mathbf{u} + N_2(R, D_x, D_y; \mathbf{u}, \mathbf{u}) + N_3(R, D_x, D_y; \mathbf{u}, \mathbf{u}, \mathbf{u}). \tag{2.16}$$

Here M and L are linear operators; N_2 and N_3 are symmetric bilinear and cubic terms (we need not consider nonlinearities of higher than third order). The listing of D_x

and D_y indicates the dependence on derivatives of $\mathbf{u}$ as well as $\mathbf{u}$ itself. The x and y variables are carried explicitly, since we want to allow arbitrary modulations in these directions; on the other hand, the z-dependence of $\mathbf{u}$ is built into the function space. The number R will serve as a bifurcation parameter; R may be any of a number of things; it could be the Reynolds number, but also, for instance, the viscosity ratio. All parameters other than R are considered fixed and given, and we shall not carry them explicitly in the equations.

Specifically,

$$M(R)\mathbf{u} = (u_{(1)}, v_{(1)}, w_{(1)}, u_{(2)}, v_{(2)}, w_{(2)}, 0, 0, 0, 0, 0, 0, 0, h), \tag{2.17}$$

where the subscripts refer to fluids 1 and 2. There are 14 components in this expression. The first 7 components come from the 3 components $\partial \mathbf{v}/\partial t$ of the Navier-Stokes equations in fluids 1 and 2, respectively, and incompressibility. The remaining 7 components come from the interface conditions; the last entry arises from $\partial h/\partial t$ in the kinematic condition. Similarly, we write $L(R, D_x, D_y)\mathbf{u}$ in components which we label f_1 through f_{14}. We have

$$\begin{aligned}
f_1 &= -U\frac{\partial u}{\partial x} - wU' + \frac{1}{R_1}\Delta u - \frac{\partial p}{\partial x},\\
f_2 &= -U\frac{\partial v}{\partial x} + \frac{1}{R_1}\Delta v - \frac{\partial p}{\partial y},\\
f_3 &= -U\frac{\partial w}{\partial x} + \frac{1}{R_1}\Delta w - \frac{\partial p}{\partial z},\\
f_4 &= -U\frac{\partial u}{\partial x} - wU' + \frac{1}{R_2}\Delta u - r\frac{\partial p}{\partial x}\\
f_5 &= -U\frac{\partial v}{\partial x} + \frac{1}{R_2}\Delta v - r\frac{\partial p}{\partial y},\\
f_6 &= -U\frac{\partial w}{\partial x} + \frac{1}{R_2}\Delta w - r\frac{\partial p}{\partial z},\\
f_7 &= \frac{\partial u}{\partial x} + \frac{\partial v}{\partial y} + \frac{\partial w}{\partial z} - [\![w]\!],\\
f_8 &= h[\![U']\!] + [\![u]\!],\\
f_9 &= [\![v]\!],\\
f_{10} &= [\![w]\!],\\
f_{11} &= [\![\frac{\mu}{\mu_1}(\frac{\partial u}{\partial z} + \frac{\partial w}{\partial x})]\!],\\
f_{12} &= [\![\frac{\mu}{\mu_1}(\frac{\partial v}{\partial z} + \frac{\partial w}{\partial y})]\!],\\
f_{13} &= [\![\frac{2\mu}{\mu_1 R_1}\frac{\partial w}{\partial z} - p]\!] - \frac{T}{mR_1}\Delta h - h[\![\frac{\partial P}{\partial z}]\!],\\
f_{14} &= -\frac{\partial}{\partial x}\Big[U(l_1)h + \int_0^{l_1} u\,dz\Big] - \frac{\partial}{\partial y}\int_0^{l_1} v\,dz.
\end{aligned} \tag{2.18}$$

The quadratic term $N_2(R, D_x, D_y; \mathbf{u}, \mathbf{u})$, written in components f_1 through f_{14}, is as follows:

$$
\begin{aligned}
f_1 &= -u\frac{\partial u}{\partial x} - v\frac{\partial u}{\partial y} - w\frac{\partial u}{\partial z}, \\
f_2 &= -u\frac{\partial v}{\partial x} - v\frac{\partial v}{\partial y} - w\frac{\partial v}{\partial z}, \\
f_3 &= -u\frac{\partial w}{\partial x} - v\frac{\partial w}{\partial y} - w\frac{\partial w}{\partial z}, \\
f_4 &= -u\frac{\partial u}{\partial x} - v\frac{\partial u}{\partial y} - w\frac{\partial u}{\partial z}, \\
f_5 &= -u\frac{\partial v}{\partial x} - v\frac{\partial v}{\partial y} - w\frac{\partial v}{\partial z}, \\
f_6 &= -u\frac{\partial w}{\partial x} - v\frac{\partial w}{\partial y} - w\frac{\partial w}{\partial z}, \\
f_7 &= \frac{\partial}{\partial x}\Big(h[\![u]\!] + \frac{h^2}{2}[\![U']\!]\Big) + \frac{\partial}{\partial y}(h[\![v]\!]), \\
f_8 &= h[\![\frac{\partial u}{\partial z}]\!] + \frac{h^2}{2}[\![U'']\!], \\
f_9 &= h[\![\frac{\partial v}{\partial z}]\!], \\
f_{10} &= h[\![\frac{\partial w}{\partial z}]\!], \\
f_{11} &= h[\![\frac{\mu}{\mu_1}(\frac{\partial^2 u}{\partial z^2} + \frac{\partial^2 w}{\partial z \partial x})]\!] - 2\frac{\partial h}{\partial x}[\![\frac{\mu}{\mu_1}(\frac{\partial u}{\partial x} - \frac{\partial w}{\partial z})]\!] \\
&\quad -\frac{\partial h}{\partial y}[\![\frac{\mu}{\mu_1}(\frac{\partial u}{\partial y} + \frac{\partial v}{\partial x})]\!], \\
f_{12} &= h[\![\frac{\mu}{\mu_1}(\frac{\partial^2 v}{\partial z^2} + \frac{\partial^2 w}{\partial z \partial y})]\!] - 2\frac{\partial h}{\partial y}[\![\frac{\mu}{\mu_1}(\frac{\partial v}{\partial y} - \frac{\partial w}{\partial z})]\!] \\
&\quad -\frac{\partial h}{\partial x}[\![\frac{\mu}{\mu_1}(\frac{\partial u}{\partial y} + \frac{\partial v}{\partial x})]\!], \\
f_{13} &= h[\![\frac{2\mu}{\mu_1 R_1}\frac{\partial^2 w}{\partial z^2} - \frac{\partial p}{\partial z}]\!] - \frac{\partial h}{\partial x}\frac{2}{R_1}[\![\frac{\mu}{\mu_1}(\frac{\partial u}{\partial z} + \frac{\partial w}{\partial x})]\!] \\
&\quad -\frac{\partial h}{\partial y}\frac{2}{R_1}[\![\frac{\mu}{\mu_1}(\frac{\partial v}{\partial z} + \frac{\partial w}{\partial y})]\!], \\
f_{14} &= -\frac{\partial}{\partial x}(hu_{(1)} + \frac{h^2}{2}U'_{(1)}) - \frac{\partial}{\partial y}(hv_{(1)}).
\end{aligned} \tag{2.19}
$$

The cubic term $N_3(R, D_x, D_y; \mathbf{u}, \mathbf{u}, \mathbf{u})$ consists of:

$$
\begin{aligned}
f_1 &= f_2 = f_3 = f_4 = f_5 = f_6 = 0, \\
f_7 &= \frac{\partial}{\partial x}\Big(\frac{h^2}{2}[\![\frac{\partial u}{\partial z}]\!] + \frac{h^3}{6}[\![U'']\!]\Big) + \frac{\partial}{\partial y}(\frac{h^2}{2}[\![\frac{\partial v}{\partial z}]\!]), \\
f_8 &= \frac{h^2}{2}[\![\frac{\partial^2 u}{\partial z^2}]\!], \\
f_9 &= \frac{h^2}{2}[\![\frac{\partial^2 v}{\partial z^2}]\!],
\end{aligned}
$$

$$
\begin{aligned}
f_{10} &= \frac{h^2}{2}[\![\frac{\partial^2 w}{\partial z^2}]\!],\\
f_{11} &= \frac{h^2}{2}[\![\frac{\mu}{\mu_1}(\frac{\partial^3 u}{\partial z^3}+\frac{\partial^3 w}{\partial z^2\partial x})]\!]-2\frac{\partial h}{\partial x}h[\![\frac{\mu}{\mu_1}(\frac{\partial^2 u}{\partial z\partial x}-\frac{\partial^2 w}{\partial z^2})]\!]\\
&\quad -h\frac{\partial h}{\partial y}[\![\frac{\mu}{\mu_1}(\frac{\partial^2 u}{\partial y\partial z}+\frac{\partial^2 v}{\partial x\partial z})]\!]-(\frac{\partial h}{\partial x})^2[\![\frac{\mu}{\mu_1}(\frac{\partial u}{\partial z}+\frac{\partial w}{\partial x})]\!]\\
&\quad -\frac{\partial h}{\partial x}\frac{\partial h}{\partial y}[\![\frac{\mu}{\mu_1}(\frac{\partial v}{\partial z}+\frac{\partial w}{\partial y})]\!],\\
f_{12} &= \frac{h^2}{2}[\![\frac{\mu}{\mu_1}(\frac{\partial^3 v}{\partial z^3}+\frac{\partial^3 w}{\partial z^2\partial y})]\!]-2\frac{\partial h}{\partial y}h[\![\frac{\mu}{\mu_1}(\frac{\partial^2 v}{\partial z\partial y}-\frac{\partial^2 w}{\partial z^2})]\!]\\
&\quad -h\frac{\partial h}{\partial x}[\![\frac{\mu}{\mu_1}(\frac{\partial^2 u}{\partial y\partial z}+\frac{\partial^2 v}{\partial x\partial z})]\!]-(\frac{\partial h}{\partial y})^2[\![\frac{\mu}{\mu_1}(\frac{\partial v}{\partial z}+\frac{\partial w}{\partial y})]\!]\\
&\quad -\frac{\partial h}{\partial x}\frac{\partial h}{\partial y}[\![\frac{\mu}{\mu_1}(\frac{\partial u}{\partial z}+\frac{\partial w}{\partial x})]\!],\\
f_{13} &= \frac{h^2}{2}[\![\frac{2\mu}{R_1\mu_1}\frac{\partial^3 w}{\partial z^3}-\frac{\partial^2 p}{\partial z^2}]\!]-(\frac{\partial h}{\partial x})h\frac{2}{R_1}[\![\frac{\mu}{\mu_1}(\frac{\partial^2 u}{\partial z^2}+\frac{\partial^2 w}{\partial z\partial x})]\!]\\
&\quad -(\frac{\partial h}{\partial y})h\frac{2}{R_1}[\![\frac{\mu}{\mu_1}(\frac{\partial^2 v}{\partial z^2}+\frac{\partial^2 w}{\partial z\partial y})]\!]+(\frac{\partial h}{\partial x})^2[\![\frac{2}{R_1}\frac{\mu}{\mu_1}\frac{\partial u}{\partial x}-p]\!]\\
&\quad +(\frac{\partial h}{\partial y})^2[\![\frac{2}{R_1}\frac{\mu}{\mu_1}\frac{\partial v}{\partial y}-p]\!]+\frac{\partial h}{\partial x}\frac{\partial h}{\partial y}\frac{2}{R_1}[\![\frac{\mu}{\mu_1}(\frac{\partial u}{\partial y}+\frac{\partial v}{\partial x})]\!]\\
&\quad -h(\nabla h)^2[\![\frac{\partial P}{\partial z}]\!]-\frac{T}{2mR_1}\frac{\partial^2 h}{\partial x^2}\Big((\frac{\partial h}{\partial y})^2-(\frac{\partial h}{\partial x})^2\Big)\\
&\quad -\frac{T}{2mR_1}\frac{\partial^2 h}{\partial y^2}\Big((\frac{\partial h}{\partial x})^2-(\frac{\partial h}{\partial y})^2\Big)+\frac{2T}{mR_1}\frac{\partial^2 h}{\partial x\partial y}\frac{\partial h}{\partial x}\frac{\partial h}{\partial y},\\
f_{14} &= -\frac{\partial}{\partial x}(\frac{h^2}{2}\frac{\partial u_{(1)}}{\partial z}+\frac{h^3}{6}U''_{(1)})-\frac{\partial}{\partial y}(\frac{h^2}{2}\frac{\partial v_{(1)}}{\partial z}).
\end{aligned}
\tag{2.20}
$$

3 Linear Stability Analysis

For the linear stability analysis, we use Squire's transformation and seek normal mode solutions proportional to $\exp(ikx+\sigma t)$, where k is the wavenumber of the disturbance and σ is the complex eigenvalue, to be computed numerically. The Navier-Stokes equations and incompressibility reduce to Orr-Sommerfeld equations in each fluid. Our results are obtained by discretizing the equations in the z variable by the Chebyshev-tau method, and computing the eigenvalues in quadruple precision. This method yields infinite order accuracy since the eigenfunctions are C^∞. The eigenvalues are discrete.

If the two fluids are identical, then the set of eigenvalues consists of the infinitely many "one-fluid" (bulk) modes and a neutral "interfacial" mode. The interfacial mode represents a perturbation which does nothing to the velocity, but makes the interface wavy:

$$\mathbf{v}=\mathbf{0},\quad h=\exp(ikx+\sigma t),\quad \sigma=-ikU(l_1). \tag{3.1}$$

When the two fluids have different properties, the interfacial mode may cause an instability even at low Reynolds numbers, with the bulk modes stable. Thus, it is of

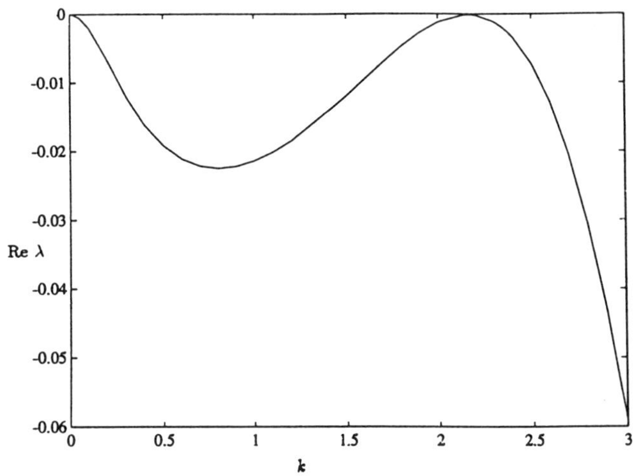

Figure 1: Growth rate $\mathrm{Re}\,\lambda$ (σ in text) versus wavenumber k for the interfacial mode. $R_1 = 10.030$, $G = 0.0043015$, $l_1 = 0.99$, $m = 1081.1$, $T = 87183$, $r = 1026.1$, $1/F^2 = 21.379$, $k_c = 2.16$.

interest to examine the interfacial mode on its own in asymptotic limits. For long-wave disturbances, k and kR_1 are assumed small and the interfacial eigenfunction is expanded in a regular perturbation in k [23]. The results show that an arrangement with the less viscous liquid in a sufficiently thin layer tends to be stable to long waves. For large k or short waves, surface tension is the dominant term and is stabilizing, while viscosity and density difference terms influence stability at higher order [11]. There are also studies of some other asymptotic limits as well as numerical investigations for general wavenumbers. We refer to Chapter IV in Volume 1 of [14] for a review. Here, we shall pursue the neutrally stable situation where the bulk modes are stable and the least stable mode is the interfacial one with a critical wavenumber k_c. The graph of the growth rate against wavenumber of the disturbance for the least stable mode in a Couette-Poiseuille flow of gas over liquid [15] is depicted in Figure 1.

A contour plot of the vertical component of the velocity perturbation is shown in Figure 2. The interface is at $z = 0.99$ so that the contours we see are those in fluid 1. The wavenumber is $k = 2.16$. For shorter waves, the contours cluster close to the interface, where the eigenfunction displays fast variation, and for longer waves, the contours are spread out more evenly.

When the Reynolds number is raised for the situation in Figure 1, the hump in the growth rate curve at k_c rises up and causes an instability, with the possibility of generation of traveling waves of that wavenumber. To see if the traveling wave solution is linearly stable with respect to disturbances of the same wavenumber as itself (supercritical bifurcation) the Stuart-Landau equation is derived and the Stuart-Landau coefficient is calculated [18]. It is known that there are many situations,

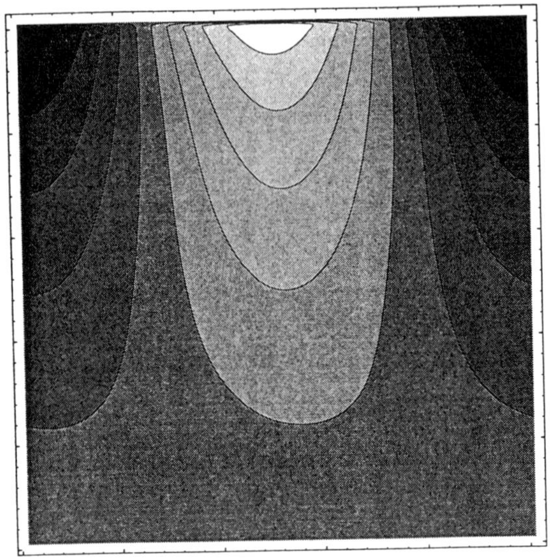

Figure 2: Contour plots for the vertical component of velocity for the critical situation in Figure 1. The horizontal size is one wavelength $2\pi/k_c$ and the vertical size of this plot is 1.

especially at moderate flow parameters, where the bifurcation is supercritical. We take a supercritical situation and then ask the question: is the traveling wave solution linearly stable with respect to perturbations with a band of wavenumbers around α_c (sideband perturbations)? It turns out that the weakly nonlinear interaction of the traveling wave mode with the sideband modes can excite the evolution of the long-wave (or mean-flow) mode. To describe this, we need to derive the appropriate amplitude equations, including the evolution of the long-wave mode, and elucidate the criteria for sideband stability.

Instability to sideband perturbations provides a mechanism for three-dimensional disturbances to grow. It is also a mechanism for the mean-flow mode to grow as a result of an instability of a traveling wave mode. This type of phenomenon has been reported by Jurman, Deutsch and McCready [15] for a pressure-driven flow of a gas over a liquid in a channel.

4 Amplitude Equations

We consider the problem in three dimensions. In the linearized problem, where the perturbation solutions are proportional to $\exp(ikx + ily + \sigma t)$, there are four eigenspaces. At $k = l = 0$, $\sigma = 0$, there is an eigenvector $\mathbf{a_0}(R_c)$ which shifts the interface up or down and an eigenvector $\tilde{\mathbf{a_0}}$ which adds a constant to the pressure field. The former is the long-wave mode, and the latter is a mode that is already present in the one-fluid problem considered by Davey, Hocking and Stewartson [8] for plane Poiseuille flow. At $k = k_0, l = 0, \sigma = i\omega_0, R = R_c$, there is an eigenvector $\mathbf{a}_1$ representing the traveling wave. Its complex conjugate is at $k = -k_0, l = 0$. The derivation of the amplitude equations is a formal calculation, based on the center manifold reduction method, although there is no center manifold theorem for the sideband problem. We assume that the dynamics close to criticality is dominated by the critical modes and their interactions.

We shall not give the fairly involved calculations leading to the amplitude equations, but instead refer to [17]. The following outlines the results. We introduce a small parameter ϵ where the bifurcation parameter R is

$$R = R_c + \epsilon^2, \quad \epsilon << 1, \tag{4.1}$$

and R_c is the critical value. We seek small amplitude traveling wave solutions with a slow modulation, where ϵ measures the smallness of the solution, and $\epsilon x, \epsilon y, \epsilon t$ describe the slow modulation. We let the solution vector be

$$\mathbf{u} = \epsilon \mathbf{w}(\zeta, \xi, \eta, \tau), \quad \zeta = k_0 x + \omega_0 t, \quad \xi = \epsilon x, \quad \eta = \epsilon y, \quad \tau = \epsilon t. \tag{4.2}$$

We expand $\mathbf{w}$ in Fourier series

$$\mathbf{w}(\zeta, \xi, \eta, \tau) = \sum_{n=-\infty}^{n=\infty} \mathbf{w}^n(\xi, \eta, \tau) \exp(in\zeta). \tag{4.3}$$

Assuming that the critical modes and their interactions dominate the dynamics, we have the following expressions for the zeroth, first and second components:

$$\mathbf{w}^0 = B\mathbf{a}_0(R_c) + P\tilde{\mathbf{a}}_0 + \epsilon P_\xi \mathbf{v}_1(R_c) + \epsilon P_\eta \mathbf{v}_2(R_c) + \epsilon \mathbf{g},$$

$$\mathbf{w}^1 = A\mathbf{a}_1 + \epsilon \mathbf{f},$$

$$\mathbf{w}^2 = \epsilon \mathbf{h}, \tag{4.4}$$

and all other Fourier components are $O(\epsilon^2)$. Here $\mathbf{v}_1(R_c)$ and $\mathbf{v}_2(R_c)$ are the modifications to the flow introduced by an additional small pressure gradient, and the interactions terms $\mathbf{f}$, $\mathbf{g}$ and $\mathbf{h}$ are given by expressions of the form

$$\begin{aligned} \mathbf{h} &= A^2\phi_1, \\ \mathbf{f} &= 2AB\phi_2 + A_\xi\psi_1 + A_\eta\psi_2, \\ \mathbf{g} &= B^2\phi_3 + 2|A|^2\phi_4 + B_\xi\psi_3 + B_\eta\psi_4 + B_\tau\psi_5 \\ &\quad +\epsilon(\chi_1 P_{\xi\tau} + \chi_2 P_{\eta\tau} + \chi_3 P_{\xi\xi} + (\chi_4 + \chi_5)P_{\xi\eta} + \chi_6 P_{\eta\eta} \\ &\quad +2\chi_7 BP_\xi + 2\chi_8 BP_\eta) + \epsilon A\overline{A}_\eta\phi_5 + +\epsilon\overline{A}A_\eta\overline{\phi}_5. \end{aligned} \tag{4.5}$$

(The coefficients ϕ_i, ψ_i and χ_i are written out in [17], and we shall not repeat them here.) We substitute these expressions substituted back into the system of equations governing each Fourier component, perform certain projections, and finally transform to a moving coordinate system. This yields the following amplitude evolution equations:

$$\begin{aligned}
A_\tau &= \epsilon\gamma A_{\xi\xi} + \epsilon\kappa A_{\eta\eta} + \epsilon\sigma A + \epsilon\beta|A|^2A + \delta AB + \epsilon\rho P_\xi A, \\
B_\tau &= (c_g - c_0)B_\xi + \epsilon\tilde{\gamma}B_{\xi\xi} + \epsilon\tilde{\kappa}B_{\eta\eta} + \epsilon^2\tilde{\rho}(BP_\xi)_\xi + \epsilon^2\tilde{\chi}(BP_\eta)_\eta \\
&\quad +\epsilon\tilde{\beta}(B^2)_\xi + \epsilon\tilde{\delta}(|A|^2)_\xi + \epsilon^2(\delta^*(A\overline{A}_\eta)_\eta + \overline{\delta^*}(\overline{A}A_\eta)_\eta), \\
\epsilon(P_{\xi\xi} &+ P_{\eta\eta}) = \epsilon r_0(|A|^2)_\xi + s_0 B_\xi.
\end{aligned} \tag{4.6}$$

Here γ, κ, σ, β, δ, δ^* and ρ are complex and the other coefficients are real, c_0 is the propagation speed for long waves, c_g is the group speed of traveling waves near the critical wave number. It is important for our analysis that these two speeds are different.

In the two-dimensional case, there is no dependence on η and we can integrate the third equation in (4.6) to obtain $\epsilon P_\xi = \epsilon r_0|A|^2 + s_0 B + \epsilon K$, where K is a constant of integration. This is inserted into the remaining equations and a coupled system of two equations is obtained:

$$\begin{aligned}
A_\tau &= \epsilon\gamma A_{\xi\xi} + \epsilon(\sigma + \rho K)A + \epsilon(\beta + \rho r_0)|A|^2A + (\delta + \rho s_0)AB, \\
B_\tau &= (c_g - c_0)B_\xi + \epsilon\tilde{\gamma}B_{\xi\xi} + \epsilon(\tilde{\beta} + \tilde{\rho}s_0)(B^2)_\xi + \epsilon\tilde{\delta}(|A|^2)_\xi.
\end{aligned} \tag{4.7}$$

We have neglected terms of order ϵ^2 in the second equation of (4.7).

If $A = 0$, then the second equation becomes Burgers' equation, which was derived by Hooper and Grimshaw [12] as an amplitude equation for long waves. They treat the situation where the basic flow is unstable to long waves, which is different from our situation where long waves are stable in the basic flow. As a consequence, the second order term is destabilizing and they include the fourth order term to derive the Kuramoto-Sivashinsky equation for weakly nonlinear interactions of long waves. Recent works concerning the K-S equation for interfacial waves are those of Tilley, Davis and Bankoff [22] for flow down an incline, and Coward, Papageorgiou and Smyrlis [7] for core-annular flow.

If we set $B = 0$ and assume there is no dependence on ξ, we have the Stuart-Landau equation which has been investigated for two-fluid flows [4],[18]. We note that if we want to allow dependence on ξ, we cannot consistently set $B = 0$, because of the presence of the term $(|A|^2)_\xi$ in the second equation of (4.6).

If we ignore the term AB in the first equation of (4.7), and scale time with an additional factor ϵ, then ϵ scales out of the equation and we arrive at the Ginzburg-Landau equation

$$A_\tau = \gamma A_{\xi\xi} + \hat{\sigma}A + \hat{\beta}|A|^2A. \tag{4.8}$$

If $\mathrm{Re}\,\gamma > 0$, $\mathrm{Re}\,\hat{\sigma} > 0$ and $\mathrm{Re}\,\hat{\beta} < 0$, this equation has traveling wave solutions of the form $A = A_0\exp(i\mu\xi + i\omega\tau)$ as long as $\mu^2 < \mathrm{Re}\,\hat{\sigma}/\mathrm{Re}\,\gamma$ (in terms of the original problem, a nonzero μ means that the wavelength of the traveling wave is not exactly the critical wavelength, but differs from it by order ϵ). Eckhaus [9] assumes that γ, $\hat{\beta}$ and $\hat{\sigma}$ are real and he shows that the traveling wave solution is unstable if $\mu^2 > \hat{\sigma}/(3\gamma)$. If the coefficients are not real, then sideband instability is possible even if $\mu = 0$. This was first observed by Benjamin and Feir [2] for an equation modeling water waves. This system is conservative, and the coefficients in the Ginzburg-Landau equation turn out purely imaginary. The analogue of the Benjamin-Feir criterion for dissipative systems was derived by Lange and Newell [16]. Stuart and DiPrima [21] study the problem for general coefficients and general μ, thus achieving a unified treatment of the Eckhaus and Benjamin-Feir instabilities. Davey, Hocking and Stewartson [8] take account of the pressure mode in three-dimensional shearing flows; their set of amplitude equations is equivalent to the first and third equation in (4.6) (without the terms involving B). The recent work of Cheng and Chang [5] stresses the analogy in the derivation of Ginzburg-Landau amplitude equations with center manifold reduction. Unfortunately, there is no version of the center manifold theorem which covers situations with a continuum of modes. The rigorous study of the validity of amplitude equations such as Ginzburg-Landau is currently an area of active research, see e.g. [6],[10],[13],[20].

Blennerhassett [3] derives a single amplitude equation for traveling interfacial waves which allows for slow modulation. His equation is a special case of (4.7) that does not allow the long waves to evolve independently. To see this, we scale B and $\partial/\partial\tau$ with an additional factor ϵ. After doing so, ϵ factors out of the first equation of (4.7) and the second equation reduces to $(c_g - c_0)B_\xi + \tilde{\delta}(|A|^2)_\xi = 0$. After integrating with respect to the spatial variable ξ, this equation is satisfied by setting $(c_g - c_0)B + \tilde{\delta}|A|^2 = 0$. Thus, the long wave mode is completely determined as a multiple of the amplitude of the traveling wave mode. Inserting this relationship between B and A into the first equation of (4.7) yields an equation for A alone:

$$A_\tau = \gamma A_{\xi\xi} + \sigma A + \left(\beta + \rho r_0 - \frac{\tilde{\delta}}{c_g - c_0}(\delta + \rho s_0)\right)|A|^2 A. \tag{4.9}$$

The approach of our paper takes into account all modes which are close to neutral stability and might therefore lead to sideband instabilities.

5 Criteria for Sideband Stability

The system (4.6) has solutions of the following form, which we refer to as traveling wave solutions:

$$A = A_0 \exp(i(\mu\xi + \tilde{\mu}\eta + \epsilon\omega\tau)),\ B = 0,\ P = P_0\xi + P_1\eta. \tag{5.1}$$

The analysis of sideband stability for the general case is extremely complicated, and we have only investigated the case $\mu = \tilde{\mu} = 0$. This corresponds to waves with exactly the critical wave number, i.e. the analogue of the situation studied by Benjamin and Feir. The lateral pressure gradient P_1 is assumed zero, and for P_0 we are interested in the cases $P_0 = 0$ or $P_0 = r_0|A_0|^2$. The first case corresponds to a situation of fixed pressure gradient (FPG), i.e. the traveling wave solution has the same average pressure drop as the unperturbed parallel flow, while the second case corresponds to fixed volume flux (FVF). By inserting (5.1) (with $\mu = \tilde{\mu} = 0$) into (4.6), we find

$$\sigma_r + \rho_r P_0 + \beta_r|A_0|^2 = 0,\ \omega = \sigma_i + \rho_i + \beta_i|A_0|^2 = 0. \tag{5.2}$$

Here and in the following the subscripts r and i denote real and imaginary parts. We can solve (5.2) uniquely for $|A_0|$ and ω as long as σ_r and β_r (for FPG) or $\beta_r + \rho_r r_0$ (for FVF) have opposite signs. If $\beta_r < 0$ ($\beta_r + \rho_r r_0 < 0$), then the bifurcation is supercritical and the bifurcated traveling wave is stable with respect to perturbations of the same wavelength. We shall now discuss stability with respect to sideband modulations; the criteria here turn out to be the same for the FPG and FVF cases.

In (4.6), we set $A = (A_0 + C)\exp(i\epsilon\omega\tau)$, $P = P_0\xi + Q$, and linearize for small C, B and Q. In the linearized equations, we separate variables in the usual way, and we find that stability is determined by the eigenvalues of the following matrix [17]:

$$\begin{pmatrix} \epsilon[-\gamma\nu^2 - \kappa\theta^2 + \hat{\beta}A_0^2] & \epsilon\hat{\beta}A_0^2 & \hat{\delta}A_0 \\ \epsilon\overline{\hat{\beta}}A_0^2 & \epsilon[-\overline{\gamma}\nu^2 - \overline{\kappa}\theta^2 + \overline{\hat{\beta}}A_0^2] & \overline{\hat{\delta}}A_0 \\ \epsilon\tilde{\delta}i\nu A_0 - \epsilon^2\overline{\delta^*}\theta^2 A_0 & \epsilon\tilde{\delta}i\nu A_0 - \epsilon^2\delta^*\theta^2 A_0 & i\nu(c_g - c_0) - \epsilon\tilde{\gamma}\nu^2 - \epsilon\tilde{\kappa}\theta^2 \end{pmatrix}. \tag{5.3}$$

In this matrix, ν and θ are the wave numbers of the perturbation in the ξ- and η-directions, and we have introduced the new notations

$$\hat{\beta} = \beta + \frac{\rho r_0 \nu^2}{\nu^2 + \theta^2}, \quad \hat{\delta} = \delta + \frac{\rho s_0 \nu^2}{\nu^2 + \theta^2}. \tag{5.4}$$

We have stability to sideband perturbations if the eigenvalues of (5.3) have negative real parts for all values of ν and θ. In [17], we have derived a set of criteria which are necessary and sufficient to assure this for small ϵ.

There are several cases to be distinguished. The simplest case is if $\epsilon\nu^2$ and/or $\epsilon\theta^2$ are large or at least of order one. In this case, (5.3) is dominated by the diagonal entries, and the resulting criteria for stability are

$$\gamma_r > 0, \ \kappa_r > 0, \ \tilde{\gamma} > 0, \ \tilde{\kappa} > 0. \tag{5.5}$$

Indeed, these are simply the conditions which are needed to insure that the unperturbed parallel shear flow is stable to sideband perturbations, i.e. that wave numbers in a neighborhood of the critical wavenumbers k_c and 0 are stable.

If $\epsilon\nu^2$ and $\epsilon\theta^2$ are small, there are two subcases: In the first case $\epsilon\theta^2$ is small relative to ν, and in the second case $\epsilon\theta^2$ is at least of the same order of magnitude as ν and $|\theta| >> |\nu|$. For $\epsilon\theta^2 << |\nu|$, we can distinguish two kinds of modes. One mode is dominated by the amplitude of the traveling wave, with the long wave component B smaller by an order of ϵ. The other mode has traveling wave and long wave components of the same order of magnitude, and it is this mode which presumably might lead to a growth of long waves. The criteria for stability of these long wave mode are

$$\frac{\tilde{\delta}\delta_r}{c_g - c_0} < 0, \quad \frac{\tilde{\delta}(\delta_r + \rho_r s_0)}{c_g - c_0} < 0. \tag{5.6}$$

To state the criteria for the stability of the traveling wave sideband mode, we introduce the following notations:

$$z_1 := \beta - \tilde{\delta}\delta/(c_g - c_0), \ z_2 := z_1 + \rho(r_0 - \tilde{\delta}s_0/(c_g - c_0)), \tag{5.7}$$

$$p : q = p_r q_r + p_i q_i. \tag{5.8}$$

Then the criteria for sideband stability are

$$(z_1)_r < 0, \ (z_2)_r < 0, \ \kappa : z_1 < 0, \ \gamma : z_2 < 0. \tag{5.9}$$

The second and fourth of these recover Blennerhassett's [3] conditions. However, even for purely two-dimensional perturbations these criteria are incomplete because they ignore the long wave mode (which yields the second condition of (5.6) in the two-dimensional case).

For the case where $\epsilon\theta^2$ is comparable to or larger than $|\nu|$, we cannot state our criteria in terms of explicit formulae. However, we can give a purely algebraic algorithm for studying sideband stability. It can be shown that stability for this case reduces to studying the eigenvalues of the matrix

$$\begin{pmatrix} -\kappa\theta^2 + \hat{\beta} A_0^2 & \hat{\beta} A_0^2 & \hat{\delta} A_0 \\ \overline{\hat{\beta}} A_0^2 & -\overline{\kappa}\theta^2 + \overline{\hat{\beta}} A_0^2 & \overline{\hat{\delta}} A_0 \\ \tilde{\delta} i\hat{\nu} A_0 - \overline{\delta^*}\theta^2 A_0 & \tilde{\delta} i\hat{\nu} A_0 - \delta^*\theta^2 A_0 & i\hat{\nu}(c_g - c_0) - \tilde{\kappa}\theta^2 \end{pmatrix}. \tag{5.10}$$

The condition for stability is that the eigenvalues have negative real parts for all real $\hat{\nu}$ and $\theta^2 > 0$. It easily follows from (5.5) that the eigenvalues have negative real parts if θ is large. Hence it suffices to rule out purely imaginary eigenvalues. For $\hat{\nu} = 0$, the eigenvalues are determined by an equation of the form

$$-\lambda^3 + p_1(\theta^2)\lambda^2 + p_2(\theta^2)\lambda + \theta^2 \tilde{p}_2(\theta^2) = 0, \tag{5.11}$$

where the coefficients are polynomials in θ^2 of the degree indicated by the index. Imaginary eigenvalues are ruled out if $\tilde{p}_2$ has no positive real roots and either $\theta^2\tilde{p}_2 + p_1 p_2$ has no positive real roots or p_2 is positive where such a root occurs.

In the general case, imaginary eigenvalues iq lead to an equation of the form

$$q^4 P_1(\theta^2) + q^2\theta^2 P_2(\theta^2) + \theta^6 \tilde{P}_2(\theta^2) = 0. \tag{5.12}$$

Moreover, if all the previous conditions for stability hold, then both P_1 and $\tilde{P}_2$ have the same sign, say positive, for $\theta^2 > 0$. To rule out imaginary roots, it is then necessary that $P_2 > 0$ or $P_2^2 < 4\theta^2 P_1 P_2$ for $\theta^2 > 0$. Sturm's criterium can be used to evaluate whether this is the case in a purely algebraic fashion.

In [17], we apply our criteria to a number of situations modeling gas-liquid flows. All the situations were found to be unstable to some kind of sideband perturbations, although it varies precisely which condition is violated.

6 Homoclinic and Heteroclinic Orbits of the Amplitude Equations

The amplitude equations also admit solutions with nonperiodic spatial behavior. We have investigated some solutions which are spatially homoclinic or heteroclinic. Throughout this section, we shall consider only two-dimensional solutions, i.e. solutions of (4.7).

For convenience, we shall write β in place of $\beta + \rho r_0$, σ in place of $\sigma + \rho K$, δ in place of $\delta + \rho s_0$ and $\tilde{\beta}$ in place of $\tilde{\beta} + \tilde{\rho} s_0$.

We first look for solutions which are of the form $A = \exp(i\epsilon\omega\tau)\hat{A}(\xi - \epsilon c\tau)$, $B = \epsilon\hat{B}(\xi - \epsilon c\tau)$. Here ω and c are a priori unknown constants which need to be found as part of the solution. Recall that A is an amplitude factor multiplying a traveling wave, which is proportional to $\exp(i(k_0 x + \omega_0 t))$. Hence $\epsilon^2\omega$ represents a frequency shift of the traveling wave. Moreover, $\xi = \epsilon(x - c_g t)$, so ϵc is a shift in the wavespeed of the modulation relative to the group speed.

Suppressing the hats on A and B, we obtain the following system of ODEs:

$$\begin{aligned} \gamma A'' + cA' + (\sigma - i\omega)A + \beta|A|^2 A + \delta AB &= 0, \\ \epsilon\tilde{\gamma}B'' + (c_g - c_0 + \epsilon c)B' + \epsilon^2\tilde{\beta}(B^2)' + \tilde{\delta}(|A|^2)' &= 0. \end{aligned} \tag{6.1}$$

We can integrate the second equation to obtain

$$\epsilon\tilde{\gamma}B' + (c_g - c_0 + \epsilon c)B + \epsilon^2\tilde{\beta}B^2 + \tilde{\delta}|A|^2 = K, \tag{6.2}$$

where K is an arbitrary constant of integration, which we henceforth regard as fixed and given. If we set $\epsilon = 0$, then (6.2) reduces to the Ginzburg-Landau equation. For small ϵ, we can show the existence of a slow manifold on which the dynamics is described by a small perturbation of the Ginzburg-Landau equation. There are many spatially homoclinic or heteroclinic orbits which are known to exist for the Ginzburg-Landau equation. One of them is the pulse or soliton solution described by

$$A = C\,\mathrm{sech}\,(k\xi)\exp(iE\ln\cosh(k\xi)), \tag{6.3}$$

with $c = 0$ in (6.2) and appropriately determined values of C, E, k and ω. This solution connects the two-dimensional unstable manifold of the origin to the two-dimensional stable manifold. Hence it should generically persist if there are two freely adjustable parameters. We have the parameters c and ω at our disposal, and the relevant transversality conditions hold in the special case of real coefficients, and hence generically. Hence, there are solutions homoclinic to the origin for small ϵ.

The Bekki-Nozaki family [1] is an explicitly known set of heteroclinic solutions to the Ginzburg-Landau equations. However, the existence of this family violates generic transversality and hence we cannot expect persistence of these solutions for nonzero ϵ. It would be of interest to consider bifurcations from the Bekki-Nozaki family.

There are other heteroclinic solutions of our amplitude equations which do not come from the Ginzburg-Landau equations. To find these solutions, we use a different scaling of A and B. We make the ansatz $A = \hat{A}(\xi + (c_g - c_0 - \epsilon\tilde{c})\tau)$, $B = \hat{B}(\xi + (c_g - c_0 - \epsilon\tilde{c})\tau)$. Note that A is the amplitude factor for a traveling wave, while B represents a mean interface shift. The solutions considered below are of heteroclinic type; they either approach different constant interface heights at $\pm\infty$, or they approach a constant interface height at one end and a traveling wave at the other.

The ansatz above leads to the system (again we suppress the hats):

$$\begin{aligned} \epsilon\gamma A'' + (c_0 - c_g + \epsilon\tilde{c})A' + \epsilon\sigma A + \epsilon\beta|A|^2 A + \delta AB &= 0, \\ \tilde{\gamma}B'' + \tilde{c}B' + \tilde{\beta}(B^2)' + \tilde{\delta}(|A|^2)' &= 0. \end{aligned} \tag{6.4}$$

If $c_0 \neq c_g$, then, for small ϵ, there is an invariant slow manifold of the form $A' = \phi(\epsilon, A, B)$; for $\epsilon = 0$ we evidently have $\phi(\epsilon, A, B) = \delta AB/(c_g - c_0)$. Moreover, we can integrate the second equation of (6.4). We thus obtain the new system

$$A' = \phi(\epsilon, A, B), \ \tilde{\gamma}B' + \tilde{c}B + \tilde{\beta}B^2 + \tilde{\delta}|A|^2 = K. \tag{6.5}$$

We have the invariance $\phi(\epsilon, A\exp(i\psi), B) = \exp(i\psi)\phi(\epsilon, A, B)$. This allows us to split the first equation in (6.5) into separate equations for the modulus and phase of A. For the following, we are interested only in the equation for the modulus. With R denoting $|A|^2$, we obtain the new system

$$R' = \tilde{\phi}(\epsilon, R, B) := 2\sqrt{R}\ \mathrm{Re}\ \phi(\epsilon, \sqrt{R}, B), \ \tilde{\gamma}B' + \tilde{c}B + \tilde{\beta}B^2 + \tilde{\delta}R = K. \tag{6.6}$$

For $\epsilon = 0$, Eq. (6.6) reads

$$R' = \frac{2\delta_r}{c_g - c_0}RB, \ \tilde{\gamma}B' + \tilde{c}B + \tilde{\beta}B^2 + \tilde{\delta}R = K. \tag{6.7}$$

The qualitative behavior of solutions to (6.7) depends on the signs of the coefficients; we consider the case where

$$\tilde{\gamma} > 0, \ \tilde{\beta} < 0, \ \tilde{\delta} < 0, \ K < 0, \ \delta_r/(c_g - c_0) < 0. \tag{6.8}$$

(A number of other combinations lead to similar qualitative behavior.) We first consider the case $\tilde{c} = 0$. In that case, (6.7) has the three equilibrium points $B = 0$, $R = K/\tilde{\delta}$ and $R = 0$, $B = \pm\sqrt{K/\tilde{\beta}}$. In the (B, R)-phase plane, the first of these points is a center; the other two are saddle points. The two saddle points are connected by two heteroclinic orbits: one is part of the line $R = 0$, the other is an arc on a parabola. The phase plane portrait is displayed in Fig. 3.

For $\tilde{c} = 0$ the quantity

$$H(R, B) = R^{1+\alpha}[\tilde{\beta}B^2 + \tilde{\delta}\frac{1+\alpha}{2+\alpha}R - K] \tag{6.9}$$

is constant along orbits of (6.7). Here we have set

$$\alpha = -1 + \frac{\tilde{\beta}(c_g - c_0)}{\tilde{\gamma}\delta_r}. \tag{6.10}$$

The two orbits connecting the saddle points lie on the level set $H = 0$. For $\tilde{c} > 0$, H increases along orbits, while for $\tilde{c} < 0$ it decreases. As a result, the phase plane

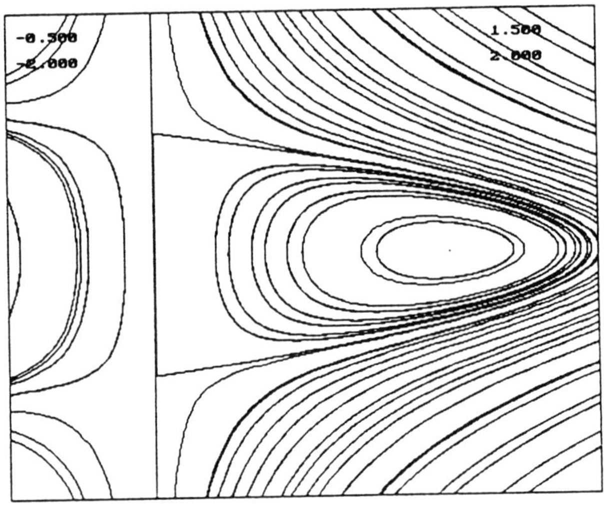

Figure 3: Phase plane portrait for the system $R' = -RB$, $B' = B^2 + R - 1$. The R axis is horizontal and the B axis is vertical. The range of R is from -0.5 to 1.5, and the range of B is from -2 to 2.

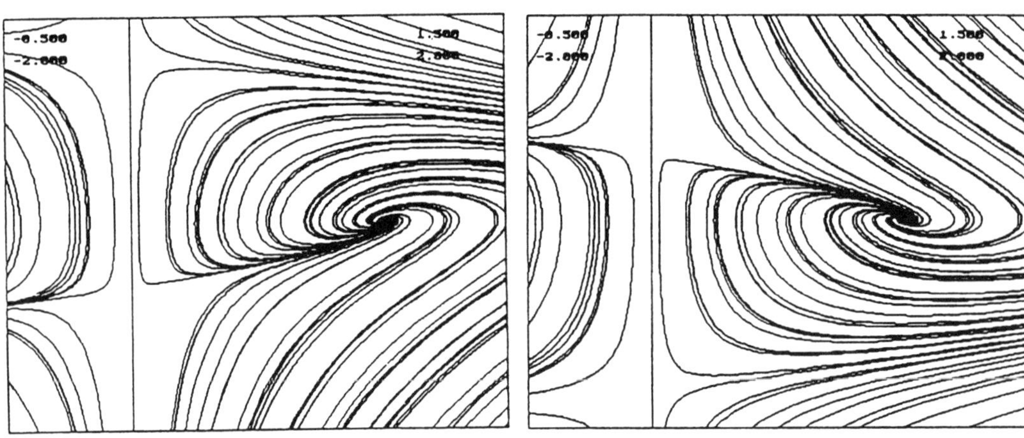

Figure 4: Phase plane portraits for the systems $R' = -RB$, $B' = \pm B + B^2 + R - 1$. The R axis is horizontal and the B axis is vertical. The range of R is from -0.5 to 1.5, and the range of B is from -2 to 2.

pictures in these cases look like Fig. 4. There is a heteroclinic orbit connecting one of the saddle points with the critical point at $B = 0$, $R = K/\tilde{\delta}$. These heteroclinic orbits are structurally stable and hence persist for $\epsilon \neq 0$. The orbit connecting the two saddle points occurs for a set of codimension one, i.e. for a specific choice of $\tilde{c}(\epsilon)$.

Even the classical Ginzburg-Landau equation allows a rich variety of solutions which is only partially understood, and the more complicated system (4.6) leads to even more possibilities. The analysis of sideband instability and the homoclinic and heteroclinic solutions studied in this section are only a beginning; we are far from an exhaustive description of physically realizable patterns. Clearly, there is much potential for further research on the existence and stability of solutions to our amplitude equations.

Acknowledgement This research is funded by the National Science Foundation under Grants CTS-9307238 and DMS-9306635, and by the Office of Naval Research under Grant N00014-92-J-1664.

References

[1] N. Bekki and K. Nozaki, Formations of spatial patterns and holes in the generalized Ginzburg-Landau equation, *Phys. Lett. A* **110** (1985), 133.

[2] T.B. Benjamin and J.E. Feir, The disintegration of wave trains on deep water, *J. Fluid Mech.* **27** (1967), 417.

[3] P.J. Blennerhassett, On the generation of waves by wind, *Phil. Trans. Roy. Soc. London A* **298** (1980), 451.

[4] K.P. Chen and D.D. Joseph, Lubricated pipelining: stability of core-annular flows. Part IV: Ginzburg-Landau equations, *J. Fluid Mech.* **227** (1991), 587.

[5] M. Cheng and H.-C. Chang, A generalized sideband stability theory via center manifold projection, *Phys. Fluids A* **2** (1990), 1364.

[6] J.P. Collet and P. Eckmann, The time dependent amplitud equation for the Swift-Hohenberg problem, *Comm. Math. Phys.* **132** (1990), 139.

[7] A.V. Coward, D.T. Papageorgiou and Y.S. Smyrlis, Nonlinear stability of oscillatory core-annular flow: A generalized Kuramoto-Sivashinsky equation with time-periodic coefficients, *Z. angew. Math. Phys.* **46** (1995), 1.

[8] A. Davey, L.M. Hocking and K. Stewartson, On the nonlinear evolution of three-dimensional disturbances in plane Poiseuille flow, *J. Fluid Mech.* **63** (1974), 529.

[9] W. Eckhaus, Problèmes non-linéaires de stabilité dans un espace à deux dimensions II, *J. Méc.* **2** (1963), 153.

[10] A. van Harten, On the validity of the Ginzburg-Landau equation, *J. Nonlin. Sci.* **1** (1991), 397.

[11] A.P. Hooper and W.G.C. Boyd, Shear flow instability at the interface between two viscous fluids, *J. Fluid Mech.* **128** (1983), 507.

[12] A.P. Hooper and R. Grimshaw, Nonlinear instability at the interface between two viscous fluids, *Phys. Fluids* **28** (1985), 37.

[13] G. Iooss, A. Mielke and Y. Demay, Theory of steady Ginzburg-Landau equation in hydrodynamic stability problems, *Euro. J. Mech. B* **8** (1989), 229.

[14] D.D. Joseph and Y.Y. Renardy, *Fundamentals of Two-Fluid Dynamics*, Parts 1 and 2, Interdisciplinary Applied Mathematics Series, Vol. 3 and 4, Springer, New York 1993.

[15] L.A. Jurman, S.E. Deutsch and M.J. McCready, Interfacial mode interactions in horizontal gas-liquid flows, *J. Fluid Mech.* **238** (1992), 187.

[16] C.G. Lange and A.C. Newell, A stability criterion for envelope equations, *SIAM J. Appl. Math.* **27** (1974), 441.

[17] M. Renardy and Y. Renardy, Derivation of amplitude equations and analysis of sideband instabilities in two-layer flows, *Phys. Fluids A* **5** (1993), 2738 (Corrigendum **6** (1994), 3502).

[18] Y. Renardy, Weakly nonlinear behavior of periodic disturbances in two-layer Couette-Poiseuille flow, *Phys. Fluids A* **1** (1989), 1666.

[19] M. Sangalli, Th. Prokopiou, M.J. McCready and H.-C. Chang, Observed transitions in two-phase stratified gas-liquid flow, *Chem. Eng. Sci.* **47** (1992), 3289.

[20] G. Schneider, Global existence via Ginzburg-Landau formalism and pseudo-orbits of Ginzburg-Landau approximations, *Comm. Math. Phys.* **164** (1994), 157.

[21] J.T. Stuart and R.C. DiPrima, The Eckhaus and Benjamin-Feir resonance mechanisms, *Proc. Roy. Soc. London A* **362** (1978), 27.

[22] B.S. Tilley, S.H. Davis and S.G. Bankoff, Nonlinear long-wave stability of superposed fluids in an inclined channel, *J. Fluid Mech.* **277** (1994), 55.

[23] C.S. Yih, Instability due to viscosity stratification, *J. Fluid Mech.* **26** (1967), 337.

A Personal Sample of Patterns in Biology

L.A. SEGEL
Department of Applied Mathematics and Computer Science, Weizmann Institute, Rehovot, Israel

Abstract

Description is given of theories concerned with examples of temporal, spatial and topological patterns in various parts of biology. Patterns in "aspect" are also described, where "aspect" is a descriptor, such as size, that is used as an independent variable in models of population dynamics.

1 Introduction

I used to be an honest patternophile. A patternophile is a person who loves patterns. An honest one investigates pattern formation in classical contexts such as fluid and solid mechanics, where known physical laws conspire to yield surprising order, e.g. see DiPrima, Eckhaus and Segel (1971). Since the underlying physical laws are well understood, the investigator can concentrate on elucidating the interactions that lead to pattern.

More than two decades ago I switched into theoretical biology. I thought to make a radical change of fields, and I have largely done so, but biology is so full of pattern formation problems that I have not been able to avoid some of them. This article is mainly a brief survey of my own encounters with these problems, concentrating almost exclusively on work of the last few years. As will be seen, in the study of biological pattern formation, often the essence of the difficulty is formulating a mathematical problem that expresses the underlying biology well enough so that the equations indeed yield the observed patterns.

Making the survey personal makes it easier to write (on occasion I will lift paragraphs bodily from my own papers). I can fairly painlessly give the reader an appropriate feel for the scope of biological pattern formation problems by showing the variety of such problems that have been met by a single investigator — whose main interests have usually resided elsewhere. A full survey of pattern formation in biology would require a book. Indeed, those whose appetites are whetted are urged to consult the books by Meinhardt (1982) and Harrison (1993), as well as the workshop proceedings edited by Othmer, Maini and Murray (1993).

The patterns to be discussed here can be put into the categories of temporal, topological, spatial and aspectal. The last category will be explained in due course.

2 Excitable Bursting: A Temporal Pattern in Neurobiology

Bursting neurons exhibit periods of oscillation separated by periods of quiesence. Here we sketch a bursting model of a new type, briefly summarizing work of Sivan, Segel and Parnas (1995).

Classical mathematical explanations of bursting behavior (see for example Rinzel and Ermentrout, 1989) consider situations (very often found in practice) where variables can be divided into two categories, fast and slow. The explanation begins with an examination of the fast variables when the slow variables are held at fixed values and hence temporarily regarded as parameters. Depending on these parameters, two types of behavior are found. Typically, for some parameter values periodic oscillations are seen, while for others the fast system always tends to a steady state. Bursting occurs when the slow parameters are allowed to vary according to their governing equations. If conditions are appropriate, oscillations in the fast variables move the slow variables out of the "parameter domain" where fast oscillations occur (when the slow variables are temporarily fixed) and into the domain where the fast system is attracted to a steady state. At steady values of the fast variables, however, the slow variables ultimately move to the "parameter domain" corresponding to fast oscillations. Bursting thus results from the alternating movement of the slow variables from the domain giving fast oscillation to the domain giving a steady state. One could term this classical burster an *on-off oscillator*.

There are four linked bursting cells (numbered 6, 7, 8, 9) in the cardiac ganglion of the lobster, a simple model nerve network. After we arrived at specific candidate models for each of these, following up on papers by Av-Ron, Parnas and Segel (1991, 1993), we began to examine the models in the spirit of the classical bursting explanation. We found to our surprise that in our cell 9 model for no relevant fixed values of the slow variables did the fast equations exhibit oscillation. There was always a globally stable steady state. See Fig. 1.

How could bursting possibly stem from such a stable system? Further examination of the equations gave a clue. When we kept the slow variables fixed at various relevant values then the two fast equations were an excitable system. That is, even though all perturbations eventually died out and the system inevitably returned to rest, there were important transient effects. Although relatively small perturbations to the steady state (the rest state) decayed monotonically, superthreshold perturbations led to an explosive increase in voltage (action potential) after which the system returned to rest. Detailed examination of the four nonlinear ordinary differential equations that comprise the model of Sivan *et al.* (1995) showed that indeed when the slow variables are permitted to vary, conditions can be such that the excitability threshold can be repeatedly crossed. Once we explained how such a spike sequence can in principal be generated, we then showed why the sequence ceases — to give the quiescent part of the cycle. Finally, we explained why the spikes start again.

The new way to obtain alternating periods of oscillation and quiesence is not via the classical on-off oscillator, but rather is an *excitable burster*. Sivan *et al.* (1995) show why the excitable burster is likely to be far more than an amusing curiosity but rather may play a central role in small neuronal networks.

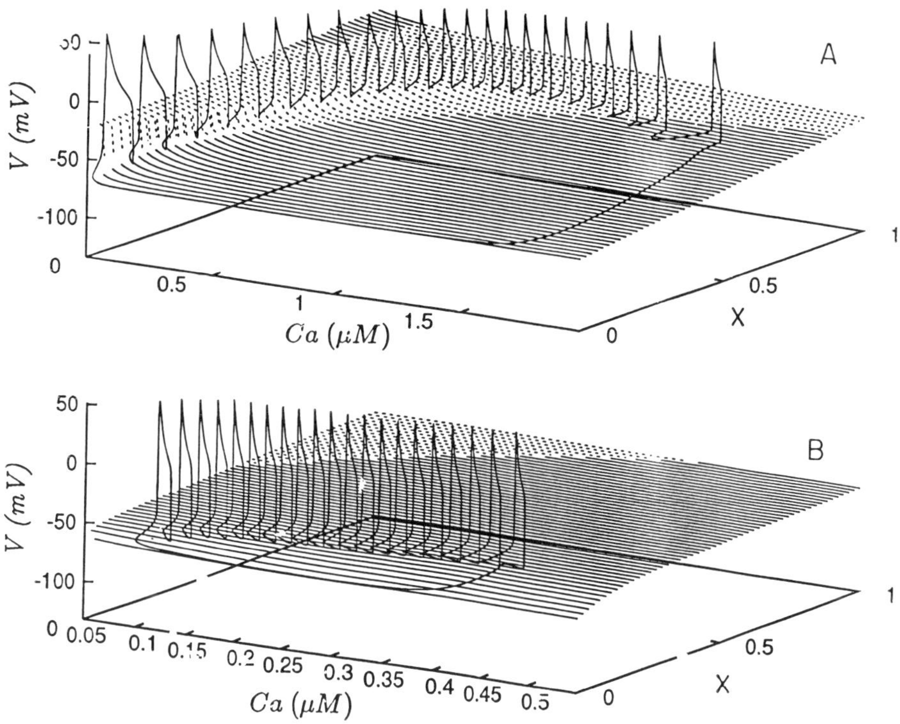

Figure 1: [Figure 2 of Sivan *et al.* (1995).] Trajectory of voltage V (a fast variable) together with calcium concentration Ca and the fraction of open calcium channels X (slow variables) for model cells. Below the trajectory is a plane each point of which describes the stability of the fast system if C and X are frozen at their instantaneous values. The plane is dashed or solid depending on whether the steady state is unstable or stable. A. Model cell 6. As expected, the oscillations take place when the trajectory is in the unstable domain of the plane. (The last two spikes are an exception; here our discussion of cell 9 is relevant.) B. Model cell 9. Contrary to expectation, even during oscillations the projected trajectory is in the stable domain of the plane.

3 Topological Pattern in Developmental Biology

A central concern of developmental biology (an extension of embryology to simpler organisms without embryos) is the study of pattern emergence. Two approaches to the study of developmental pattern can be discerned. One approach is on a supracellular level; the most famous example is the suggestion of Turing (1952) that pattern may be determined by the reaction and diffusion of chemical "morphogens". A second approach concentrates on interactions between the fundamental building blocks of tissue, the cells.

It is essential to keep track of cell geometry when the relation between force and deformation must be tracked in morphogenesis problems (Weliky and Oster, 1990). But there could well be circumstances wherein what is of critical importance is not geometrical but topological — cell adjacency and the concomitant possibility of specific cell-cell communication.

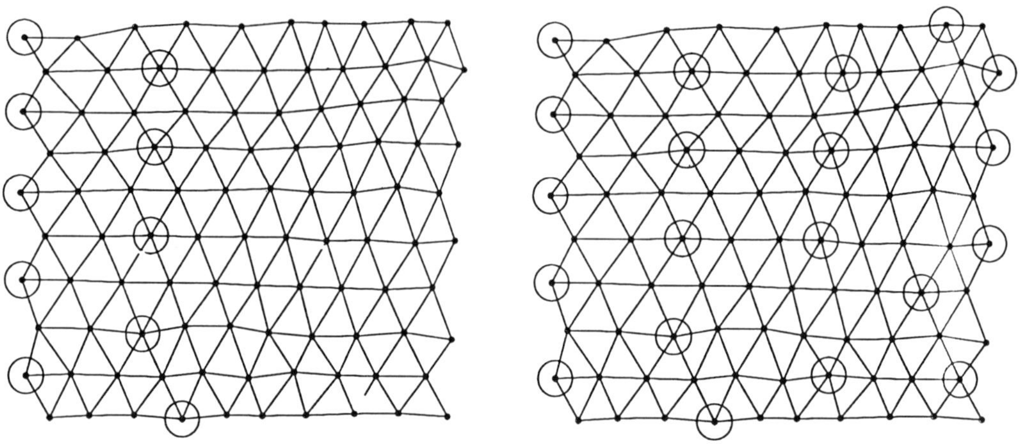

Figure 2: [Fig. 11(c,d) of Duvdevani-Bar and Segel (1994).] Simulation of two stages in the differentiation of R8 cells (represented by circles) in the *Drosophila* retina. Activator-inhibitor (nonlinear) reaction-diffusion equations are assumed, with the property that a suitable activator gradient is automatically formed. R8 cells have a probability of differentiating when the activator concentration exceeds a certain threshold. Differentiation is blocked if a second inhibitor, secreted by R cells, surpasses its threshold. See Duvdevani-Bar and Segel (1994) for details of the surprisingly complex model that seemed necessary to provide a simulation of the observations.

There have been a number of computer simulations of development at the cellular level. Most of these have employed a rigid two-dimensional square or hexagonally tesselated space of "cells"; see for example Ede and Law (1969). By contrast, a novel topological approach to the representation of cells on the computer was proposed by Matela and Fletterick (1980). This approach suggests that a simulation of a cell system can be based on a planar graph and its dual structure, a planar geometric map. In the graph, the vertices are the cells and the adjacency of these cells is represented by the edges of the graph. The dual representation is a finite set of closely packed n-gons, $n \geq 3$. Each of these represents an individual cell, and the connections between the cells are represented by the polygon edges. The topological model framework has been used to simulate cell sorting (Matela and Fletterick, 1980) and cell division (Matela and Ransom, 1985). Duvdevani-Bar and Segel (1994) showed how the topological approach can also be employed for the study of several other processes that are considered to be fundamental ingredients of development: cell growth and death, cell motion, cell differentiation and chemical diffusion.

As a case study of the topological approach, Duvdevani-Bar and Segel (1994) considered the compound eye pattern in the fruitfly *Drosophila*. The first major event in the pattern formation is a wave of morphogenesis that results in successive rows of rather regularly spaced differentiated "R8 cells". Zak and Shilo (1992) hypothesized that the spacing is due to the secretion of an inhibitor. Duvdevani-Bar and Segel (1994) used their formalism to study this hypothesis, with the aid of a reaction-diffusion activator-inhibitor model of the Meinhardt-Gierer type (Meinhardt, 1982). Figure 2 shows the results of typical computer simulations.

Although it has many advantages, there is a difficulty with the topological model, namely, visualizing the results. Visualization is necessary, since a changing list of cell identification numbers, neighbours, and descriptors conveys virtually no information. But trying to locate the nodes and edges so as to form a clear and pleasing picture of the graph is not an easy task. In fact, it turns out that, technically speaking, attaining an optimal solution is computationally infeasible (NP-hard) (Davidson and Harel, 1995).

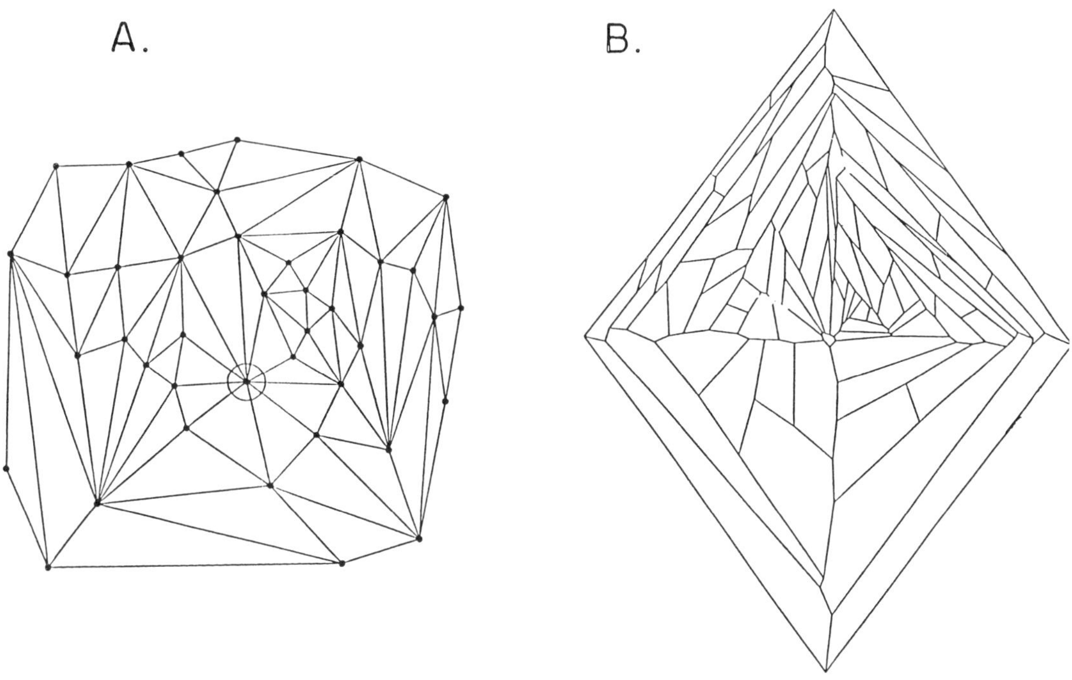

Figure 3: [Figs. 6 and 7 of Duvdevani-Bar and Segel (1994).] A. The topological representation of 7 generations of cleavage, which result in approximately 40 cells. In accord with observations, cell division has been made more rapid closer to the animal pole, toward the top of the picture. The large circle represents an internal cavity, now bounded by 10 cells instead of the 4 bounding cells with which the simulation started. B. The dual representation of 8 generations of cleavage. Compared to A, it is clearer that there are more cells near the animal pole.

Duvdevani-Bar and Segel (1994) used a home-made algorithm to produce adequate graphs for their "toy" simulations (Fig. 3). But, alas, the need to determine a fixed point in space for elements of a drawing of a graph stands in complete contrast to the essence of the topological approach to the modelling of cell pattern. Nonetheless progress by computer scientists in drawing "nice graphs" (Davidson and Harel, 1995) indicates that there will be a role for topological approaches to patterning.

4 Inflammatory Patterns: Rash Theory

Rashes are spatial patterns of skin abnormalities that are reproducibly associated with various diseases, thereby providing a major diagnostic tool. Nonetheless, almost nothing is known about why various particular patterns constitute signatures of particular diseases.

There are two principle bases for patterns of rashes, which we have termed *allogenesis* and *autogenesis*. In allogenesis, the reason for the pattern is in essence separate from the underlying mechanisms of the disease. For example, a rash in shingles is due to the attack of a virus on a nerve, so that the rash pattern reflects the underlying nerve pattern. By contrast, there are numerous patterns of rashes that seem to have no obvious counterpart in underlying anatomy. They are generated autonomously; hence the term autogenesis. We will be concerned with autogenic patterns. Of major interest to us, and to physicians, are rashes that are associated with inflammatory processes or with other manifestations of the interaction of the immune system with pathogens. Just as galactic patterns give clues to underlying astrophysical events, so rash patterns can give us clues to the subdermal struggle between the immune system and its foes (which are our foes).

We shall search for pattern formation mechanisms in the framework of interaction-transport equations (Levin and Segel, 1985). One type of pattern that can be exhibited by this broad class of equations consists of stationary stripes, hexagonal networks, or regular arrays of spots. We term these *Type I patterns*. *Type II patterns* are spiral and "target" waves and combinations thereof. Unlike the stationary Type I patterns, Type II patterns always move.

Segel *et al.* (1992) constructed illustrative examples of models that would give Type I and Type II patterns, based on previously studied pairs of nonlinear reaction-diffusion partial differential equations. In the context of rashes, the dependent variables are interpreted as the level $V(x,y,t)$ of a pathogen population (e.g., viruses) that are presumed to cause the rash and the level $I(x,y,t)$ of some product of the immune system that attacks the pathogen.

Upon considering the literature and clinical experience of dermatologists, Segel *et al.* (1992) found that there seem to be no clear instances of Type I patterns such as more or less regular, stationary, stripes of inflammation or regular arrays of inflamed spots or hexagonal lattices. Nonetheless the appearance of spots in a number of common diseases might not be random, as it seems at first sight, but rather the rashes may exhibit an order that is consistent with a Type I pattern-forming mechanism. It would be worthwhile for dermatologists to make serial photographs of the emergence of the spots and to examine the pattern for randomness by various statistical procedures (Bartlett, 1976).

"An exception that proves the rule" is the hexagonal network stemming from stresses generated by an abnormal drying of the skin and concomitant cracking. This is a Type I pattern, but not one generated directly by a pathogen-immune system interaction.

Dermatologists are aware of moving Type II patterns (although they are rare), which are hypothesized to be concomitants of the inflammation process. One such pattern has been described as possessing a "distinct erythematous border which may be annular, arcuate, gyrate, polycyclic, serpiginous or reticulate in overall pattern. This border typically demonstrates relatively rapid expansion or migration over a period of hours or days..." (Willis, 1978). In a brief letter to the editor, Moore (1982) called attention to the "striking morphological similarity" of what we have termed Type II patterns to chemical waves seen in the Belousov-Zhabotinski reaction. Segel *et al.* (1992) took this observation further (at first without knowledge of Moore's remarks) by examining a simple reaction-diffusion model for the genesis of rash patterns that is of the same mathematical class as models for the Belousov-Zhabotinski reaction and other pattern forming mechanisms. They suggested that the apparent lack of Type I patterns might stem from the absence of the requirements (in a reaction-diffusion model) of positive feedback of pathogen on its own proliferation, and negative feedback of the immune factor on *its* production. The presence of suitable excitability presumably underlies the Type II patterns.

The next stage in constructing "rash theories" is to pass beyond the tentative first steps mentioned here to far more profound modelling of the physiology.

5 Patterns in Aspect

Patterns are traditionally regarded as spatial or spatiotemporal, where the "space" in question is that of ordinary Euclidean geometry. Recently, modellers of biological events and processes have shown the relevance of patterns in nontraditional spaces. These spaces are abstract in that they deal with variables that are often not nearly as tangible as geometric space, but concrete in that they refer to specific properties of biological organisms. Levin and Segel (1985) surveyed a number of examples of patterns in *aspect space*, where "aspect" is a descriptive variable of a population. A major example concerned predator-prey interactions wherein the probability of a predator attacking a certain prey was influenced by the latter's appearance with respect to such variables as color, spot density and configuration. Other examples cited by Levin and Segel of aspects of biological entities that influence their interactions include age, size or maturity of organisms, and chemical composition that effect an organism's palatability to its prey. Later Levin, Segel and Adler (1990) discussed coevolution of plant defensive chemicals and herbivore detoxification ability — an example of how such aspect variables influence population dynamics.

We now briefly mention some other instances of patterns in various types of aspect spaces. To introduce our first example we note that the dynamics of the immune system can be formulated in terms of what is called *shape space*. This possibility stems from the fact that a fundamental aspect of a receptor-ligand interaction is the shapes of the two molecules that participate in such an interaction, in particular the extent to which they are complementary. Thus, for example, an invading bacteria will trigger an attack by the immune system to the extent that the shapes of key bacterial molecules, or parts of them, are such that they ligate (bind) certain receptor molecules on lymphocytes. Precise description of the relevant "shapes" requires an infinite dimensional space. More than geometrical dimensions are needed, for such properties as charge distribution and hydrophobicity influence the complementarity of shapes and hence their binding affinity. Patterns in a two-dimensional shape space are illustrated in Fig. 4. This figure is taken from de Boer, Segel and Perelson (1992) wherein references can be found to other approaches to shape space.

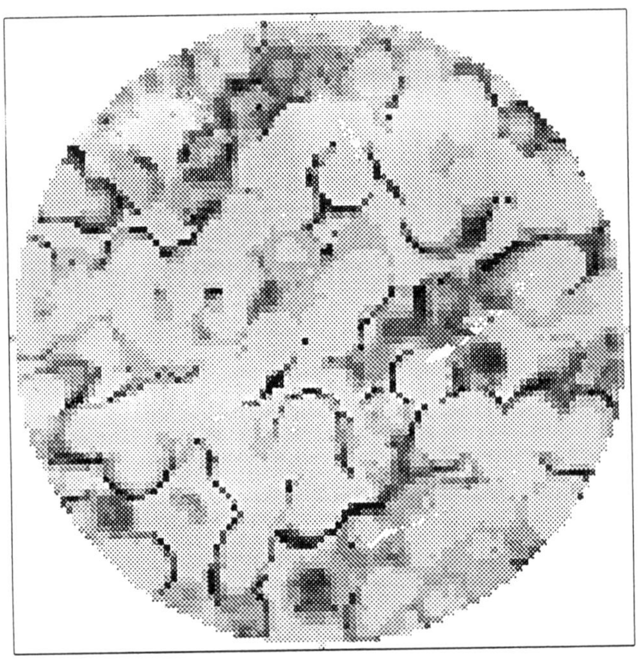

Figure 4: [Fig 8a of Deboer, Segel, and Perelson (1992).] Higher intensities of black correspond to higher population levels of cells in a simulated immune system, based on a two-dimensional shape space. The proliferation rate of a cell of a given shape is a log bell-shaped function of field strength, which in turn is an increasing function of the number of cells present of complementary or near-complementary shapes.

A biological context in which a low dimensional aspect space may be relevant is the interaction of anonymous social organisms such as bees. There is evidence that bees possess a scalar character called *dominance*, and that this character changes as a result of "fights" between pairs of organisms. The more dominant organism has a higher probability of winning; the winners dominance is augmented and that of the loser is decremented. The less probable the result of a fight, the bigger the consequent change in dominance (Hogeweg and Hesper, 1983). Simulations show that for certain specific rules, the population divides into two groups, one with high dominance and one with low. It is possible that the two groups perform different tasks in the hive, so that dominance serves as an automatic organizing principle (Hogeweg and Hesper, 1983).

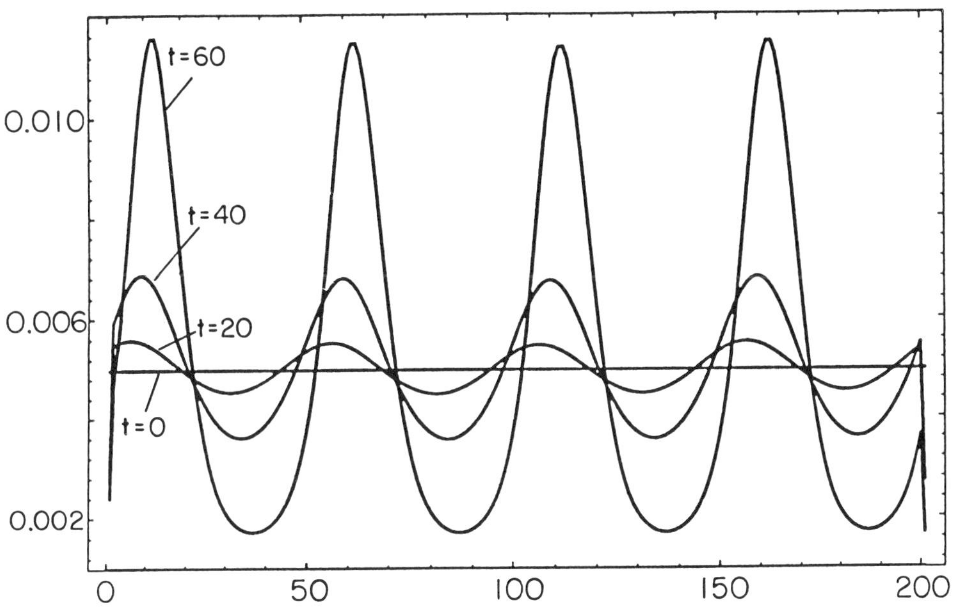

Figure 5: [Fig. 6.7 of Ladizhansky (1994).] Simulations of the development in time t of the distribution $f(y,t)$ of actin filaments at angles y, $-\pi < y < \pi$, where

$$\frac{\partial f(y,t)}{\partial t} = -f(y,t)\int_{-\pi}^{\pi} f(u,t)du + \int_{-\pi}^{\pi}\int_{-\pi}^{\pi} \psi(y-x, y-u) f(x,t) f(u,t) dx du.$$

The kernel ψ models the effect of a molecular "hinge" that tends to move interfilament angles toward a right angle:

$$\psi(r,s) = \begin{matrix} \frac{1}{2\delta} & , & -\delta \le r - g(r-s) \le \delta; \\ 0 & , & \text{otherwise.} \end{matrix} \qquad g(w) = \begin{matrix} -kw & , & 0 \le w \le \frac{\pi}{2}; \\ -k(w-\pi) & , & \frac{\pi}{2} < w < \pi. \end{matrix}$$

Jaeger and Segel (1992) derived a nonlinear integro-differential equation for the distribution of dominance in a population. They analyzed emerging patterns of dominance numerically, and also by means of a Fokker-Planck equation that emerges from approximations in the case of weak interactions. Significant extensions to the Jaeger-Segel paper were made in the M.Sc. dissertation of Ladizhansky (1994), under my supervision and with the decisive collaboration of A. Mogilner (University of British Columbia). Among the results are the following. As suggested by Mogilner, the concept of aspect can be used to examine the bundling of actin fibers. (The dynamic organization and re-organization of such fibers plays a central role in forming the cytoskeleton, the labile "skeleton" of a cell.) Here the distribution of aspect in the population corresponds to the distribution of angles of actin fibers. Analytic approximations can be obtained in the form of interacting sharp peaks of aspect. Numerical simulations back up the analytic findings. A sample result is shown in Fig. 5. The four peaks correspond to tight bipolar orthogonal bundles of actin that are seen under certain circumstances.

The ratings of chess players is another example where the aspect concept can be employed. Ladizhansky (1994) showed that an ever-spreading distribution is obtained if rules for calculating chess ratings are introduced into the aspect formalism. This unrealistic result is converted into a satisfactory approach to a steady bell-like state if the model is realistically augmented by entry of new players at a low dominance value and by a finite "lifetime" of chess players as participants in tournaments.

6 Summary

We have indicated something of the variegated settings for pattern formation research in the biological sciences. The contexts range widely; cell and molecular biology (actin fiber orientation), developmental biology (retinal morphogenesis), dermatology (rashes), ecology and evolution (predator-prey patterns), ethology (dominance), immunology (shape space) and even human behavior (chess ratings). Biology brings new patterns (temporal periodic bursts) and new contexts for old patterns (spiral rashes). Biology suggests that topological patterns may be relevant, not just geometric patterns, with concomitant quandries concerning the pleasing geometric representation of topological graphs. The independent variables are not restricted to time and geometric space but are expanded to include aspect variables such as shape, orientation and dominance. Not just differential equations are involved, but integro-differential equations, with interesting consequences of non-locality of interaction. And nonlinearity permeates everything.

Acknowledgements. Supported is acknowledged from the US-Israel Binational Science Foundation, Grant 92-00171. Reproduced figures are with the kind permission of the publishers.

References

Av-ron, A, Parnas, H and Segel, L (1991) A minimal biophysical model for an excitable and oscillatory neuron, *Biol. Cybern.* **65**, 487–500.

Av-ron, E, Parnas, H and Segel, L (1993) A basic biophysical model for bursting neurons, *Biol. Cybern.* **69**(1), 87–95.

Bartlett, M (1976) *The Statistical Analysis of Spatial Pattern*, Chapman and Hall, London.

Davidson, R and Harel, D (1995) Drawing graphs nicely using simulated annealing, *Comm. Assoc. Comput. Mach.*, in press.

de Boer, R J, Segel, L A and Perelson, A S (1992) Pattern formation in one- and two-dimensional shape-space models of the immune system, *J. theor. Biol.* **155**, 295–333.

DiPrima, R C, Eckhaus, W and Segel, L A (1971) Nonlinear wave-number interaction in near-critical two-dimensional flows, *J. Fluid Mechanics* **49**, 705–744.

Duvdevani-Bar, S and Segel, L (1994) On topological simulations in developmental biology, *J. theor. Biol.* **166**, 33–50.

Ede, D A and Law, J T (1969) Computer simulation of vertebrate limb morphogenesis, *Nature* **221**, 244–248.

Harrison, L G (1993) *Kinetic Theory of Pattern Formation*, Cambridge University Press, Cambridge.

Jäger, J and Segel, L (1992) On the distribution of dominance in a population of interacting anonymous organisms, *SIAM J. Appl. Math.*, **52**(5), 1442–1468.

Hogeweg, P and Hesper, P (1983) The ontogeny of the interaction structure in bumble bee colonies, *Behav. Ecol. Sociobiol.* **12**, 271–283.

Ladizhansky, K (1994) Distribution of generalized aspect with applications to actin fibers and social interactions, M.Sc. Thesis, Dept. of Appl. Math. and Computer Sci., Weizmann Institute, Rehovot, Israel.

Levin, S A and Segel, L A (1985) Pattern generation in space and aspect by interaction and redistribution, *SIAM Rev.* **27**, 45–67.

Levin, S, Segel, L A and Adler, F (1990) Diffuse coevolution in plant-herbivore communities, *Theor. Population Biol.* **37**(1), 171–191.

Matela, R J and Fletterick, R J (1980) Computer simulation of cellular self-sorting — a topological exchange model, *J. theor. Biol.* **84**, 673–690.

Matela, R J and Ransom, R (1985) A topological model of cell division: structure of the computer program, *BioSystems* **18**, 65–78.

Meinhardt, H (1982) *Models of Biological Pattern Formation*, Academic Press, London.

Moore, H J (1982) Does the pattern of erythema gyratum repens depend on a reaction-diffusion system? *British J. Dermatol.* **107**, 723.

Othmer, H G, Maini, P K and Murray, J D (1993) *Experimental and Theoretical Advances in Biological Pattern Formation*, Nato ASI Series, Vol. 259, Plenum Press, New York.

Rinzel, J and Ermentrout G B (1989) Analysis of neural excitability and oscillations, in: Koch, C and Segev, I (eds.), *Methods in Neuronal Modeling: From Synapses to Networks*, MIT Press, Bradford, MA.

Segel, L A, Perelson, A S, Hyman, J M and Klaus, S N (1992) Rash theory, in: Perelson, A S and Weisbuch, G (eds.), *Theoretical and Experimental Insights into Immunology*, NATO ASI Series, pp. 333–352, Springer-Verlag, Berlin.

Sivan, E, Segel, L A and Parnas, H (1995) Modulated excitability: a new way to obtain bursting neurons, *Biol. Cybern.*, in press.

Turing, A M (1952) The chemical basis of morphogenesis, *Phil. Trans. Roy. Soc. London B* **237**, 37–72.

Weliky, M and Oster, G (1990) The mechanical basis of cell rearrangement, *Development* **109**, 373–386.

Willis W F (1978) The gyrate erythemas, *Intern. J. Dermatol.* **17**, 698–702.

Zak, N B and Shilo, B (1992) Localization of DER and the pattern of cell divisions in wild-type and Ellipse eye imaginal discs, *Developmental Biol.* **149**, 448–456.

Invitation to river morphodynamics

G. SEMINARA

Instituto di Idraulica, Università di Genove, Via Montallegro, 1, 16145 Genoa, Italy

Abstract

This review is intended to provide an introduction to river morphodynamics for a mathematically oriented audience. Some as yet unpublished results are also presented. By no means we pretend to cover the subject exhaustively. Our aim is simply to show that some progress in our understanding of extremely complex phenomena has been possible by attempting to construct rational theories though our modeling still has to rely vastly on often uncertain empirical inputs. The applied mathematician is then warned that it is dangerous to build skyscrapers on weak foundations.

1 Morphological patterns

1.1 Introductory remarks

The natural environment often displays a high degree of order and regularity detectable in the context of natural patterns that may be highly complex. We are concerned herein with patterns originating from the interaction of sediments with the flow of natural fluids. The main characteristic of rivers, coasts and deserts is the fact that they are *self-formed* features, the self-forming mechanism being activated by the process of sediment transport. We will deal with the major agent of sediment transport in the terrestrial environment, i.e. rivers, where hydrodynamic forces drive particle motion. Mass flows of sediments, dominated by the action of gravity and occurring on the upper portions of river basins (debris flows, pyroclastic flows, mud flows) will not be considered in the present review (see Takahashi, 1981). Similarly other morphological environments distinct from rivers will not be covered, though analogies can often be drawn with mechanisms controlling the formation of patterns observed in the coastal and estuarine environments.

The self-forming processes are typically *unsteady* as both the flow conditions and the production of sediment available for transport are controlled by the climate. Furthermore the evolution of the *container shape* is itself a cause of variations of the flow field, i.e. unsteadiness is intrinsically associated with the self-adjusting nature of the process. As a result one can hardly speak of *equilibrium configurations* of rivers, estuaries, coasts or deserts; equilibrium may only be temporary. Nonetheless it is often useful to think of an evolving pattern as a perturbed state relative to some ideal equilibrium state of the system viewed on some particular spatial and temporal scales. Taking the latter viewpoint allows one to interpret morphological patterns as the outcome of inherent instabilities of the above basic equilibrium states. A hierarchy of unstable modes then arises, corresponding to the variety of morphological features observed in nature. Let us briefly review their classification. The reader interested in the descriptive aspects of bedforms is referred to Allen (1984).

1.2 Transverse bedforms

The smallest spatial scale relevant to morphodynamics is the typical size of sediments. For cohesionless sediments the size may range from values of the order of .1 mm (fine sand) to values of the order of meters (large boulders). Finer sizes are typical of cohesive materials containing silt and clay.

The smallest self-formed pattern involving a distinct deformation of the boundary is that of *ripples* observed in rivers and estuaries. Ripples are asymmetrical bedforms observed in sediments not coarser than medium-coarse sand, their wavelengths range about few tenth of centimeters, their heights do not exceed few centimeters. They migrate downstream through a continuous process of erosion on the stoss side and deposition on the lee side, their migration speed being much smaller than the average speed of the flow. They may be fairly two-dimensional (long-crested) or strongly three-dimensional (short-crested).

The next step in the spatial scale of river patterns includes forms whose geometrical characteristics scale on flow depth: they comprise *dunes*, *antidunes* and *gravel sheets* (or *bedload sheets*). River *dunes* are perhaps the most common among the water formed features. They are usually fairly two-dimensional with long, moderately sinuous, crests orthogonal to the flow direction and nearly triangular, strongly asymmetric shape. They migrate downstream in sand bed rivers through the mechanism of accretion of the avalanche slip face. Oblique two-dimensional as well as strongly three dimensional dunes of lunate type are also observed (fig. 1). Dunes in gravel are rarely observed. River dunes are of considerable practical importance as their presence affects the overall resistance of the stream which may increase by a factor of two or three.

As the stream Froude number F increases dunes disappear and the bed recovers its flat configuration. Further increase of F leads to the formation of *antidunes*, symmetrically shaped, nearly sinusoidal, bed undulations in phase with large amplitude surface waves. Antidunes may migrate either upstream or downstream. Antidunes are also observed in gravel bed rivers.

When the bed consists of poorly sorted sediments it has been observed that sediment motion may organize itself in the form of *sorting waves*, called *gravelsheets* (or *bedload sheets*), coherent migrating patterns consisting of periodic sequences of coarse and fine material with distinct coarse fronts. Such waves cannot be considered actual bedforms as the bottom keeps essentially plane.

A peculiar type of aeolian dune observed in rivers, deserts and coastal areas is the *barkhan dune* which forms in regions where sand accumulates on a stony, rocky or cohesive surface while the supply of sand is insufficient. Aeolian barkhan dunes are often spectacular.

The next larger spatial scale of rivers is their *width*. Bedforms characterized by wavelengths of the order of a few channel widths are called *bars*. They may occur in single rows (*alternate bars*) or in many parallel rows (*multiple row bars*). Alternate bars (fig. 2) occur both in straight and in meandering channels, with gravel or sandy bottoms. They are three dimensional bedforms characterized by fairly steep

Figure 1: Three dimensional dunes of lunate type (reproduced from Allen, 1984).

consecutive diagonal fronts, deep pools at the downstream face of each front, followed by relatively gentler riffles along the upstream face of the next front. Alternate bars are of great practical importance as they migrate with speeds of the order of several metres/day and carry scours of the order of the average flow depth of the stream.

Multiple row bars are observed in *rivers* characterized by fairly wide cross sections.

1.3 Longitudinal bedforms

The type of bedforms considered in sect. 1.2 were all characterized by crests either orthogonal or oblique relative to the direction of fluid flow. In all morphological environments bedforms also exist with crests roughly aligned with the flow. They are called *ridges* (or sometimes *ribbons*).

Current ridges observed both in the laboratory and in the field, have spanwise wavelengths roughly twice the flow depth and very small amplitudes ranging about few grain diameters. They are usually observed in the initial stage of motion before more important bedforms like ripples and dunes develop. Much larger longitudinal bedforms are the so called *tidal ridges*, observed in tidal shallow marine environments with transverse spacing of the order of few kilometers, heights of the order of 10 m with crests commonly reaching to within a metre or so of the free surface.

1.4 Planimetric forms

Natural rivers are very seldom straight, more often their planimetric form is either *meandering* or *braiding. Meandering* rivers are characterized by a single channel (typically wandering in alluvium but observed also in rocky environments) the axis of which exhibits some periodicity, which is often quite distinct (fig. 3) but may sometimes be vague or even absent.

Braiding rivers are charaterized by the presence of a network of interconnecting channels which persist both at low and at high stage (fig. 4). Braiding is typically observed in flat plains, glacial valleys, outwash deposits and alluvial fans.

Figure 2: Alternate bars in the Alpine Rhine, Switzerland (reproduced from Jaeggi, 1984).

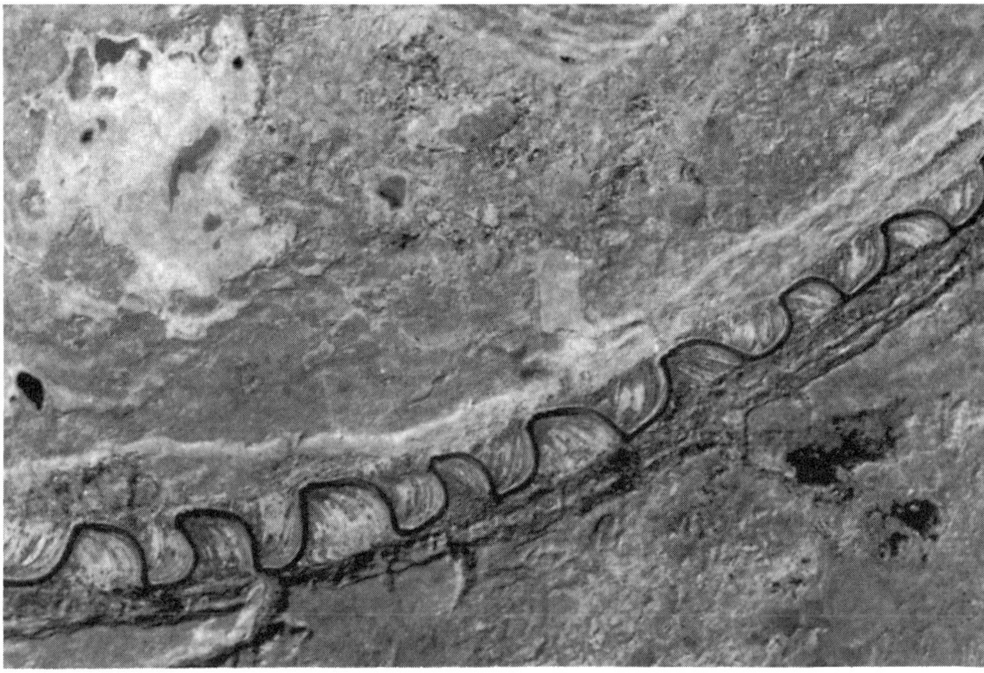

Figure 3: Regular sequence of confined meanders of Beaver river, Canada. Flow from right to left (reproduced from Allen, 1984).

2 Formulation of the basic problem of river morphodynamics

2.1 Statement of the problem

The basic problem of morphodynamics may be simply stated as that of determining the shape of a cohesionless boundary molded by the flow of a fluid able to mobilize the sediment. In the following we will focus our attention on river morphodynamics, though the formulation is readily extended to other morphological environments.

2.2 Sediment motion: mass conservation

The first mathematical ingredient needed to formulate the above problem is a global statement of mass conservation for sediment motion. Let $\eta(x_1, x_2, t)$ be the elevation of the cohesionless boundary relative to a horizontal plane referred to cartesian coordinates x_1 and x_2, with t time. Furthermore let $\mathbf{q}(x_1, x_2, t)$ be the volumetric sediment discharge per unit width. Sediment continuity then requires:

$$(HC)_{,t} + (1 - n)\eta_{,t} + \nabla_h \cdot \mathbf{q} = 0 \;, \qquad \nabla_h \equiv \left(\frac{\partial}{\partial x_1}, \frac{\partial}{\partial x_2}\right) \tag{2.1}$$

Figure 4: Braided reach of Sandur river, Iceland (courtesy of G. Parker).

where n is sediment porosity $H(x_1, x_2, t)$ is local flow depth and C is the depth averaged volumetric concentration of sediment. Notice that the effect of temporal variations of sediment concentration is usually neglected in the literature. The one-dimensional form of (2.1) was first derived by Exner (1925). In order to make any progress we then need to model the dynamics of sediment motion i.e. express $\mathbf{q}$ in terms of the flow characteristics. In spite of considerable efforts of hydraulic engineers and geomorphologists this subject is still vastly empirical and one has to rely on physical intuition and qualitative reasoning to overcome an insufficient knowledge of the fluid mechanics of the process. It is common to distinguish two main mechanisms of sediment transport: *bed load transport* and *transport in suspension.*

2.3 Sediment motion: bed load

Uniform sediment

Bed load transport occurs by rolling, sliding and saltating particles. Particles have an irregular shape and move unsteadily interacting with an irregular wall and with neighbouring particles being subject to the action of a fluid which is highly sheared, fluctuating and intermittently unstable. Though attempts to predict the average particle dynamics have been proposed (see for instance Sekine & Kikkawa (1992)) a completely reliable theory of bed load transport will have to await a better knowledge of such complex near wall processes. However in the sediment transport literature it is taken as well established on the basis of experimental observations and qualitative theoretical arguments that the bed load transport rate per unit width $|\mathbf{q}_b|$ is a function of some dimensionless measure of the average bottom stress that the fluid exerts on the boundary. In general, for *uniform* sediments, such correlation may be written in the following dimensionless form

$$\Phi = \frac{|\mathbf{q}_b|}{\sqrt{(s-1)gd_s^3}} = \Phi(\theta, \theta_c) \tag{2.2}$$

where θ is the dimensionless *Shields stress* defined as:

$$\theta = \frac{|\boldsymbol{\tau}_o|}{(\varrho_s - \varrho)gd_s} \tag{2.3}$$

and Φ is a rapidly increasing function of θ for which a large number of expressions have been proposed (see for instance Brownlie, 1981, for an assessment of their reliability). In (2.2) and (2.3) ϱ_s and d_s are density and diameter of the sediment respectively, $\boldsymbol{\tau}_o$ is the average bottom stress, ϱ is fluid density, $s \equiv \varrho_s/\varrho$ is relative density and θ_c is the threshold value of the Shields stress above which the sediment is set into motion. The value of θ_c varies roughly in the range $.03 \div .06$ depending on the value of particle Reynolds number. The above results refer to the *equilibrium* conditions occurring in uniform streams flowing over a plane bottom. They have been extended to cover the *non equilibrium* conditions associated with the presence of spatial variations of the gradient of bottom elevation. For relatively small values of the latter it has been shown that the effect of gravity acting on sediment particles leads to a deviation of the direction of sediment motion relative to that of bottom stress by an amount proportional to the local gradient of bed elevation. The following vectorial form of $\mathbf{q}_b$ arises on pure dimensional ground:

$$\mathbf{q}_b = \sqrt{(s-1)gd_s^3}\,\Phi(\theta, \theta_c)\left[\frac{\boldsymbol{\tau}_o}{|\boldsymbol{\tau}_o|} - \mathbf{R}\cdot\nabla_h\eta\right] \tag{2.4}$$

where Φ is the equilibrium relationship (2.2) and $\mathbf{R}$ is a 2-D second order tensor. The diagonal components of $\mathbf{R}$ have been empirically estimated while the cross components have so far been ignored. Further sources of non equilibrium may be associated with spatial variations of the Shields stress. They are usually neglected though their role may be significant for coarse sediments. In river morphodynamics temporal variations of the flow field are usually associated with flood propagation, a very slow process whose non equilibrium effects on sediment transport are also usually neglected.

Hence equation (2.4) is the form of the dynamic bed load transport relationship appropriate for uniform sediments with θ evaluated in terms of the local and instantaneous value (averaged over turbulence) of the bottom stress.

Sediment mixtures

Recently the description of sediment motion has been extended to the more realistic case of sediment mixtures characterized by a continuous spectrum of grain sizes. Space does not allow us to cover this subject. The interested reader is referred to the recent review by Parker (1991).

2.4 Suspended load

Particle entrainment into suspension requires a sufficiently strong turbulent intensity. The role played in particle pick up by coherent structures associated with the intense eruption events occurring in the near wall region is currently under investigation. In the engineering literature the condition for bottom particles to be suspended by a uniform stream is empirically based and reads:

$$\frac{u_*}{W_s} > N(Rp) \tag{2.5}$$

with u_* *friction velocity* of the stream, W_s particle *fall velocity* and $R_p \equiv \sqrt{(s-1)gd_s^3}/\nu$ *particle Reynolds number* and N is a function which is experimentally determined (see Van Rijn, 1984).

The classical theory of motion of suspended particles in a turbulent shear flow is based on early fundamental work by Barenblatt (1953a,b) and relies on the assumption that the size of suspended particles be small in comparison with the length scales of turbulence and the concentration c be small enough for particle interactions (both hydrodynamic and collisional) to be negligible. Under the further assumption that the Richardson number of the suspension be small, one is led to the so called *diffusion approximation* consisting of neglecting the effect of suspended particles on the dynamics of the flow. Though the above assumptions are not always completely justified in the context of river dynamics most of the engineering literature on the effect of transport in suspension on river morphodynamics is based on the above framework. Its output is the classical convection-diffusion equation for the mean volumetric concentration of particles (averaged over turbulence) c in the form

$$\frac{\partial c}{\partial t} + \frac{\partial c u_j}{\partial x_j} - W_s \frac{\partial c}{\partial x_3} = -\frac{\partial \overline{c' u_j'}}{\partial x_j} \tag{2.6}$$

where u_j is the $j-th$ component of the local fluid velocity averaged over turbulence, while c' and u'_j are turbulent fluctuations.

In order to complete the formulation one needs closure assumptions for the turbulent diffusion fluxes $\overline{u'_j c'}$, usually expressed in Fickian form, along with appropriate boundary conditions. In particular the issue of formulating an appropriate condition at the bottom raises the fundamental problem of modeling the interaction of the flowing suspension with the cohesionless bed. This is an extremely complex matter as in the vicinity of the bottom all the assumptions discussed above fail. In particular the concentration is often very high as to induce strong hydrodynamic and collisional interactions; furthermore there occurs a continuous exchange of particles between the suspension and the bottom as a result of particle deposition and particle entrainment. The process is even more complicated if small scale bedforms are present along the bed as, under these conditions, flow separates in the lee of the crests producing intense free shear layers which interact with the bed significantly enhancing the entrainment process.

Again the available experimental knowledge refers to the case of uniform suspensions. Under the latter circumstances the bottom condition may be written in the general form

$$c|_{x_3=\eta+a} = c_e\left(\frac{\theta}{\theta_c}, R_p\right) \tag{2.7}$$

$$\frac{a}{H} = f\left(\theta, \theta_c, \frac{d_s}{H}\right) \tag{2.8}$$

with H constant flow depth. The latter condition stipulates that the concentration must attain some equilibrium value c_e at some reference distance from the bottom a. Garcia & Parker (1991) have recently reviewed the available knowledge on the dependence of c_e and a on the flow and sediment parameters.

Under non-equilibrium conditions (2.7) and (2.8) are replaced by a statement whereby the flux of sediment normal to the bottom is assigned in terms of some entrainment coefficient α. If the flux at the bottom is nearly vertical this can be stated in the following form

$$\left[-\overline{u'_3 c'} - W_s c\right]_{x_3=\eta} = \alpha\left(c_e - c|_{x_3=\eta}\right) \tag{2.9}$$

A commonly used form of (2.9) (called *gradient boundary condition*) assumes α to be equal to W_s, which is equivalent to assuming that the entrainment flux $-\overline{u'_3 c'}$ equals its equilibrium value $(W_s c_e)$.

Finally the boundary condition to be imposed at the free surface stipulates that the normal component of sediment flux must vanish.

2.5 Coupling between fluid motion and sediment motion

The volumetric sediment discharge per unit width $\mathbf{q}$ consists of the bed load contribution $\mathbf{q}^b$ and the suspended load contribution $\mathbf{q}^s$ defined such that

$$q_i^s = \int_{\eta+a}^{h} (u_i c + \overline{u_i' c'})\, dx_3 \qquad (i = 1, 2) \tag{2.10}$$

where $h(x_1, x_2, t)$ denotes free surface elevation.

The evaluation of $\mathbf{q}$ involves the knowledge of the flow field which determines the bottom stress (hence the Shields stress θ) on which both $\mathbf{q}^b$ (see eqn. 2.4, c_e and a (see eqns. 2.7 and 2.8) depend. Furthermore the velocity field directly affects the convection of suspended sediment in (2.6) and the suspended load through (2.10).

The formulation of the general problem of morphodynamics is then completed through the equations and boundary conditions governing the hydrodynamics. The appropriate form of the flow equations to be employed in each particular context will mainly depend on the spatial scale of bottom perturbations being examined. In river morphodynamics the flow is invariably turbulent, hence the degree of complexity of the models ranges widely. The investigation of the formation and development of ripples, dunes, antidunes, and sand ridges has been based on 2-D (or 3-D) turbulent models, while the shallow water equations are usually sufficient to investigate bars and, with some quasi 3-D corrections, meandering. A 1-D de Saint-Venant equation is suitable to analyze morphodynamical changes at the spatial scale of the longitudinal profile of the river.

3 Small scale bedforms

Space does not allow us to review this topic. The interested reader is referred to Engelund & Fredsøe (1982) for the linear aspects. Various theoretical and experimental efforts have recently been directed to investigate finite amplitude effects on transverse bedforms. The recent works of Colombini (1993) and Colombini & Parker (1994) provide the first predictive theories of longitudinal befdforms. As for the effects of grain sorting the reader is referred to the recent review by Seminara (1994).

4 Large scale bedforms

The formation of river bars has been traditionally investigated with the aim of ascertaining the origin of meandering initiation. However recent developments (Blondeaux & Seminara, 1985) have led to the recognition that alternate bars are essentially *free* migrating modes, while meanders are associated with the *forced* quasi-steady bars induced by channel curvature. Free bars are reviewed in section 4.1 while forced bars are considered in section 4.2.

4.1 Free bars

Formulation

Free bars develop as a result of an instability of uniform turbulent flow in a straight channel with erodible bottom and non erodible banks, to three-dimensional perturbations in the form of sand waves characterized by oblique alternating fronts and wavelengths which scale by the channel width ($2B^*$). (Hereafter a star apex will denote dimensional quantities). Stability analyses have been formulated based on two main assumptions, namely: i) the channel is wide enough for side wall effects to be negligible outside relatively ineffective side boundary layers, ii) flow separation downstream the oblique bar fronts does not play a major role.

Since bed develops on a time scale much larger than that characteristic of flow adjustment to the deforming boundary, the mathematical problem is readily formulated in the 2-D form

$$UU_{,s} + VU_{,n} = -h_{,s} - \beta\frac{\tau_s}{H} \tag{4.1}$$

$$VV_{,n} + UV_{,s} = -h_{,n} - \beta\frac{\tau_n}{H} \tag{4.2}$$

$$(VH)_{,n} + (UH)_{,s} = 0 \tag{4.3}$$

$$\eta_{,t} + Q_o(q_{n,n} + q_{s,s}) = 0 \tag{4.4}$$

$$V = q_n = 0 \quad n = \pm 1 \tag{4.5a, b}$$

where (U, V) are depth-averaged axial and transverse velocity components, (τ_s, τ_n) are bottom stresses, h is water surface elevation, H is local flow depth, (q_s, q_n) are sediment transport rates. Furthermore three dimensionless parameters arise, namely β, Q_o and F_o, defined in the form

$$\beta = \frac{B^*}{H_o^*} \quad F_o = \frac{U_o^*}{\sqrt{gH_o^*}} \quad Q_o = \frac{[(s-1)gd_s^{*3}]^{1/2}}{(1-n)H_o^*U_o^*} \ . \tag{4.6a - c}$$

$\rightarrow$Variables have been made dimensionless in the form

$$(U^*, V^*) = U_o^*(U, V) \qquad (h^*, H^*, \eta^*) = H_o^*(F_o^2 h, H, \eta)$$
$$(s^*, n^*) = B^*(s, n) \qquad (\tau_s^*, \tau_n^*) = \varrho U_o^*(\tau_s, \tau_n) \tag{4.7}$$
$$(q_s^*, q_n^*) = \left\{(s-1)gd_s^{*3}\right\}(q_s, q_n) \qquad t^* = \frac{B^*}{U_o^*}t$$

with H_o^* and U_o^* undisturbed uniform flow depth and average speed. Moreover the sediment has been assumed to be uniform.

It should be noticed that Q_o is a small parameter, ranging about 10^{-3} which confirms that η develops slowly relative to the flow field. The aspect ratio β is the *crucial parameter* controlling the development of bar instability. Two further parameters arise in the 'constitutive' relationships for bottom stresses and sediment transport rates which must be associated with (4.1-5) to close the formulation of the problem, namely Shields stress θ and relative roughness $d_s (\equiv d_s^*/H^*)$ which determines the local value of the friction coefficient C_f appearing in the bottom stress closure:

$$\boldsymbol{\tau} \equiv (\tau_s, \tau_n) = C_f(U, V)\sqrt{U^2 + V^2} \tag{4.8}$$

$$C_f = 6 + 2.5 \ln(H^*/2.5 d_s^*) \tag{4.9}$$

However notice that both F_o and Q_o may be expressed in terms of θ and d_s.

Linear stability

The stability problem may then be posed in the following form: for given undisturbed flow and sediment transport properties, namely for given values of θ_o and d_{so}, is there a threshold value of the aspect ratio β above which free bars grow? The answer to this question was given through a series of papers published in the seventies and early eighties (see Seminara & Tubino, 1989 for a review) pointing out various competing effects. The basic idea is to perform a classical normal mode analysis setting

$$(U, V, h, H) = (1, 0, h_o, 1) + (S_m u_1, C_m v_1, S_m h_1, S_m H_1) E_1 \tag{4.10}$$

where

$$E_n = \exp ni(\lambda s - \omega t) \qquad S_m = \sin\left(\frac{\pi}{2} mn\right) \qquad C_m = \cos\left(\frac{\pi}{2} mn\right) \tag{4.11}$$

In (4.10) λ and ω are wavenumber and complex growth rate of the bar perturbation while the order of the mode m gives the number of rows characterizing it.

With the help of (4.10), (4.11) the mathematical problem can be linearized and leads to a dispersion relationship which allows to draw the following conclusions (Colombini, Seminara & Tubino, 1987) referring to the case of dominant bedload transport:

i) For given θ_o and d_{so} infinitesimal single row bars ($m = 1$) grow for values of β exceeding a critical value $\beta_{c1}(\theta_o, d_{so})$

ii) The critical value β_{cm} for multiple row bars satisfies the simple condition

$$\beta_{cm} = m\beta_{c1} \tag{4.12}$$

iii) Bar perturbations migrate downstream at critical conditions.

Comparison of theoretical predictions with observations regarding single row bars is fairly satisfactory (see also Garcia & Niño, 1993).

The physical mechanism underlying bar growth is somewhat similar to that controlling dune growth: the destabilizing role is played by bottom friction which makes the longitudinal sediment transport lag behind bottom elevation, while the stabilizing effect is associated with the role of transverse sediment transport which is significantly affected by gravity. The latter gives rise to a diffusion type stabilizing contribution in the sediment continuity equation which increases as the order m of the bar mode increases.

Finite amplitude effects

A weakly nonlinear stability analysis of finite amplitude bars was first developed by Colombini & al. (1987). The latter authors restricted their attention to the growth of monochromatic waves characterized by the critical wavenumber λ_c in a neighbourhood of the threshold condition for bar growth. The classical multiple scale type expansion was employed. Denoting by $A(T)$ the amplitude function describing the growth of bars on the *slow* timescale described by the *slow* timevariable T defined as ϵt with ϵ small parameter measuring the distance of the parameter β from its critical value β_c, the output of the analysis is a classical Landau type amplitude equation for $A(T)$.

Colombini & al. (1987) showed that the latter describes a supercritical bifurcation and were able to derive an expression for the maximum bar height which compared satisfactorily with experimental observations even for values of ϵ of order one. Further interesting feature emerged from the finite amplitude solution, namely its tendency to form diagonal fronts and to give rise to a nonlinear reduction of the bar wavespeed.

Recently Schielen, Doelman & de Swart (1993) have extended the above analysis to allow for the growth of the whole group of waves in a narrow spectrum centred around the critical wavenumber. The technique used goes back to similar analyses performed in a variety of other nonlinear instability phenomena and leads to an amplitude equation of →Ginzburg-Landau type. The latter admits of periodic solutions as found by Colombini & al. (1987). However conditions exist such that the latter solutions are unstable as shown by Stuart & Di Prima (1978) in a different context. Numerical solutions showed that, provided the channel length exceeds some critical value (of the order of few hundred channel widths) quasi-periodic bed patterns may arise with 2 or 3 fundamental frequencies. No chaotic solutions were found under realistic conditions.

Strongly nonlinear effects

As the bifurcation parameter β attains values much larger than β_c strongly nonlinear effects may be expected which are coupled with the destabilization of higher order modes ($m > 1$). Some investigation of the latter effects was performed by Colombini & Tubino (1991) who obtained fully nonlinear periodic solutions of the problem by means of spectral methods. As expected the process was described whereby, as β increases, the bar topography shifts towards a 'central bar' configuration (fig. 5). The occurrence of transition from the single-alternating bar mode to the central-bar mode was also found to depend on the initial conditions.

Unsteady effects

All the above works were based on assuming the basic state to be steady. Such assumption is acceptable and instructive as a first step in the analysis. However, particularly in gravel bed rivers, the formative conditions of alternate bars are typically

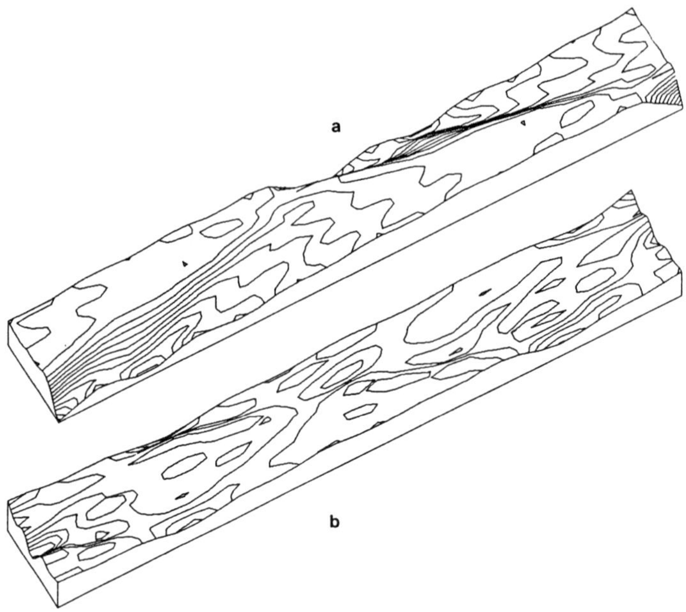

Figure 5: Strongly nonlinear bar topography predicted by Colombini & Tubino (1991). a) $\beta = 12$ (alternate bar type) b) $\beta = 30$ (central bar type).

unsteady as they develop at flood stage and undergo minor variations at lower stage.

A modification of Colombini & al.'s (1987) theory to account for weak unsteady effects was proposed by Tubino (1991) and leads to the following picture of bar growth: bars undergo a quasilinear growth during the rising stage of the flood, then a modest decay if the depth reaches values such that $\beta(T) < \beta_c$, and finally a growth dominated by nonlinear effects during the longer falling stage which is the formative part of the flood. The assumptions of weak nonlinearities and unsteadiness are not wholly met in practice; however the above qualitative behaviour is likely to occur also under conditions of less weak nonlinearity and unsteadiness.

The fundamental assumption that unsteadiness is characterized by the same time scale as bar growth can be stated in the form

$$U = \frac{B^*}{\epsilon \bar{U}_o^* T_f \alpha_{1r}} \sim O(1) \tag{4.13}$$

where T_f is the characteristic period of the flood and α_{1r} is the real part of α_1. If $U \gg 1$ bar development is much slower than the time evolution of the basic flow which is unable to affect it. If $U \ll 1$ bars develop on a time scale much shorter than that of the basic unsteadiness hence they are able to reach their steady *equilibrium amplitudes* corresponding to the instantaneous flow characteristics. In nature usually $U \sim O(1)$.

Curvature effects

Rivers are seldom straight and often meander. It has been observed (Kinoshita & Miwa, 1974) that curvature tends to suppress free migrating bars. The latter effect was explained theoretically by Tubino & Seminara (1990) in terms of nonlinear interactions free-forced bars (see sect. 4.2).

Effects of suspended load

The effect of suspended load on bar formation has been treated by Fredsøe (1978), Kuroki (1988) and, more recently, by Watanabe & Tubino (1992). The latter authors show that suspended load is stabilizing for large values of θ and leads to increasing bar wavelengths. These results disagree with those found by Fredsøe (1978) and Kuroki (1988). The reason of discrepancy is likely to be due to the treatment of suspended load which is quite sensitive to the boundary condition adopted in the analysis.

4.2 Forced bars

Spatial variations of channel alignment give rise to *curvature induced secondary flows* which tend to move sediment in the transverse direction leading to the establishement of zones of sediment deposition called *forced bars.* The development of spatial variations of bed elevation gives then rise to *topographically induced secondary flows* similar to those characteristic of free bars.

The simplest forced bar: fully developed flow and bed topography in constant curvature channels

It has long been recognized (Rozovskij, 1957) that in uniformly curved channels circular streamlines are impossible as centrifugal effects increase from the bottom to the free surface while the transverse pressure gradient is practically uniform in the vertical direction. Such unbalance requires the establishment of an outward circulation close to the surface and an inward one close to the bottom. As a result particles tend to be displaced from the outer towards the inner bank until a uniform transverse slope is established such that the transverse bottom stress acting on the particle is balanced by the downslope component of gravity. Cross sectional deepening then pushes the thread of high velocity towards the outside bank. Notice that if the bed were fixed the thread of high velocity would be initially located along the inner bank due to the 'metric' increase of longitudinal slope from the outside to the inside bank.

Ignoring the side wall layers (an assumption appropriate for the case of shallow bends) and assuming that, as it is common in nature, curvature effects are *weak*, namely

$$\nu \equiv \frac{B^*}{R^*} \ll 1 \tag{4.14}$$

with R^* constant curvature of the channel axis, the leading order fully developed solution for the flow field-bed topography may be written in the form

$$(u, h, H) \quad = \quad (u_o(z), h_o, 1) + \nu(u_1(z), h_1, H_1)n + O(\nu^2) \tag{4.15}$$

$$v = \nu v_1(z) + O(\nu^2) \tag{4.16}$$

where quantities have been made dimensionless according to (4.7), (u, v) are local (mean) velocities s, n, z are orthogonal curvilinear dimensionless coordinates with s longitudinal, n transverse and z vertical.

Notice that, the flow being fully developed in the longitudinal direction, the distribution $v_1(z)$ must satisfy the condition of vanishing transverse flux. With the help of simple closures various authors (see in particular →Kalkwijk & de Vriend, 1980) were able to obtain the structure of flow and bottom perturbations (u_1, h_1, H_1, v_1) from linearized forms of Reynolds and sediment continuity equations.

Forced bars in weakly meandering channels

Channel alignment of meandering rivers can often be described by the so called *sine generated curve* (Langbein & Leopold, 1964), i.e. a line the curvature of which exhibits a sinusoidal dependence on the longitudinal coordinate. It has thus become customary to investigate the latter configuration as a prototype of the regular sequences of meanders often encountered in nature. Let us then define the dimensionless curvature of channel axis $\mathcal{C}$ in the form:

$$\mathcal{C}(s) = \nu(e_1 + c.c.) \tag{4.17}$$

where

$$e_1 = \exp i\lambda s \qquad \nu = \frac{B^*}{R_o^*} \ll 1 \tag{4.18a, b}$$

Hence $R_o^*/2$ is the radius of curvature at the bend apex.

The mathematical formulation of the flow problem in this case must account for the role played by the curvature induced secondary flow described above. On the other hand it would be desirable to formulate the problem in terms of the depth averaged components of the velocity vector (U, V). The shallow water equations written in the curvilinear coordinates s and n accomplish the latter goal but not the former one. Kalkwijk & de Vriend (1980) have suggested a procedure able to overcome this deficiency. The idea is to decompose the local velocity (u, v) in the form

$$u = u_o(z)U(s, n) \tag{4.19}$$

$$v = u_o(z)V(s, n) + \nu[v_1(z)e_1 + c.c.]U(s, n) \tag{4.20}$$

where $v_1(z)$ is the transverse component with zero depth-average associated with the centrifugally induced secondary flow. This 'local' assumption is an approximate way to keep memory of the 3-D centrifugally induced secondary flow within a 2-D formulation of the problem. Depth averaging the 3-D Reynolds equations with the help of →(4.19) and (4.20) one ends up with a differential problem identical with the free bar problem except for new forcing terms associated with curvature, directly through the metric coefficients of the curvilinear coordinate system or indirectly through the depth-averaging procedure accounting for the curvature dependent secondary flow.

The assumption of weak channel curvature then suggests to set up the following regular expansion

$$(U, V, h, H) = (1, 0, h_o, 1) + \nu[(U_1, V_1, h_1, H_1)e_1 + c.c.] + O(\nu^2) \tag{4.21}$$

Substituting (4.21) into the governing differential problem one finds

$$L \begin{bmatrix} U_1 \\ V_1 \\ h_1 \\ H_1 \end{bmatrix} = \begin{bmatrix} -n\beta C_{fo} \\ 1 - i\lambda_m k_1 - \beta C_{fo} k_4 \\ 0 \\ 0 \end{bmatrix} \tag{4.22}$$

$$V_1|_{n=\pm 1} = 0 \tag{4.23}$$

$$(F_o^2 h_1 - H_1)_{,n}|_{n=\pm 1} = k_4 \tag{4.24}$$

where L is the linear homogeneous ordinary differential operator governing free bars except for λ_m replacing λ and for the free bar growth rate vanishing in (4.22). Furthermore k_1 and k_4 are redistribution parameters originating from the effect of the secondary flow with zero depth-average on the average longitudinal and trasverse momentum transport.

The latter problem has been solved in closed form by Blondeaux & Seminara (1985). (See also Seminara & Tubino (1986) for the case when suspended load is significant). The solution allows one to determine the topographic pattern of forced bars. The main feature of the latter is a phase shift of the peak of the forced bar relative to the bend apex. Such shift depends on meander wavenumber (fig. 6).

Resonance

The structure of the linear problem immediately suggests that resonance may be induced by curvature provided the forcing wavenumber λ_m and the aspect ratio β coincide with the values λ_R, β_R characterizing, for given θ_o and d_{so} marginal free bars with vanishing migration speed. Such conditions do indeed occur as shown by Blondeaux & Seminara (1985). A peak in the response of flow and bed topography to the forcing effect of curvature occurs for values of λ and β falling in a neighbourhood of (λ_R, β_R). In other words weakly meandering channels behave like a linear resonator leading to an infinite response if the resonance conditions are exactly met. Such conditions are dependent on the mode and intensity of sediment transport and on bottom roughness. Of course the latter conditions are fairly ideal as, in nature, one does never observe an infinite sequence of identical meanders. However the theoretical existence of resonance conditions is of great interest to understand the possible mechanisms of meandering initiation and development.

Near resonant meanders: weakly nonlinear effects

Seminara & Tubino (1992) have theoretically investigated finite amplitude effects on near resonant meanders. A weakly nonlinear expansion for the solution in powers of $\nu^{1/3}$ was set up whithin an $O(\nu^{2/3})$ neighbourhood of the resonant conditions.

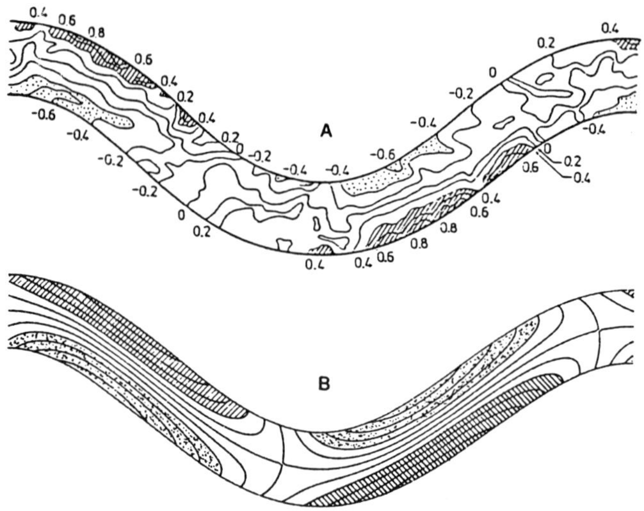

Figure 6: Forced bar topography predicted by Seminara & Tubino (1986) compared with Ikeda & Nishimura's (1985) experimental findings ($\lambda = .31$, $\beta = 2.8$, $\theta = .22$, $d_s = .0028$, $R_p = 7.5$, $\nu = 0.1$, contour intervalrs = .2).

Substituting this expansion in the nonlinear differential problem led at $O(\nu)$ to a non-homogeneous problem the solvability of which imposed the following amplitude equation for the amplitude A of the fundamental

$$(\beta_1 a_1^R + i\lambda_1 a_o^R)A + a_2^R A^2 \bar{A} + a_3^R = 0 \tag{4.25}$$

with $a_o^R, a_1^R, a_2^R, a_3^R$ complex coefficients. The equation (4.25) may be solved for $|A|$ and in general exhibits one real and two (physically meaningless) complex conjugate solutions. Narrow ranges of λ may however exist where three real solutions are possible. Seminara & Tubino (1992) showed that the upper branch of the solution is always stable, the lower branch is invariably unstable and bifurcates into a time periodic solution, while the loop joining upper and lower branches is invariably unstable and bifurcates into the steady upper branch solution.

Comparison with the experimental results of Colombini, Tubino & Whiting (1990) was fairly good as regards the amplitude of the fundamental component of the bed elevation, though the response of the channel exhibited →a harmonic content richer than that predicted by the weakly nonlinear analysis.

Recently →(S, T & W, 1992) obtained a fully nonlinear numerical solution of the problem by means of finite difference approximation. A comparison with the experimental results of Colombini & al. (1992) both for the maximum scour and for the harmonic content of bottom elevation was satisfactory.

Suppression of free bars

In an important paper (Kinoshita & Miwa, 1974), it was experimentally shown that, for a given set of flow and sediment characteristics a critical 'sinuosity' (say ν_c for a sequence of sine generated meanders) exists above which free bars are unable to grow and migrate. Tubino & Seminara (1990) provided some explanation of this phenomenon in terms of the free-forced interactions of a critical *free* bar with the *forced* bars induced by curvature, an interaction significant if

$$\nu = n\epsilon^{1/2} \qquad n \sim O(1) \tag{4.26}$$

where ϵ is the small parameter $(\beta - \beta_c)/\beta_c$.

The above analysis restricts the attention to free bars at critical conditions. A more general approach able to investigate the stability of flow in weakly meandering channels with respect to *free bars of arbitrary wavelength* has not been performed yet. A mathematical model which may be relevant to the latter problem has been analyzed by Schielen & Doelman (1994).

5 Planimetric forms

Rivers evolve planimetrically through the process of bank erosion. We will concentrate our attention on meandering rivers though in braided rivers the main process of lateral shift of the channels composing the network is also controlled by bank erosion. The latter process is extremely complex and far from being understood while some progress has been made in the understanding of meander development.

5.1 Planimetric evolution equation of meandering rivers

The process of meander formation has been long associated with the development of free bars in the originally straight channel (*bar theory*). However this viewpoint is disputable as free bars often migrate at a speed much larger than that of meander migration. An alternative viewpoint was taken by Ikeda, Parker & Sawai (1981) who investigated the stability of a straight channel configuration to planimetric perturbations of the channel axis (*bend theory*). Though the latter approach was still fairly idealized as flow was assumed steady and the initial channel infinitely long and straight, however the basic idea, namely associating meandering initiation with a planimetric instability appears to be sound and conceptually appealing.

In order to proceed in this direction we need a planimetric evolution equation. Let ζ^* be the speed of transverse migration of channel axis (a quantity which takes values around metres/year). Experimental observations (Hasegawa, 1989) suggest that it is reasonable to assume ζ^* proportional to a measure of the perturbation of the longitudinal speed at the bank. In other words in dimensionless form we may set

$$\zeta = e(u|_{n=-1} - u|_{n=1}) \tag{5.1}$$

with e dimensionless effective erosion coefficient ($e \sim O(10^{-7})$) which describes in a continuous fashion the bank erosion process smoothing out its intermittency associated with flood events. Notice that any transversely symmetric components of the perturbation, which would lead to channel widening but would not produce any lateral shift of the channel, do not affect ζ as defined in (5.1).

The statement (5.1) can be transformed into a planimetric evolution equation. This was first done by Parker, Sawai & Ikeda (1982) who derived an integro-differential equation in cartesian coordinates based on the flow model of Ikeda & al. (1981). Seminara, Tubino & Zardi (1994a) have recently proposed an alternative approach which allows one to overcome some limitations of the original model of Ikeda & al. (1981) (see Blondeaux & Seminara, 1985 and Johannesson & Parker, 1989) and is readily extended to the analysis of finite amplitude meander development.

Let $s(s_o, t)$ be the curvilinear coordinate which identifies the location at time t of the cross section that at time $t = 0$ was located at s_o. Furthermore let $\vartheta(s,t)$ be the angle that the local tangent to the channel axis forms with some chosen direction in the flood plain.

The evolution equation in integro-differential form for $\vartheta(s,t)$ derived by Seminara & al. (1994a) reads

$$\frac{\partial \theta}{\partial t} - \theta_{,s} \int_o^s \zeta \theta_{,s} ds = \zeta_{,s} \tag{5.2}$$

with ζ given by (5.1). The knowledge of the solution for the flow field in the channel reach considered is required to close (5.1) and (5.2). Equation (5.2) contains a nonlinear integro differential term which accounts for the *history* of the process of straining of the channel axis due to the erosion mechanism. The latter is intrinsically nonlinear on pure geometric ground as pointed out by Parker & al. (1982).

5.2 Meandering initiation

Consider the ideal case of an infinitely long channel with constant width and uniform sediment characteristics and investigate the stability of an initially uniform and stable configuration to infinitesimal harmonic perturbations of channel axis. Let us then assume

$$\vartheta_1 = |\vartheta_1(t)| \exp i\lambda s \tag{5.3}$$

with $|\vartheta_1| \ll 1$. The linearized form of (5.2) may then be readily reduced to the form

$$\frac{d\vartheta_1}{dt} = G\vartheta_1 \tag{5.4}$$

which admits of exponential solutions for ϑ_1 with complex growth rate G, a quantity depending on meander wavenumber λ_m for given flow and sediment characteristics. This quantity may be obtained using the solution of Blondeaux & Seminara (1985) for the flow field in weakly meandering channels. The typical behaviour of $Re(G)$ and $Im(G)$ under non resonant conditions shows that meanders grow and migrate downstream if λ_m does not exceed a threshold value λ_{mc}.

Notice that (5.3) is an exact linear solution of the planimetric evolution equation. This shows that Langbein & Leopold's (1964) *sine generated curve is not simply an empirical correlation but rather the necessary form that a periodic sequence of meanders must take* at least within the context of the present theoretical scheme.

It also emerges that bend instability selects the resonant wavenumber as the most unstable perturbation for given flow and sediment characteristics.

The above results rely on several simplifying approximations. In particular the presence of free alternating bars in the initially straight channel may affect the selection mechanism in various ways as yet unexplored. Furthermore the initial configuration from which the actual patterns observed in nature originates is not necessarily straight and usually unknown, while previous configurations may have evolved into the observed one through sequences of cutoffs.

A related viewpoint taken by Struiksma & al. (1986) associated meandering initiation with the occurrence of spatially growing steady bars developing from an initial disturbance in space.

5.3 Nonlinear planimetric development of meandering channels to cutoff

Kinoshita (1961) pointed out that mature meandering rivers can often be given a planimetric description slightly modified with respect to that proposed by Langbein & Leopold (1964), namely:

$$\mathcal{C}^* = \mathcal{C}_o^*[\cos\lambda_s - C_F\cos 3\lambda_s - C_S \sin 3 \sin 3\lambda_s] \tag{5.5}$$

where $\mathcal{C}_s$ and $\mathcal{C}_f$ are dimensionless coefficient of *skewing* and *fattening* which reproduce characteristic features of natural meanders. This simple fundamental observations pose some interesting questions for theoretical investigation. Firstly, why does (5.5) not contain even harmonics? And why are higher harmonics not significant? Furthermore can we predict the structure (5.5) or is it a purely empirical correlation?

These questions have already received some answers in a series of papers of Parker and his school (Parker & al., 1982). Recently Seminara & al. (1994b) have derived a solution of the evolution equation (5.2) coupled with the weakly nonlinear model for the flow field of Seminara & Tubino (1992) which provides a fairly complete answer to the above questions.

In fact the general periodic solution of (5.2) may be written in the form:

$$\vartheta = \sum_{k=1} \vartheta_k(t)\exp[i(2k-1)\lambda_k(t)s] \tag{5.6}$$

with

$$\lambda_k = (2k-1)\lambda(t) \tag{5.7}$$

and $\lambda(t)$ time-dependent wavenumber which accounts for the continuous straining of channel axis. Notice that only odd harmonics are present in (5.6) as a result of the cubic nonlinearity of (5.2). Furthermore, using the model for the near resonant flow field developed by Seminara & Tubino (1992) and carried to higher order by Seminara & al. (1994b) one readily finds

$$\zeta = (\zeta_{1,1/3}|\vartheta_1|^{1/3} + \zeta_{11}|\vartheta_1|)e_1^R + (\zeta_{31}|\vartheta_1| + \zeta_{33}|\vartheta_3|)e_3^R + c.c. \tag{5.8}$$

Substituting from (5.5, 6, 7, 8) into (5.2) we end up with a nonlinear system of ordinary differential equations for $\vartheta_k(t)(k = 1, 2, \ldots)$ and $\lambda(t)$. Attention can be restricted to the first and third harmonics as it will appear that higher order components cannot develop due to the occurrence of meander 'cutoff'. Seminara & al. (1994a) solved numerically the system for ϑ_1, ϑ_3 and $\lambda(t)$.

Their results show that the planimetric evolution starting from a sequence of small amplitude *near resonant* meanders is characterized by distinct stages (fig. 7).

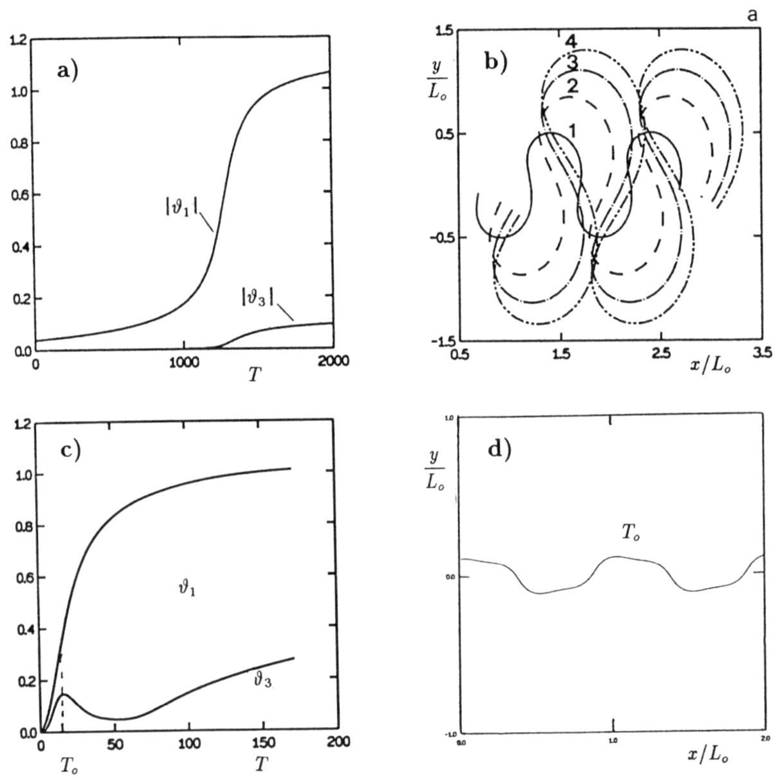

Figure 7: The planimetric evolution of non near-resonant meanders (a,b) from incipient formation to cutoff is compared with the near-resonant case (c,d). ($\theta = 0.2$, $\beta = 10$, $d_s = 0.001$). T is a dimensionless time and L is the cartesian meander wavelength.

i) an initial stage dominated by near resonant flow nonlinearities where both the first and the third harmonics grow rapidly with significant downstream migration;

ii) an intermediate stage dominated by the linear growth of the fundamental with the third harmonics keeping fairly constant;

iii) a final stage proceeding till cutoff where the growth of the fundamental is slowed down, the third harmonics undergo a growth controlled by geometric non linearities and a rapid reduction of migration speed is observed.

Stage i) which is characterized by a degree of fattening and skewing more pronounced than stage iii) appears to explain the shape of confined meanders like those of Beaver river (fig. 3), while stage iii) applies to large amplitude unconfined meanders.

5.4 Braided rivers

In spite of the great interest of the geomorphological community (Best & Bristow, 1993) very little is as yet known on the mechanics of braided rivers. Research developed in the seventies concentrated on predictions of their occurrence. The basic idea underlying most of the analyses was that of associating braiding with a multiple bar pattern. Hence theories of formation of free bars were set as the basis of morphological criteria (see Fredsøe (1978)). Though the importance and the relative success of the latter theories should not be undermined however they appear to be obviously relevant to understand what might be called *multiple bar braiding* rather than *multiple channel braiding*. The former case is that of rivers which display an apparently braiding character only at low stage: they are such that the pattern observed at low stage is generated by an *altimetric* instability of the *free bar* type occurring at high stage when the cross section is completely submerged. Multiple channel braiding is characteristic of rivers such that the distinction among different branches does not disappear even at high stage. In this case the pattern may more appropriately be considered as an interconnected network of channels.

Braiding phenomena display a degree of complexity which is typical of systems characterized by a large number of degrees of freedom. This has recently led some researchers to try and make use of tools developed for the analysis of complex systems. The interesting achievements of Murray and Paola (1994) in spite of the relative simplicity of their cellular model appears to suggest that this may turn out to be the right avenue to the general understanding of such complex patterns. Descriptive tools, like techniques able to analyze the fractal properties of river networks, have also been recently employed though they do not seem to add much to the physics of the processes controlling the mechanics of braiding.

Acknowledgements I wish to thank M. Tubino for some useful comments on a first draft of the present review.

References

Allen, J R L (1984) *Sedimentary structures*, Elsevier.

Barenblatt, G (1953a) Motion of suspended particles in a turbulent flow, *Prikl. Matem. Mekh.*, 17(3), 261-274.

Barenblatt, G (1953b) Motion of suspended particles in a turbulent flow occupying a half-space or plane channel of finite depth, *Prikl. Matem. Mekh.*, 19(1), 61-68.

Best, J L and Bristow, C S Eds. (1993) *Braided rivers*, Geol. Soc. London.

Blondeaux, P and Seminara, G (1985) A unified bar-bend theory of river meanders, *J. Fluid Mech.*, 157, 449-470.

Brownlie, W R (1981) Prediction of flow depth and sediment discharge in open channels, *Report No. Kh-R-43A*, Keck Laboratory, California Institute of Technology.

Colombini, M (1993) Turbulence-driven secondary flows and formation of sand ridges, *J. Fluid Mech.*, 254, 701-719.

Colombini, M and Tubino, M (1991) Finite amplitude free-bars: a fully nonlinear spectral solution, Euromech 262 Colloquium on *Sand Transport in Rivers, Estuaries and the Sea*, Wallingford 26-29 June 1990, A.A. Balkema, 163.

Colombini, M and Parker, G (1994) Longitudinal streaks, submitted for publication on *J. of Fluid Mech.*.

Colombini, M, Seminara, G and Tubino, M (1987) Finite amplitude alternate bars, *J. Fluid Mech.*, 181, 213-232, 1987.

Colombini, M, Tubino, M and Whiting, P (1992) Topographic expression of bars in meandering channels, *Dynamics of Gravel-bed Rivers*, Firenze, 24-29 settembre 1990, J. Wiley, Chichester, 457.

Engelund F and Fredsøe, J (1982) Sediment ripples and dunes, *Ann. Rev. of Fluid Mech.*, 14, 13-37.

Exner, F M (1925) Uber die wechselwirkung zwischen wasser und geschiebe in flussen, it Sitzer Akad. Wiss., Wien, 165-180.

Fredsøe, J (1978) Meandering and braiding of rivers, *J. Fluid Mech.*, 74, 609-624, 1978.

Garcia, M and Niño, Y (1993) Dynamics of sediment bars in straight and meandering channels: experiments on the resonance phenomenon, *J. Hydraul. Res.*, 31(6), 739-761.

Garcia, M and Parker, G (1991) Entrainment of bed sediment into suspension, *J. of Hydraul. Eng. ASCE*, 117(4), 414-435.

Hasegawa, K (1989) Studies on qualitative and quantitative prediction of meander channel shift, in *River Meandering*, (eds. S. Ikeda & G. Parker), AGU, Water Resources Monograph, 12.

Ikeda, S and Nishimura, T (1986), Flow and bed profle in meandering sand-silt rivers, *J. Hydraul. Eng. ASCE*, 112(7), 562-579.

Ikeda, S, Parker, G and Sawai, K (1981) Bend theory of river meanders, Part 1. Linear development, *J. Fluid Mech.*, 112, 363-377.

Jaeggi, T (1984) Formation and effects of alternate bars, *J. Hydraul. Eng. ASCE*, 110(2), 142-156.

Johannesson, H and Parker, G (1989) Linear theory of river meanders, in *River Meandering* (eds. S.Ikeda & G.Parker), AGU, Water Resources Monograph, 12, 181-213.

Kalkwijk, J P Th and de Vriend, H J (1980) Computation of the flow in shallow river bends, *J. Hydr. Res.*, 18(4), 327-342.

Kinoshita, R (1961) Investigation on channel deformation in Ishikari River, *Rpt. Bureau of Resources*, Dept. Science and Technology, Japan, 1-174.

Kinoshita, R and Miwa, H (1974) River channel formation which prevents downstream translation of transverse bars, *Shinsabo 94*, 12-17 (in Japanese).

Kuroki, M (1988) *Study on flow and bedforms of rivers in alluvial*, Ph.D. Dissertation, Hokkaido University (in Japanese).

Langbein, W B and Leopold, L B (1964) Quasi-equilibrium states in channel morphology, *Amer. J. of Science*, 262, 782-794.

Murray, B and Paola, C (1994) A cellular model of braided rivers, *Nature*, 371, September 1st, 54-57.

Parker, G (1991) Some random notes on grain sorting, *Proc. Grain Sorting Seminar, IAHR*, Ascona, Switzerland, 20-76.

Parker, G, Sawai, K and Ikeda, S (1982) Bend theory of river manders. Part. 2. Nonlinear deformation of finite-amplitude bends. *J. Fluid Mech.*, 115, 303-314.

Rozovskij, I L (1957) Flow of water in bends of open channels, Kiew: *Acad. Sci. Ukranian SSR.*

Schielen, R and Doelman, A (1994) Nonlinear stability analysis in unbounded, weakly oscillating, domains, Paper presented at *ICPF'94*, Noordwijkerhout, July 4-7,

Schielen, R, Doelman, A and de Swart, H (1993) On the nonlinear dynamics of free bars in straight channels, *J. Fluid. Mech.*, 252, 325-356.

Sekine, M and Kikkawa, H (1992) Mechanics of saltating grains. II, *J. of Hydraul. Eng. ASCE*, 118(4), 536-558.

Seminara, G (1994) The effect of grain sorting on the formation of bedforms, Invited paper at Euromech 310 on *Sediment Transport Mechanism in Coastal Environment and Rivers*, To appear in Appl. Mech. Review Journal.

Seminara, G and Tubino, M (1986) Further results on the effects of transport in suspension on flow in weakly meandering channels, Colloquium on *The Dynamics of Alluvial Rivers*, Genova, June.

Seminara, G and Tubino, M (1992) Weakly nonlinear theory of regular meanders, *J. Fluid Mech.*, 244.

Seminara, G and Tubino, M (1989) Alternate bars and meandering: free, forced and mixed interactions, in *River Meandering*, (eds. S.Ikeda & G.Parker), AGU, Water Resources Monograph, 12.

Seminara, G, Tubino, M and Zardi, D (1994a) Evoluzione planimetrica dei corsi d'acqua meandriformi dall'incipiente formazione al cutoff, *XXIV Convegno di Idraulica e Costr. Idrauliche*, Napoli.

Seminara, G, Tubino, M and Zardi, D (1994b) Flow and geometric nonlinearities in meander development from incipient formation to cutoff, To be submitted for publication on JFM.

Shimizu, Y, Tubino, M and Watanabe, Y (1992) Numerical calculation of bed deformation in a range of resonant wavenumber of meandering channel, *Proc. of Hydr. Eng. JSCE*, 36, (in Japanese).

Stuart, J T and Di Prima, R C (1978) The Eckhaus and Benjamin-Feir resonance mechanisms, *Proc. Roy. Soc. London* Ser. A 362, 27-41.

Struiksma, N, Olesen, K W, Flokstra, C and de Vriend, H J (1986) Bed deformation in curved alluvial channels, *J. Hydr. Res.*, 23(1), 57-79.

Takahashi, T (1981) Debris flow, *Ann. Rev. Fluid Mech.*, 13, 57-77.

Tubino, M (1991) Growth of alternate bars in unsteady flow, *Water Res. Research*, 27(1), 37-52.

Tubino, M and Seminara, G (1990) Free-forced interactions in developing meanders and suppression of free bars, *J. Fluid Mech.*, 214.

Van Rijn, L C (1984) Sediment transport. Part II. Suspended load transport, *J. Hydraul. Eng. ASCE* 110(11), 1613-1641.

Watanabe, Y and Tubino, M (1992) Influence of bed load and suspended load on alternate bars, *Proc. of Hydr. Eng. JSCE*, 36.

Periodic orbits in singularly-perturbed systems

C. SOTO-TREVIÑO and T.J. KAPER
Department of Mathematics, Boston University, Boston, MA 02215 U.S.A.

Abstract

In this paper, we discuss a class of finite-dimensional singularly-perturbed ordinary differential equations. We present a technique that simultaneously yields the existence and local uniqueness of periodic orbits, as well as detailed information about their location in phase space and their stability type. In order to illustrate the method, and due to space constraints, we focus here on two simple example problems. The orbits include resonant multiple-pulse subharmonics and multiple-pulse periodic orbits, which have finitely many rapid transition layers in between successive passages (either long or short) near slow manifolds. The results for general systems, including systems with small or large dissipation, will be presented in [11]. One of the main technical tools is version of the Exchange Lemma with Exponentially Small Error, developed recently in [8], modified to treat the case of periodic orbits.

1 Introduction

In this work, we are concerned with finite-dimensional systems of singularly perturbed ordinary differential equations of the form:

$$\begin{aligned}\dot{\mathbf{x}} &= \mathbf{f}(\mathbf{x},\mathbf{z},\varepsilon),\\ \dot{\mathbf{z}} &= \varepsilon\mathbf{g}(\mathbf{x},\mathbf{z},\varepsilon),\end{aligned} \tag{1.1}$$

where $0<\varepsilon\ll 1$, $\mathbf{x}\in\mathbb{R}^{\mathbf{k}}$, $\mathbf{z}\in\mathbb{R}^{\mathbf{m}}$, $k\geq 2$, $m\geq 1$, $\mathbf{f}$ and $\mathbf{g}$ are smooth, and the dot denotes differentiation with respect to the independent variable t. Due to the smallness of ε, equations (1.1) possess two time scales (or two length scales), and the $\mathbf{x}$ and $\mathbf{z}$ variables are referred to as fast and slow variables, respectively.

We are interested in those systems (1.1) for which there exists an open set $U\subset\mathbb{R}^{\mathbf{m}}$ of $\mathbf{z}$ values such that the equation $\mathbf{f}(\mathbf{x},\mathbf{z},0)=0$ defines an m–dimensional manifold $\mathcal{M}_0$ of equilibria that can locally be expressed as a graph $(\mathbf{x},\mathbf{z})=(\mathbf{X}(\mathbf{z}),\mathbf{z})$ over the $\mathbf{z}$ variables. Moreover, we assume that the eigenvalues of the Jacobian of the function $f(\mathbf{x},\mathbf{z},0)$ at $\mathbf{x}=\mathbf{X}(\mathbf{z})$ have nonzero real part so that $\mathcal{M}_0$ is normally hyperbolic and so that when $0<\varepsilon\ll 1$, the Fenichel theory (see in particular Theorem 11.1 of [3]) guarantees the existence of a slow invariant manifold $\mathcal{M}_\varepsilon$ $\mathcal{O}(\varepsilon)$ close to $\mathcal{M}_0$.

Under the above assumptions, equations (1.1) model a diverse array of physical and biological phenomena. Moreover, special orbits, such as heteroclinic orbits and periodic orbits, play central roles in the system dynamics. We discuss a few of these applications here, although our list is by no means exhaustive.

The first example concerns resonant sloshing of shallow water in a tank:

$$\ddot{q}-q^2+1+c+\cos z=0,\qquad \dot{z}=\varepsilon \tag{1.2}$$

see [18] and [5]. Here $0 < \varepsilon \ll 1$, $k = 2$, $m = 1$, $\mathbf{x} = (q, \dot{q})$, and $c > 0$ is a parameter. These equations arise as an idealized model when a rectangular tank of length πL filled with a shallow layer of water is oscillated horizontally over a distance $a \ll L$ with $\mathcal{O}(1)$ frequency. Considering the motion as inviscid and irrotational, the full governing equation is a partial differential equation for the velocity potential along with the nonlinear free surface condition. In the limit of small surface displacements, small dispersion, neglecting surface tension, and under the proper scaling, the full equations can be reduced to the above form for q. Solutions of period $2\pi/\varepsilon$ are sought. The surface elevation can then be recovered from q and other parameters, see [18].

With both numerical methods and matched asymptotic expansions, the existence of a large class of spiked periodic orbits in (1.2) was discovered and analyzed in [18]. These solutions exhibit both outer regimes, in which they are near the slowly-varying saddle of (1.2), and inner (or fast) regimes, when they make excursions away from the saddle near a separatrix of the system (1.2) with $\varepsilon = 0$ (so that z is held constant). Solutions have either one or no spikes near separatrices that enclose a minimum area, and may have finitely many spikes near maximum separatrices. [18] also continued many of these orbits into a regime where small viscous damping is added to (1.2). The presence of damping results in the undamped solutions becoming asymmetric, as well as in a shift in the location of the spikes.

A mathematically rigorous demonstration of the existence of the spiked periodic orbits of (1.2) was recently given in [5] using a shooting technique. In [5], it was also shown that a kneading theory could be developed for these periodic orbits and subharmonic orbits. Finally, the existence of an entire Smale horseshoe of periodic orbits was discussed in [13]. We shall return to equation (1.2) and all of these periodic orbits in Section 3.

A second set of problems modeled by equations of the form (1.1) arises in perturbed N-degree-of-freedom Hamiltonian systems. Particular cases include systems subjected to external forces that resonate one of the modes as well as systems in which the frequencies of two of its degrees of freedom are in resonance. In these problems, there exist two-dimensional resonance manifolds; and, it is of fundamental importance to understand the dynamics on the resonance manifolds, as well as in the directions transverse to them. See [12] and the references therein. The structure of the governing equations reveals that the resonance manifolds and the directions transverse to them naturally separate the phase space variables into fast and slow types:

$$
\begin{aligned}
\dot{\mathbf{x}} &= J D_{\mathbf{x}} H(\mathbf{x}, I_0 + \sqrt{\varepsilon} h) + \varepsilon g^{\mathbf{x}}(\mathbf{x}, I_0 + \sqrt{\varepsilon} h, \theta, \lambda), \\
\dot{h} &= \sqrt{\varepsilon} g^{I}(\mathbf{x}, I_0 + \sqrt{\varepsilon} h, \theta, \lambda), \\
\dot{\theta} &= D_I H(\mathbf{x}, I_0 + \sqrt{\varepsilon} h) + \varepsilon g^{\theta}(\mathbf{x}, I_0 + \sqrt{\varepsilon} h, \theta, \lambda),
\end{aligned}
$$

Here, $I \in \mathbb{R}$ and $\theta \in S^1$ are the action-angle coordinates for the resonant degree-of-freedom and $\mathbf{x}$ with $k = 2(N-1)$ denotes the canonical position and momentum variables for the remaining $N-1$ degrees of freedom, J denotes the standard $(N-1) \times (N-1)$ symplectic matrix, H is the Hamiltonian for the unperturbed integrable system, and $\lambda \in \mathbb{R}^p$ is a vector of parameters. Furthermore, I_0 is the value of I corresponding to the resonance. Therefore, the standard rescaling $I = I_0 + \sqrt{\varepsilon}h$ is used to analyze the resonance; and, on the resonance manifold, both θ and h are slow variables. They evolve on a time scale of $\mathcal{O}(\frac{1}{\sqrt{\varepsilon}})$. Moreover, the $\mathbf{x}$ variables that govern the transverse directions change on the rapid $\mathcal{O}(1)$ time scale.

It has recently been shown that there exist heteroclinic orbits consisting of slow segments, when the orbit is near a resonance manifold, interspersed with fast segments, in which the orbit makes one or more excursions away from a resonance manifold and then approaches another (or the same) resonance manifold. One of the most interesting examples of such orbits is a new multiple-pulse Silnikov orbit discovered in a problem consisting of a Duffing oscillator coupled to an anharmonic oscillator in the presence of small damping in [12]. The difficulties inherent in working with the two time scales of these problems are overcome using techniques from geometric singular perturbation theory, see [8].

Numerical evidence indicates that there are also many periodic orbits consisting of alternating fast and slow segments (transverse to and near the resonances, respectively) both when the perturbations are Hamiltonian and dissipative. The method outlined in this work was designed in part to rigorously establish their existence. Full details will be given in [11].

Thirdly, systems of the form (1.1) occur in plasma physics and oceanography. Frequently, in plasma physics, the motion of particles in large-amplitude coherent wavepackets is characterized by two time scales. There is the short scale corresponding to the bounce time of the particle in the wave and the long time scale during which the wave amplitude varies significantly. The presence of two time scales in these problems naturally leads to model equations of the form (1.1). Particles in large regions of phase space repeatedly trap into and detrap from the wavepacket. In oceanography, simplified one-dimensional models have been derived for the interaction of small-scale internal waves with packets of large-amplitude inertial waves that have the form (1.1). In terms of ξ, the nondimensionalized phase of the internal wave in the frame moving with the large-amplitude inertial waves, $\tilde{\omega}$, the nondimensionalized local frequency of the internal wave, and q, the nondimensionalized vertical distance, the governing equations are:

$$\begin{aligned} \frac{d\xi}{dq} &= 1 - [\tilde{\omega} - A(\varepsilon q)\cos\xi]^{-2} \\ \frac{d\tilde{\omega}}{dq} &= A(\varepsilon q)\sin\xi\,[\tilde{\omega} - A(\varepsilon q)\cos\xi]^{-2}. \end{aligned}$$

These equations have a slowly-varying Hamiltonian $[\tilde{\omega} - A(\varepsilon q)\cos\xi]^{-1} + \tilde{\omega}$ that is the absolute frequency of the internal wave in the moving frame.

In the plasma problem, particles in large regions of phase space repeatedly trap into and detrap from the wavepacket; and in the oceanographic problem, the internal waves also trap into and detrap from the packet of large-amplitude near-inertial waves, leading to an average transport to smaller wavenumbers. Adiabatic separatrix-crossing theory and the adiabatic Melnikov function theory have been the main tools used in quantifying this phase space flux in the resonance (or stochastic) regions. A more detailed analysis of the fine structure in these regions has not yet been carried out, and depends on knowledge of the various elliptic and hyperbolic periodic orbits contained in them. Quantities associated to these orbits (such as their actions) can be used in a careful local analysis of the resonances and islands of stability on the Poincare map.

A fourth application of (1.1) comes from a model of nerve impulse transmission. The FitzHugh-Nagumo equations, as a system of partial differential equations, describe the propagation of membrane action potentials on nerve axons. Traveling wave solutions of the partial differential equations are also solutions of the ordinary differential equations:

$$\begin{aligned} \dot{v} &= w, \\ \dot{w} &= cv - f(v) + z, \\ \dot{z} &= \varepsilon/c(v - \gamma z), \\ \dot{c} &= 0, \end{aligned}$$

where γ is a constant, $0 < \varepsilon \ll 1$, v represents voltage across the membrane and z denotes a voltage-dependent conductance, and, finally, the function f is generally cubic-shaped. These FitzHugh-Nagumo equations are of the form (1.1) with fast variables $\mathbf{x} = (v, w)$ and slow variables z and c. If the effects of additional fast and slow currents are incorporated into the model, then the dimensions of the dependent variables increase, as for example is the case in the Hodgkin-Huxley equations.

Of particular interest are homoclinic and periodic solutions that have alternating fast and slow segments, reflecting the two time scales in the equations. Topological methods to demonstrate the existence of these orbits were established in [2]. Geometric singular perturbation theory, was used in [10] and [9] to show both existence and uniqueness for homoclinic orbits. In particular, a technique known as the Exchange Lemma was established that enables tracking of invariant manifolds while orbits on them spend long intervals of time near the slow manifolds.

The fifth and final application of (1.1) we shall cite concerns slowly, parametrically excited pendulums. The governing equations have the form

$$\ddot{q} + +\varepsilon\delta\dot{q} + A(z = \varepsilon h(q, t)) \sin q = 0 \tag{1.3}$$

where q represents the angle as measured counterclockwise from the downward vertical and h is a smooth function. The slowly varying amplitude function $A(z)$ can be derived from either a slowly varying length or from a situation in which the point of suspension is slowly, vertically oscillated. These equations also arise as the normal form equation in certain resonance bands. See also [6].

It has been shown that, when $A(z)$ varies periodically or quasiperiodically with z, there exist heteroclinic orbits that lie in the transverse intersections of particular stable and unstable manifolds, see [13] for an exposition. Hence, it is known that these systems possess Smale horseshoes, and in particular hyperbolic periodic orbits of every possible period. Section 2 is devoted to a particular realization of (1.3).

For the examples mentioned above, and in general for systems of the form (1.1), there are two reduced equations associated to (1.1), namely the fast system:

$$\dot{\mathbf{x}} = \mathbf{f}(\mathbf{x}, \mathbf{z}, 0) \qquad \text{and} \qquad \dot{\mathbf{z}} = 0 \tag{1.4}$$

in which $\mathbf{z}$ acts as a parameter, and the slow system:

$$0 = \mathbf{f}(\mathbf{x}, \mathbf{z}, 0) \qquad \text{and} \qquad \mathbf{z}' = \mathbf{g}(\mathbf{X}(\mathbf{z}), \mathbf{z}, 0) \tag{1.5}$$

which governs the leading order dynamics on the slow manifold. Singular periodic orbits are constructed by piecing together segments of solutions of (1.4) with segments of solutions of (1.5). The central question in this geometric approach to singular perturbation theory is therefore: which of the singular periodic orbits persist in the full system (1.1) when ε is nonzero, but small?

In this work we establish a technique to answer this question. In particular, we derive a technique that: (i) yields simultaneously existence and local uniqueness of the periodic orbits, as well as explicit information about their locations; and, (ii) relies only on geometrical information from the reduced system, and therefore has explicitly verifiable hypotheses.

To illustrate the method, we apply it to two slowly-modulated Hamiltonian systems. The general technique and its application to systems with damping, and to higher-dimensional singularly-perturbed systems, including those with more than one center (or slow) direction such as in some of the examples cited above, are given in [11].

Other techniques used on the problem of finding periodic orbits in time-dependent Hamiltonian vector fields include variational methods (see for example [4], [17] [19] [22] and the references therein), shooting methods [6], the method of sub- and super-solutions [1], and matched asymptotic expansions [14]. Techniques for singularly-perturbed boundary value problems are given in [16] and [21], and the references therein.

2 First example

2.A. The adiabatic pendulum.

In this section, we consider the parametrically-excited pendulum:

$$\dot{q} = p, \qquad \dot{p} = -(1 + \gamma \sin z) \sin q, \qquad \dot{z} = \varepsilon, \tag{2.1}$$

where $\gamma \in (0, 1)$ is a parameter, and the modulation frequency satisfies $0 < \varepsilon \ll 1$. This corresponds to a special choice of the amplitude function A discussed in the Introduction. The total (kinetic plus potential) energy for (2.1) written in terms of the slowly-varying parameter z is $H(q,p,z) = p^2/2 + (1+\gamma \sin z)(1-\cos q)$. Now, without altering the equations of motion, we may add $-2(1+\gamma \sin z)$ to the energy so that the Hamiltonian:

$$H(q,p,z) = p^2/2 - (1+\gamma \sin z)(1+\cos q) \tag{2.2}$$

vanishes at $(q_0^z = \pm\pi, p_0^z = 0, z)$ and on the separatrices connecting these saddles for each z.

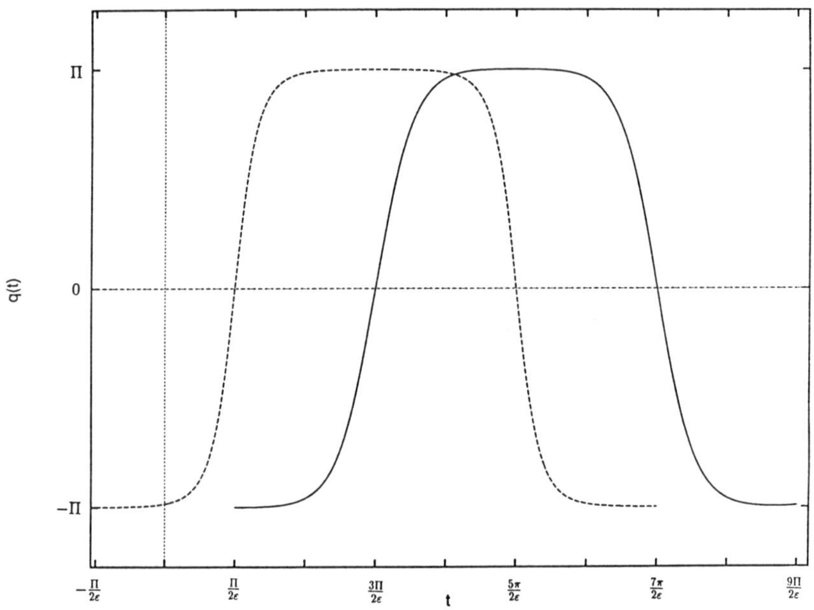

Figure 1: A pair of 2 : 1 librational periodic orbits of (2.1). One (dashed line) makes fast excursions near maximum instantaneous separatrices and the other (solid line) makes fast excursions close to minimum instantaneous separatrices. The initial conditions are $(q = -3.137013892601834, p = 0, z = -\frac{\pi}{2})$ and $(q = -3.138944064264264405087, p = 0, z = \frac{\pi}{2})$, respectively, and $\varepsilon = 0.4$ and $\gamma = 0.5$ for both. The dashed orbit corresponds to $k = -1$ and $m = 1$ in Theorem 1. A Runge-Kutta 4th order scheme was used for this and all other numerical figures.

We look for periodic orbits whose periods are rational multiples of the modulation period $T = 2\pi/\varepsilon$. These resonant orbits may either be librational (see figure 1), rotational, or of mixed type (see figure 2). For clarity of exposition and due to space limitations, we focus on librational $2m : 1$ subharmonics in this work. Letting $Z = 2\pi$, we establish:

Theorem 1 For each positive integer m there exist (at least) two locally unique $2m : 1$ subharmonic periodic orbits, $q^+(z), q^-(z)$, whose periods are $2mZ$. $q^+(z)$ has the following properties: for any given odd integer k, (i) as $\varepsilon \longrightarrow 0$, $q^+(z)$ tends uniformly

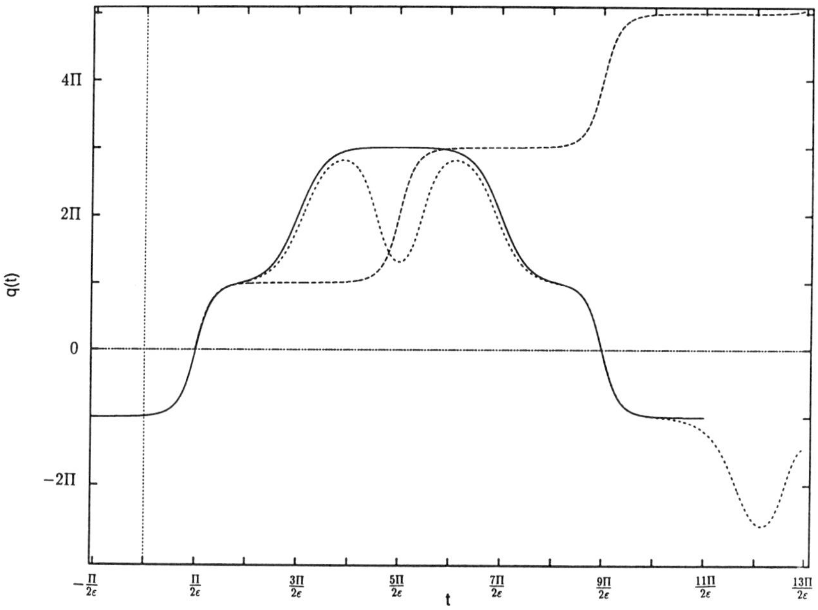

Figure 2: Three different types of resonant periodic orbits of (2.1). A rotational 1 : 1 resonance (long dashes) is shown on the z interval $[\frac{\pi}{2}, \frac{13\pi}{2}]$ with initial condition $(q = 0, p = 2.418497315201805, z = \frac{\pi}{2})$. A 3 : 1 mixed rotational-librational resonance (solid line) is shown for z in $[-\frac{\pi}{2}, \frac{11\pi}{2}]$ with initial condition $(q = -3.1373527193251982, p = 0, z = -\frac{\pi}{2})$. A segment of a 4 : 1 mixed rotational-two-pulse-librational resonance (short dashes) is shown on the interval $[\frac{\pi}{2}, \frac{13\pi}{2}]$ with initial condition $(q = 0, p = 2.4233907, z = \frac{\pi}{2})$. All three orbits have $\varepsilon = 0.4$ and $\gamma = 0.5$.

to π on any compact subset of $(\frac{k\pi}{2}+\frac{mZ}{2}, \frac{k\pi}{2}+\frac{3mZ}{2})$; (ii) $\dot{q}^+(z = \frac{k\pi}{2}+mZ) = 0$; (iii) $q^+(z)$ makes a fast excursion from the neighborhood of $q = \pi$ to the neighborhood of $q = -\pi$ on open intervals centered at $z = \frac{k\pi}{2} + \frac{3mZ}{2}$; (iv) as $\varepsilon \longrightarrow 0$, $q^+(z)$ tends uniformly to $q = -\pi$ on compact subsets of $(\frac{k\pi}{2} + \frac{3mZ}{2}, \frac{k\pi}{2} + \frac{5mZ}{2})$; (v) $\dot{q}^+(z = \frac{k\pi}{2} + 2mZ) = 0$; (vi) $q^+(z)$ makes a fast excursion from the neighborhood of $q = -\pi$ to the neighborhood of $q = \pi$ on open intervals centered at $z = \frac{k\pi}{2} + \frac{5mZ}{2}$. Moreover, $q^-(z)$ has the same properties as $q^+(z)$, with $\frac{k\pi}{2}$ replaced by $\frac{k\pi}{2} + \frac{Z}{2}$ in (i)-(vi) above.

This theorem is proven for $m = 1$ in subsections 2.B and 2.C. The generalization of Theorem 1 to the case of $m > 1$ follows from the same techniques and is given in [11]. There we also address the existence of multiple-pulse resonant harmonic orbits for (2.1) of the types shown in figure 2 and show that the results obtained here continue to hold when the system (2.1) is subjected to $\mathcal{O}(\varepsilon)$ damping. Finally, in subsection 2.D, we go beyond the present example and briefly discuss some of the main steps involved with the general version of the present technique.

2.B. Singular periodic orbits.

The geometric structure in $q-p-z$ space determined by the dynamics of (2.1) with $\varepsilon = 0$, *i.e.* the classical nonlinear pendulum for which the potential energy depends on the now constant value of the parameter z, guides our study of the $0 < \varepsilon \ll 1$ problem. For each $z \in [0, 2\pi)$, $(2.1)_{\varepsilon=0}$ possesses saddle equilibria at $(q_0^z = -\pi, p_0^z = 0, z)$ and at $(q_0^z = \pi, p_0^z = 0, z)$ and a center equilibrium at $(0, 0, z)$. Moreover, the two saddles are connected to each other by upper and lower heteroclinic orbits, that we denote $\Gamma_{\pm}^z$, respectively, and solutions on the heteroclinic orbits $\Gamma_{\pm}^z$ are parameterized by

$$(q_{\pm}^z(s), p_{\pm}^z(s), z) = (\pm 2 \arcsin(\tanh \sqrt{1+\gamma \sin z}s), \pm 2\sqrt{1+\gamma \sin z}\,\mathrm{sech}(\sqrt{1+\gamma \sin z}s), z). \tag{2.3}$$

Finally, the phase portrait for q outside the interval $[-\pi, \pi]$ is determined using the periodicity of (2.1) in q.

In the $q-p-z$ phase space of $(2.1)_{\varepsilon=0}$, we define the following curves formed by the unions of the saddle equilibria that shall play a central role in the analysis. Let z_1 and z_2 be real numbers satisfying $z_1 < z_2$, then:

$$\gamma_0^{-\pi}(z_1, z_2) \equiv \bigcup_{z \in [z_1, z_2]} (0, -\pi, z) \qquad \text{and} \qquad \gamma_0^{\pi}(z_1, z_2) \equiv \bigcup_{z \in [z_1, z_2]} (0, \pi, z)$$

Definition 1 For any $z_1 \in [0, 2\pi)$ and for any $z_2 \in [2m\pi, (2m+1)\pi)$, a singular $2m:1$ periodic orbit $\mathcal{Q}^{2m:1}(z_1, z_2)$ is the union:

$$\mathcal{Q}^{2m:1}(z_1, z_2) \equiv \Gamma_+^{z_1} \cup \gamma_0^{\pi}(z_1, z_2) \cup \Gamma_-^{z_2} \cup \gamma_0^{-\pi}(z_2, z_1 + 2\pi). \tag{2.4}$$

A singular periodic orbit is not the orbit of one system, rather it is the concatenation of orbits and equilibria of various different systems. Theorem 1 may therefore be viewed as an answer to the question: For which pairs of values z_1 and z_2 do the $2m:1$ singular periodic orbits $\mathcal{Q}^{2m:1}(z_1, z_2)$ persist as real $2m:1$ periodic orbits of (2.1) when $0 < \varepsilon \ll 1$?

2.C. Existence of 2m:1 subharmonics

The detailed geometrical information about the reduced system, (2.1) with $\varepsilon = 0$, given in the previous subsection is extremely useful for determining the geometry and dynamics of the full system (2.1) with $0 < \varepsilon \ll 1$. It may be seen from inspection that the full system possesses slow saddle periodic orbits $\gamma_\varepsilon^{-\pi} = \cup_{z \in \mathbb{R}}(q = -\pi, p = 0, z)$ and $\gamma_\varepsilon^{\pi} = \cup_{z \in \mathbb{R}}(q = \pi, p = 0, z)$. These slow orbits coincide with the curves of saddle equilibria of the system (2.1) with $\varepsilon = 0$. We shall refer to $\gamma_\varepsilon^{-\pi}$ and γ_ε^{π} as slow manifolds, and remark that they are one-dimensional. In addition, the theory in [3] guarantees that these slow manifolds have two-dimensional C^r smooth (for any finite $r > 0$) stable and unstable manifolds. We denote these by $W^s(\gamma_\varepsilon^{-\pi})$, $W^u(\gamma_\varepsilon^{-\pi})$, $W^s(\gamma_\varepsilon^{\pi})$, and $W^u(\gamma_\varepsilon^{\pi})$, respectively.

With this information about the full system, we proceed to prove Theorem 1. However, rather than prove it for all m, we consider only one of the $m = 1$ librational orbits for definiteness as indicated above. Central to the technique is the demonstration that a pair of two-dimensional manifolds intersect transversely in three-dimensional $q-p-z$ phase space, and, as curves, solutions in this intersection lie close to the singular periodic orbit

$$\gamma_0^{-\pi}\left(z_1=-\frac{\pi}{2}, z_2=\frac{\pi}{2}\right)\cup\Gamma_+^{\frac{\pi}{2}}\cup\gamma_0^{\pi}\left(z_1=\frac{\pi}{2}, z_2=\frac{5\pi}{2}\right)\cup\Gamma_-^{\frac{5\pi}{2}}\cup\gamma_0^{-\pi}\left(z_1=\frac{5\pi}{2}, z_2=\frac{7\pi}{2}\right).$$

Let Π_{z_0} be the plane in $q-p-z$ space with $z=z_0$ constant. We will find an appropriate one-dimensional set of initial conditions $I_\varepsilon \subset \Pi_{-\frac{\pi}{2}}$ and let it flow in forward for a finite amount of time to obtain a two-dimensional manifold $\mathcal{M}_\varepsilon$. We will take another set of initial conditions, $J_\varepsilon \subset \Pi_{\frac{3\pi}{2}}$, which we will let flow in backwards time to produce another two-dimensional manifold $\mathcal{N}_\varepsilon$. By showing that $\mathcal{M}_\varepsilon$ and $\mathcal{N}_\varepsilon$ intersect transversely, we obtain an orbit segment. Finally, we shall show that I_ε and J_ε can be chosen so that this orbit segment can be extended in time for $z\in[3\pi/2, 7\pi/2]$ to form a periodic orbit.

The proof is divided into three main steps. First, we construct the intervals I_ε and J_ε:

Step 1 *Let $\mathcal{B}$ be a tubular neighborhood of the slow manifold $\gamma_\varepsilon^{-\pi}$. There exists an interval $I_\varepsilon=\{r=(q,p,z)|q\in[q_2,q_1], p=0, z=-\pi/2\}$ of points such that the solution through $r_1=(q_1,0,-\frac{\pi}{2})$ leaves $\mathcal{B}$ at time $z=\frac{\pi}{2}-\Delta z_1$, and the solution through $r_2=(q_2,0,-\frac{\pi}{2})$, with $-\pi<q_2<q_1<-\pi/2$, leaves $\mathcal{B}$ at time $z=\frac{\pi}{2}+\Delta z_2$. Here $0<\Delta z_1,\Delta z_2<\pi$.*

Similarly, let $\mathcal{B}'$ be a neighborhood of γ_ε^{π}. There exists an interval $J_\varepsilon=\{r=(q,p,z)|q\in[q_3,q_4], p=0, z=3\pi/2\}$, such that the solution through r_3 leaves $\mathcal{B}'$ in backwards time at $z=-\frac{\pi}{2}+\Delta z_3$, and the solution through r_4 with $\pi/2<q_3<q_4<\pi$ leaves $\mathcal{B}'$ in backwards time at $z=-\frac{\pi}{2}-\Delta z_4$, where $0<\Delta z_3,\Delta z_4<\pi$.

Proof of Step 1. We prove the existence of the interval I_ε; that for J_ε is similar. Fix a small constant $\Delta>0$. Let $q\equiv-\pi+\tilde{q}$ and $p\equiv\tilde{p}$. The components of the eigenvectors of (2.1) with $\varepsilon=0$ at the saddle $(\tilde{q}=0,\tilde{p}=0,z)$ for each $z\in[0,2\pi)$ enable us to define new variables:

$$\xi\equiv(1+\gamma\sin z)^{1/2}\,\tilde{q}+\tilde{p} \qquad \text{and} \qquad \eta\equiv(1+\gamma\sin z)^{1/2}\,\tilde{q}-\tilde{p}.$$

Then, in a sufficiently small neighborhood of $\gamma_\varepsilon^{-\pi}$, (2.1) may be rewritten as:

$$\begin{aligned}\dot{\xi} &= (1+\gamma\sin z)^{1/2}\,\xi+\tilde{g}_1(\xi,\eta,z;\varepsilon)\\ \dot{\eta} &= -(1+\gamma\sin z)^{1/2}\,\eta+\tilde{g}_2(\xi,\eta,z;\varepsilon)\\ \dot{z} &= \varepsilon,\end{aligned} \tag{2.5}$$

where the scalar functions $\tilde{g}_i, i=1,2$, represent terms that are nonlinear in ξ and η, as well as terms that are linear in ξ and η that are $\mathcal{O}(\varepsilon)$. These terms satisfy the estimates

$$|\tilde{g}_i(\xi,\eta,z;\varepsilon)|\leq K\,(|\xi|+|\eta|)\left[\varepsilon+(|\xi|+|\eta|)^2\right]$$

for some constant $K > 0$ and $i = 1, 2$.

Having written the linear terms in nice coordinates, we now also transform the nonlinear terms. The theory of [3] cited above specifically guarantees the existence of C^r smooth functions $h^s(\eta, z; \varepsilon)$ and $h^u(\xi, z; \varepsilon)$, for any $r > 0$, defined for $|\xi|, |\eta| \leq \Delta$, such that the sets

$$\{\xi = h^s(\eta, z; \varepsilon)\} \qquad \text{and} \qquad \{\eta = h^u(\xi, z; \varepsilon)\}$$

are the invariant manifolds $W^s_{\text{loc}}(\gamma_\varepsilon^{-\pi})$ and $W^u_{\text{loc}}(\gamma_\varepsilon^{-\pi})$, respectively. Thus, defining final variables:

$$a \equiv \xi - h^s(\eta, z; \varepsilon) \qquad \text{and} \qquad b \equiv \eta - h^u(\eta, z; \varepsilon),$$

the local manifolds $W^s_{\text{loc}}(\gamma_\varepsilon^{-\pi})$ and $W^u_{\text{loc}}(\gamma_\varepsilon^{-\pi})$ are straightened out. In particular, they are given by the sets $\{a = 0\}$ and $\{b = 0\}$, respectively, for $|a|, |b| \leq \Delta$. Moreover, since these sets are invariant, it follows that $\dot{a} = 0$ on $\{a = 0\}$ and $\dot{b} = 0$ on $\{b = 0\}$. Hence, the normal form of (2.1) in $\mathcal{B}$ is:

$$\begin{aligned} \dot{a} &= (1 + \gamma \sin z)^{1/2} a + g_1(a, b, z; \varepsilon) a \\ \dot{b} &= -(1 + \gamma \sin z)^{1/2} b + g_2(a, b, z; \varepsilon) b \\ \dot{z} &= \varepsilon \end{aligned} \tag{2.6}$$

where g_1 and g_2 are C^r smooth functions. Note also that without loss of generality, we may take the neighborhood $\mathcal{B}$ to be the set: $\mathcal{B} \equiv \{(a, b, z) \| |a|, |b| \leq \Delta\}$, where $\Delta > 0$ is the fixed constant chosen initially, independent of ε, and sufficiently small.

This normal form (2.6) directly yields the following estimates for a trajectory in $\mathcal{B}$:

$$|b(t)| \leq C_b |b(s)| e^{-\lambda_0 (t-s)} \qquad \text{and} \qquad |a(t)| \geq C_a |a(s)| e^{\lambda_0 (t-s)} \tag{2.7}$$

where the constants $C_a, C_b > 0$, $s \leq t$, and λ_0 (which can be taken to be $\sqrt{1-\gamma} - \delta$ for some small $\delta > 0$) is a lower bound on the positive eigenvalue of all the systems (2.1) with $\varepsilon = 0$. We reiterate that the above estimates are valid as long as the trajectory stays in $\mathcal{B}$.

Since $\gamma_\varepsilon^{-\pi}$ is a hyperbolic slow manifold, all trajectories in $\mathcal{B}$ not on the local stable and unstable manifolds $W^s_{\text{loc}}(\gamma_\varepsilon^{-\pi})$ and $W^u_{\text{loc}}(\gamma_\varepsilon^{-\pi})$ will cross either the $q-$axis or the $p-$axis precisely one time between the times they enter and exit $\mathcal{B}$. So, consider an initial condition $r = (q, p = 0, z = -\frac{\pi}{2})$ on the $q-$axis of the plane $\Pi_{z=-\pi/2}$. Let $s = -\frac{\pi}{2\varepsilon}$ in (2.7). Then, (2.7) yields: $|a(t)| \geq C_a |a(-\frac{\pi}{2\varepsilon})| e^{\lambda_0 (t + \frac{\pi}{2\varepsilon})}$ for $t \geq -\frac{\pi}{2\varepsilon}$ as long as the forward orbit of r stays in $\mathcal{B}$. Therefore, the choice of the initial a coordinate $|a(-\frac{\pi}{2\varepsilon})|$ such that $(C_a + 1)|a(-\frac{\pi}{2\varepsilon})| e^{\lambda_0 \frac{\pi}{\varepsilon}} = \Delta$ implies that the forward orbit of r has crossed the exit face $\{|a| = \Delta\}$ of the neighborhood $\mathcal{B}$ at some some time t satisfying $-\frac{\pi}{2\varepsilon} < t < \frac{\pi}{2\varepsilon}$. In addition, there exists a value of $|b(-\frac{\pi}{2\varepsilon})|$ such that the initial conditions $(q(-\frac{\pi}{2\varepsilon}), p(-\frac{\pi}{2\varepsilon}))$ satisfy $p(-\frac{\pi}{2\varepsilon}) = 0$. Finally, by continuous dependence of solutions on initial data, $|a(-\frac{\pi}{2\varepsilon})|$ may be increased (while a value of $b(-\frac{\pi}{2\varepsilon})$ is maintained so that $p(-\frac{\pi}{2\varepsilon}) = 0$) until one has an initial condition $r_1 \equiv (a(-\frac{\pi}{2\varepsilon}), b(-\frac{\pi}{2\varepsilon}), z = -\frac{\pi}{2})$

such that the forward orbit through r_1 exits $\mathcal{B}$ when $z = \frac{\pi}{2} - \Delta z_1$ for the prescribed $0 < \Delta z_1 < \pi$. In fact, due to the symmetry of (2.1), $b(-\frac{\pi}{2\varepsilon}) = -a(-\frac{\pi}{2\varepsilon})$.

In a similar fashion, we can find $r_2 = (a'(-\frac{\pi}{2\varepsilon}), b'(-\frac{\pi}{2\varepsilon}))$ on the q-axis and close to $(a = 0, b = 0, z = \frac{\pi}{2})$, *i.e.*, close to $(q = -\pi, p = 0, z = -\frac{\pi}{2})$ such that the solution through r_2 leaves $\mathcal{B}$ at some finite value of z after $z = \frac{\pi}{2}$. Together, the initial conditions between r_1 and r_2 form the desired one-dimensional set I_ε of initial conditions on the q-axis. □

Next, flow the initial conditions of I_ε forward in time to obtain a two-dimensional manifold $\mathcal{M}_\varepsilon$, and those of J_ε backward in time to obtain another two-dimensional manifold $\mathcal{N}_\varepsilon$. The goal of step 2 is to accurately determine the location of those portions of $\mathcal{M}_\varepsilon$ and $\mathcal{N}_\varepsilon$ near the boundaries of $\mathcal{B}$:

Step 2 *For any point $r \in I_\varepsilon$, let $\overline{r}$ denote the point on the boundary of the neighborhood $\mathcal{B}$ where the orbit with initial condition at r exits $\mathcal{B}$ in forward time. Then, at $\overline{r}$, $\mathcal{M}_\varepsilon$ is $C^1 - \mathcal{O}(e^{-\frac{c}{\varepsilon}})$ close to $W^u_{\text{loc}}(\gamma_\varepsilon^{-\pi})$, for some constant $c > 0$. Similarly, for any point $r' \in J_\varepsilon$, let $\overline{r}'$ denote the point on the boundary of $\mathcal{B}'$ where the orbit with initial condition at r' exits the neighborhood $\mathcal{B}'$ in backward time. Then, at $\overline{r}'$, $\mathcal{N}_\varepsilon$ is $C^1 - \mathcal{O}(e^{-\frac{c}{\varepsilon}})$ close to $W^s_{\text{loc}}(\gamma_\varepsilon^{\pi})$, for some constant $c > 0$.*

Proof of Step 2 for $\mathcal{M}_\varepsilon$. The result for $\mathcal{N}_\varepsilon$ is obtained in a similar fashion, with the only difference being the reversal of time and of the stable and unstable manifolds.

The desired exponential closeness estimates must be obtained for both the manifolds and their tangent planes. A useful tool, known as Exchange Lemma with Exponentially Small Error (referred to as ELESE for short), for establishing these types of estimates was presented in [8]. The statement of ELESE adapted to the present normal form (2.6) is:

Modified Exchange Lemma with Exponentially Small Error. Let M be a two-dimensional invariant manifold obtained by flowing forward a strip of initial conditions K_ε inside $\mathcal{B}$. Let $s \in K_\varepsilon$ be a point whose trajectory exits $\mathcal{B}$ at $\overline{s}$ after a time $\mathcal{O}(\frac{1}{\varepsilon})$. If the projection of the tangent plane T_sM onto the tangent plane of the local unstable manifold $\{b = 0\}$ is $\mathcal{O}(1)$, then, for ε sufficiently small, the manifold M is $C^1 - \mathcal{O}(e^{-\frac{c}{\varepsilon}})$ close to $\{b = 0\}$ at $\overline{s}$ for some $c > 0$.

The standard version of ELESE requires that the tracked manifold transversely intersects the center-stable manifold of the slow manifold on the entrance face of the neighborhood $\mathcal{B}$. This requirement is needed to obtain initial estimates. However, in the search for periodic orbits for (2.1), this requirement is not met by the tracked manifold $\mathcal{M}_\varepsilon$, and so we make use of the above modified formulation of the Lemma. Moreover, given the conclusion of the modified Lemma, it suffices to show that its hypotheses are satisfied by the tracked manifold $\mathcal{M}_\varepsilon$ of equation (2.1) in order to complete the proof of this step.

Consider the two-dimensional tangent planes $\mathcal{L}_M$ of the manifold $\mathcal{M}_\varepsilon$ in $q - p - z$ space. At a point $r \in I_\varepsilon$ obtained in Step 1, the plane $\mathcal{L}_M$ is spanned by the vector field $(0, -(1-\gamma)\sin q, \varepsilon)$ and the unit vector $\mathbf{i}$. Also, at r, the unit normal vector to $\mathcal{L}_M$ is:

$$\hat{n}_M = \frac{\varepsilon \mathbf{j} + (1-\gamma)\sin q \mathbf{k}}{\sqrt{(1-\gamma)^2 \sin^2 q + \varepsilon^2}} \tag{2.8}$$

Also, let $\mathcal{L}_W$ denote the plane in $q-p-z$ space spanned by the vectors $(1, (1-\gamma), \varepsilon)$ and the unit vector $\mathbf{k}$. By definition, $\mathcal{L}_W$ coincides with the tangent plane of $W^u_{\text{loc}}(\gamma_0^{-\pi})$ at the point $(-\pi, 0, -\pi/2)$. The unit normal to $\mathcal{L}_W$ is:

$$\hat{n}_W = \frac{-\sqrt{1-\gamma}\mathbf{i} + \mathbf{j}}{\sqrt{2-\gamma}}, \tag{2.9}$$

and the angle θ between the *unit* normals $\hat{n}_M$ and $\hat{n}_W$ is given by:

$$\hat{n}_M \cdot \hat{n}_W = \cos\theta = \frac{\varepsilon}{\sqrt{2-\gamma}\left[\varepsilon^2 + (1-\gamma)^2 \sin^2 q\right]^{1/2}}. \tag{2.10}$$

Now, since q is $\mathcal{O}(e^{-\frac{c}{\varepsilon}})$ close to $-\pi$ by Step 1, the expression on the right hand side of (2.10) is to leading order $\frac{1}{\sqrt{2-\gamma}}$. Thus, the projection of a square patch of unit area on $\mathcal{L}_M$ onto $\mathcal{L}_W$ has area equal to $\frac{1}{\sqrt{2-\gamma}} + \mathcal{O}(\varepsilon)$. Moreover, (2.10) also implies that the projection of this same patch on $\mathcal{L}_M$ onto the tangent plane of the *perturbed* manifold $W^u_{\text{loc}}(\gamma_\varepsilon^{-\pi})$ also has area equal to $\frac{1}{\sqrt{2-\gamma}} + \mathcal{O}(\varepsilon)$, as $\varepsilon \to 0$, since the perturbed and unperturbed local unstable manifolds are $C^r - \mathcal{O}(\varepsilon)$ close.

Finally, the fact that the trajectory through r spends an $\mathcal{O}(\frac{1}{\varepsilon})$ amount of time in $\mathcal{B}$ follows from the normal form (2.6) in Step 1. Thus, the hypotheses of the modified Exchange Lemma are satisfied, and its application completes the proof of Step 2. □

Step 3 *$\mathcal{M}_\varepsilon$ and $\mathcal{N}_\varepsilon$ intersect transversely $\mathcal{O}(\varepsilon)$ close to the point $(0, \sqrt{1+\gamma}, \pi/2)$ and at an angle of $\mathcal{O}(\varepsilon)$.*

Proof of Step 3. The adiabatic Melnikov function for (2.1):

$$M_A(z) = \int_{-\infty}^{\infty} \left(\frac{\partial H}{\partial z}(q^z_\pm(s), p^z_\pm(s), z) - \frac{\partial H}{\partial z}(-\pi, 0, z) \right) ds = \frac{4\gamma \cos z}{\sqrt{1+\gamma \sin z}} \tag{2.11}$$

measures the $\mathcal{O}(\varepsilon)$ leading order term of the distance between $W^u(\gamma_\varepsilon^{-\pi})$ and $W^s(\gamma_\varepsilon^{\pi})$ along the normal to the unperturbed separatrix Γ^z_+ at the point $(q^z_\pm(0), p^z_\pm(0), z)$ (obtained by setting $s = 0$ in (2.3)). It has simple zeroes at $z_n = (2n+1)\pi/2$ for every integer n. Therefore, using results of [20] or [13], on Π_{z_n}, and in particular on $\Pi_{\pi/2}$, $W^u(\gamma_\varepsilon^{-\pi})$ and $W^s(\gamma_\varepsilon^{\pi})$ intersect transversely $\mathcal{O}(\varepsilon)$ close to the point $(0, \sqrt{1+\gamma \sin z_n}, z_n)$ and at an angle of $\mathcal{O}(\varepsilon)$. Let s_ε denote this principal intersection point near $(0, \sqrt{1+\gamma}, \pi/2)$ on $\Pi_{\pi/2}$.

Next, we recall that the Modified Exchange Lemma used in the previous step and the original Exchange Lemma (see [9]) are established by analyzing the equations of motion for the differential two-forms dual to the two-dimensional tangent planes of the tracked manifold. Applying a straightforward, finite-time estimate based on Gronwall's inequality to the integral form of these equations in $q-p-z$ coordinates, one readily extends the $C^1 - \mathcal{O}(e^{-\frac{c}{\varepsilon}})$ closeness of the manifolds $\mathcal{M}_\varepsilon$ and $W^u_{\rm loc}(\gamma_\varepsilon^{-\pi})$ to portions of $W^u_{\rm loc}(\gamma_\varepsilon^{-\pi})$ outside of $\mathcal{B}$. In particular, $\mathcal{M}_\varepsilon$ remains $C^1 - \mathcal{O}(e^{-\frac{c}{\varepsilon}})$ close to $W^u_{\rm loc}(\gamma_\varepsilon^{-\pi})$ also in a neighborhood of the principal intersection point s_ε.

Similarly, it may be shown that $\mathcal{N}_\varepsilon$ remains $C^1 - \mathcal{O}(e^{-\frac{c}{\varepsilon}})$ close to $W^s_{\rm loc}(\gamma_\varepsilon^{\pi})$ in a neighborhood of the principal intersection point s_ε. Putting these two extended closeness results together with the conclusion from the adiabatic Melnikov theory immediately yields the desired conclusion that $\mathcal{M}_\varepsilon$ and $\mathcal{N}_\varepsilon$ intersect $\mathcal{O}(\varepsilon)$ close to the point $(0, \sqrt{1+\gamma}, \pi/2)$ and at an angle of $\mathcal{O}(\varepsilon)$. □

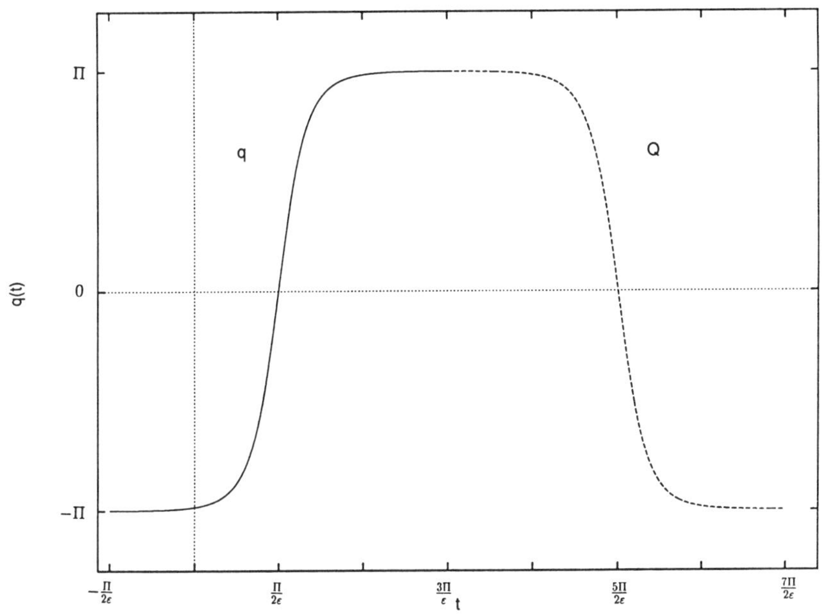

Figure 3: The 2 : 1 librational resonant subharmonic orbit for (2.1), with fast excursion at $z = \frac{\pi}{2}, \frac{5\pi}{2}$. The segment labeled "$q$" (solid) is obtained as the intersection of the two-dimensional manifolds $\mathcal{M}_\varepsilon$ and $\mathcal{N}_\varepsilon$, and the segment labeled "Q" (dashed) is obtained from the symmetries of equations (2.1).

Completion of proof of Theorem 1. The transverse intersection of $\mathcal{M}_\varepsilon$ and $\mathcal{N}_\varepsilon$ obtained in Step 3 yields a locally unique orbit segment with $q(z=\pi/2) = \mathcal{O}(\varepsilon)$ and whose p coordinates satisfy $p(z=-\pi/2) = 0$ and $p(z=3\pi/2) = 0$. Now, we use the symmetries of (2.1) to extend this orbit segment into the complete 2 : 1 periodic orbit.

Let $'$ denote derivative with respect to z. Then, (2.1) may be rewritten as:

$$q' = \frac{p}{\varepsilon}, \qquad p' = -\frac{1}{\varepsilon}(1+\gamma \sin z)\sin q, \qquad z' = 1. \tag{2.12}$$

First, the equation (2.12) stays invariant under the coordinate change $Q = -q, P = -p, Z = z$, and hence if $q(z)$ is a solution of (2.12), then so is $Q(z) = -q(z)$. Second, the equation (2.12) stays invariant under the coordinate change $Q = -q, P = p, Z = k\pi - z$, for odd integers k, and hence if $q(z)$ is a solution of (2.12), then there exists a second solution $Q(z)$ satisfying $q\left((2m+1)\pi/2 - w\right) = Q\left((2m+1)\pi/2 + w\right)$ for all $w \in \mathbb{R}$ and for any integer m. Putting these two symmetries together, one readily sees that if $q_0(z)$ is a solution of (2.12) with $p_0\left([2m+1]\pi/2\right) = 0$ for some integer m, (*i.e.*, the solution has zero momentum when the time z is an odd multiple of $\pi/2$) then $q_0\left((2m+1)\pi/2 - w\right) = q_0\left((2m+1)\pi/2 + w\right)$ for all $w \in \mathbb{R}$, see figure 3.

Therefore, we may take the orbit segment with z between $-\pi/2$ and $3\pi/2$ obtained in Step 3 and extend it using the symmetries to the interval of z values between $3\pi/2$ and $7\pi/2$ to obtain the desired periodic orbit. □

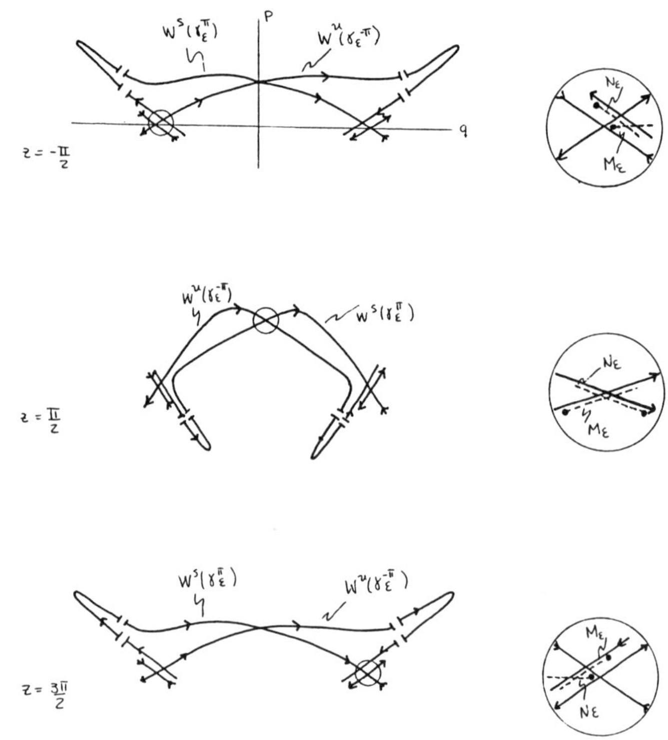

Figure 4: A sequence of phase portraits for the adiabatic pendulum (2.1), at various values of the slow variable z. Left panels show the transverse intersection of the unstable manifold of $\gamma_\varepsilon^{-\pi}$ with the stable manifold of γ_ε^{π}. Right panels show enlarged picture of the region enclosed by a circle in the corresponding left panel, and, in particular, the location of the tracked manifolds $\mathcal{M}_\varepsilon$ and $\mathcal{N}_\varepsilon$ (dashed) with respect to the stable and unstable manifolds (solid) mentioned above.

Figure 4 shows a sequence of phase portraits showing the relevant geometric structures for three different values of z. We illustrate the transverse intersection of $W^u(\gamma_\varepsilon^{-\pi})$ and $W^s(\gamma_\varepsilon^{\pi})$ (note these intersections correspond to heteroclinic connections from $\gamma_\varepsilon^{-\pi}$ to γ_ε^{π}). The regions enclosed by a circle, and enlarged in the right panels, show where $\mathcal{M}_\varepsilon$ and $\mathcal{N}_\varepsilon$ are situated, as well as their transverse intersection, which corresponds to our periodic orbit.

The location of $\mathcal{M}_\varepsilon$ and $\mathcal{N}_\varepsilon$ with respect to the manifolds $W^{u,s}(\gamma_\varepsilon^{-\pi})$, $W^{u,s}(\gamma_\varepsilon^{\pi})$, together with the variational equations of (2.1) are important tools to determine the stability type of this 2:1 periodic orbit. A full discussion of the instability of this orbit and other resonant orbits will be given in [11]. We also remark that, although the context is slightly different, the notion of tracking vectors and measuring their rotation as set forth in [15] was influential both for this figure as well as for the development of the original exchange lemmas. The subharmonics appear to be the same as those found in a Smale horseshoe, but the multiple-pulse periodic orbits, such as the ones shown in figure 2, do not.

2.D. The general periodic orbit problem.

For general systems of the form (1.1) with the assumptions stated in the Introduction,othere is an analog for each of the three steps performed above. In particular, the dynamics of (1.1) near the slow manifolds $\mathcal{M}_\varepsilon$ is given by the full Fenichel normal form:

$$\begin{aligned} \dot{a} &= \Lambda(a, b, z; \varepsilon)a, \\ \dot{b} &= \Gamma(a, b, z; \varepsilon)b \\ \dot{z} &= \varepsilon \mathbf{g}_1(\mathbf{z}, \varepsilon) + \varepsilon \mathbf{h}(a, b, \mathbf{z}, \varepsilon)ab \end{aligned} \tag{2.13}$$

where a and b are vectors in $\mathbb{R}^{\mathbf{k_1}}$ and $\mathbb{R}^{\mathbf{k_2}}$, respectively where $k_1 + k_2 = k$, the matrices Λ and Γ have k_1 (k_2) eigenvalues with positive (negative) real parts when $a, b, \varepsilon =$ (*i.e.*, on $\mathcal{M}_0$), and h is a rank three tensor. See also the recent exposition [7]. Moreover, the analogs of the estimates (2.7) are precisely those given by applying Lemma 1 in [8] to the full Fenichel normal form so that the existence of good initial conditions may be shown as in step 1 above. In many cases, one finds that the shooting set is the union of the points :$\{|a| = c\} \cap \{|b| = c\}$ over $c \in [0, c_0]$.

The analogs of steps 2 and 3 for (1.1) are based upon the full ELESE from [8] and an analysis of the fast foliations of the slow manifolds and their transverse intersections. Furthermore, these intersections may be at either $\mathcal{O}(\varepsilon)$ angles such as may be established via a Melnikov type calculation or $\mathcal{O}(1)$ angles such as in the examples discussed, for example, in [15], [1], [10], and [9]. A full account is contained in [11].

3 Second example

In this section, we show the existence and local uniqueness of both subharmonic and multiple-pulse resonant periodic orbits for a free particle in a slowly-varying cubic potential:

$$\dot{q} = p, \qquad \dot{p} = q^2 - 1 - c - \cos z, \qquad \dot{z} = \varepsilon, \tag{3.1}$$

where c is a positive constant, and the modulation frequency ε again satisfies $0 < \varepsilon \ll 1$. As discussed in the Introduction, this model equation arises in the analysis of resonant sloshing in a square tank that contains a shallow layer of water and that is oscillated horizontally near the resonant frequency of certain surface waves, see [18] and [5].

As in Section 2, an analysis of the structure of (3.1) with $\varepsilon = 0$ in the $p - q - z$ space guides our study of the full problem (3.1). When $\varepsilon = 0$, (3.1) has one saddle equilibrium at $(p = 0, q = \sqrt{1 + c + \cos z}, z)$ and a center equilibrium at $(p = 0, q = -\sqrt{1 + c + \cos z}, z)$ for each real number z. We denote the unions of the saddles over the interval of values $z_1 \leq z \leq z_2$ by $\gamma_0(z_1, z_2)$. Each saddle equilibrium is connected to itself by a homoclinic orbit, that we label Γ^z, containing the center equilibrium. So, in the same spirit as in the previous section, we make:

Definition 2 For any $z \in [0, 2\pi)$, a singular $1:1$ periodic orbit of (3.1) with $\varepsilon = 0$ is the union:

$$\gamma_0(0, z) \cup \Gamma^z \cup \gamma_0(z, 2\pi).$$

Definition 3 For any $z \in [0, 2\pi)$ and any positive integer k, a singular $k:1$ subharmonic periodic orbit of (3.1) with $\varepsilon = 0$ is the union:

$$\gamma_0(0, z) \cup \Gamma^z \cup \gamma_0(z, 2k\pi).$$

Definition 4 For any $z_1, z_2, ..., z_n \in [0, 2\pi)$, a singular $n-$pulse harmonic periodic orbit of (3.1) with $\varepsilon = 0$ is the union:

$$\gamma_0(0, z_1) \cup \Gamma^{z_1} \cup \gamma_0(z_1, z_2) \cup \Gamma^{z_2} \cup \gamma_0(z_2, z_3) \cup \Gamma^{z_3} \cup \cdots \cup \Gamma^{z_{n-1}} \cup \gamma_0(z_n, 2\pi).$$

where the z_i need not be distinct.

Remark. These definitions do not exhaust all of the possible singular periodic orbits.

In what follows, we will concentrate, for definiteness and due to space limitations, on the existence and uniqueness of a 1:1 resonance that, again, will be constructed as the transverse intersection of two two-dimensional manifolds, close to the singular periodic orbit $\gamma_0(0, \pi) \cup \Gamma^\pi \cup \gamma_0(\pi, 2\pi)$, see figure 5. These correspond to the (1,0) and (0,1) orbits in [5]. The orbits of Theorems 3 and 5 in [5] are also obtained with the present geometric technique. See [11] for the proofs and for a discussion of which of these are the same as the periodic orbits known to exist from the Smale horseshoe construction and which are not.

The general ideas are the same as the ones presented in section 2.C for the resonances of the adiabatic pendulum; so we will keep the exposition concise, and point out only the main differences. $\gamma_0(0, 2\pi)$ is a hyperbolic periodic orbit for $(3.1)_{\varepsilon=0}$ that, for $\varepsilon > 0$ small, will persist as a hyperbolic one-dimensional manifold of (3.1), denoted by γ_ε, satisfying $\gamma_\varepsilon = \gamma_0 + \mathcal{O}(\varepsilon)$, see [3]. Let $\mathcal{B}$ be a neighborhood of γ_ε (recall that in a, b, z coordinates, $\mathcal{B}$ is a tube of square cross-section). Let $I_\varepsilon = \{r = (q, p, z) | q \in [q_1, q_2], p = 0, z = 0\}$, with $0 < q_1 < q_2 < \sqrt{2 + c}$, be our set of initial conditions such that the solution through $r_1 = (q_1, 0, 0)$ leaves the box $\mathcal{B}$ at $z = \pi - \Delta z_1$

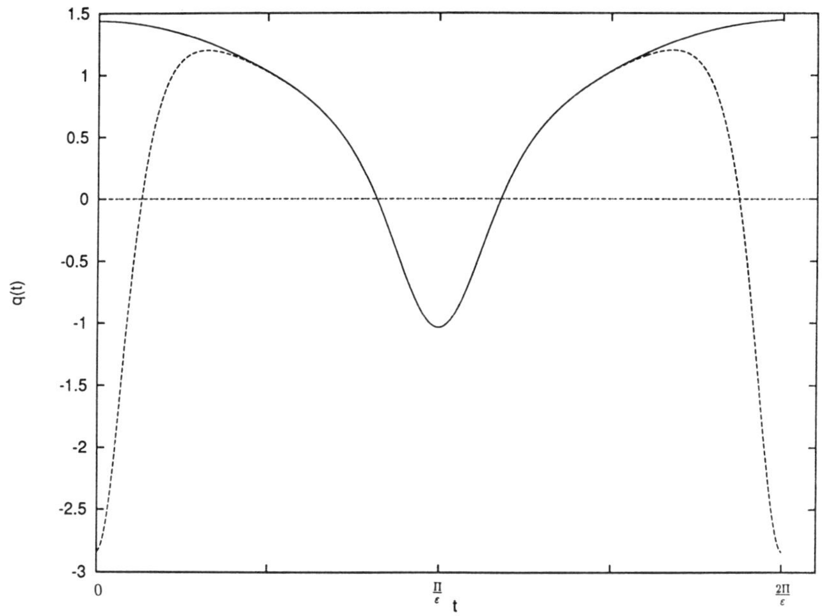

Figure 5: A pair of 1 : 1 resonances for (3.1). The solid one makes one fast excursion near the minimum separatrix. The dashed one corresponds to a two-pulse periodic orbit, making excursions at the maximum and minimum separatrices. The respective initial conditions, for $\varepsilon = 0.3$,$c = 0.1$, and 1500 time steps per period are $(q = 1.438398632065, p = 0, z = 0)$ and $(q = -2.8374234947, p = 0, z = 0)$.

and the one through $r_2 = (0, q_2, 0)$ exits $\mathcal{B}$ at $z = \pi + \Delta z_2$, for $\Delta z_1, \Delta z_2 > 0$. Let $J_\varepsilon = \{r' = (q, p, z) | q \in [q_1, q_2], p = 0, z = \pi\}$. (Note that the only difference from $r \in I_\varepsilon$ and $r' \in J_\varepsilon$ is their z coordinate).

Consider now $\mathcal{M}_\varepsilon$, a two-dimensional invariant manifold that is the union of forward orbits with initial conditions in I_ε. $\mathcal{N}_\varepsilon$ is obtained similarly, by letting J_ε flow in negative time. It follows, from the modified version of the Exponentially Small Exchange Lemma, that $\mathcal{M}_\varepsilon$ and $W^u_{\text{loc}}(\gamma_\varepsilon)$ (respectively $\mathcal{N}_\varepsilon$ and $W^s_{\text{loc}}(\gamma_\varepsilon)$) are $\mathcal{O}(e^{-\frac{c}{\varepsilon}}) - \mathcal{C}^1$ close when $\mathcal{M}_\varepsilon$ ($\mathcal{N}_\varepsilon$, respectively) exits $\mathcal{B}$, for some $c > 0$. Gronwall-type estimates guarantee that the exponential closeness of $\mathcal{M}_\varepsilon$ to $W^u_{\text{loc}}(\gamma_\varepsilon)$ (respectively $\mathcal{N}_\varepsilon$ to $W^s_{\text{loc}}(\gamma_\varepsilon)$) is maintained up to $z = \pi$.

From the simple zeroes of the Melnikov function,

$$M_A(z) = 6\sqrt{2}(1 + c + \cos z)^{1/4} \sin z$$

computed at the reference point $(q^z(0), p^z(0), z) = (-2c, 0, z)$, we know that $W^u_{\text{loc}}(\gamma_\varepsilon)$ and $W^s_{\text{loc}}(\gamma_\varepsilon)$ intersect transversely at an $\mathcal{O}(\varepsilon)$ angle when $z = \pi$, and this intersection is $\mathcal{O}(\varepsilon)$ close to the reference point. Thus, as in Section 2.C, step 3, these last two facts imply that $\mathcal{M}_\varepsilon$ and $\mathcal{N}_\varepsilon$ intersect transversely at $z = \pi$, which yields a locally unique orbit $q_0(z)$, defined on $[0, 2\pi]$, that satisfies $p_0(0) = p_0(z = 2\pi) = 0$.

The periodicity of $q_0(z)$ will be a consequence on the following symmetry of equations (3.1): Let $z = \varepsilon t$ be the new slow time variable, then (3.1) becomes

$$q' = \frac{p}{\varepsilon}, \qquad p' = -\frac{1}{\varepsilon}\left(q^2 - 1 - c - \cos z\right), \qquad z' = 1. \tag{3.2}$$

where $' \equiv \frac{d}{dz}$. Under the change of coordinates $\tau = k\pi - z$, $P = -p$, $Q = q$, (3.2) remains invariant for all even integers k. As a consequence, if $q(z)$ is a solution of (3.2) defined on the z interval $[0, \frac{k}{2}\pi]$, then there exists another solution of (3.2), $\widehat{q}(z)$, defined on $[\frac{k}{2}\pi, k\pi]$ such that $q(\frac{k}{2}\pi - w) = \widehat{q}(\frac{k}{2}\pi + w)$ for all $w \in [0, \frac{k}{2}\pi]$, see figure 6.

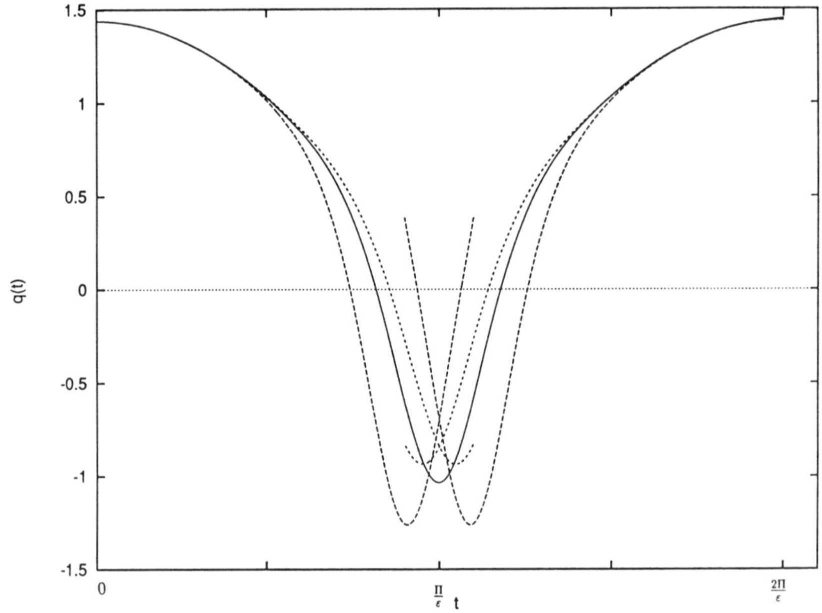

Figure 6: The symmetries of equations (3.2) force the orbit obtained by the transverse intersection of $\mathcal{M}_\varepsilon$ and $\mathcal{N}_\varepsilon$ (solid line) to have $p = 0$ for $z = \pi$. The dashed curves starting at $z = 0$ are orbit segments of (3.1) or (3.2) that lie on $\mathcal{M}_\varepsilon$; the dashed curves ending at $z = 2\pi$ are the corresponding symmetric ones, lying on $\mathcal{N}_\varepsilon$.

This symmetry implies that, in the plane $\Pi_{z=\pi}$, $W^s(\gamma_\varepsilon) \cap \Pi_\pi$ is the reflection of $W^u(\gamma_\varepsilon) \cap \Pi_\pi$ about the q axis, which immediately forces the p coordinate of $W^u(\gamma_\varepsilon) \cap W^s(\gamma_\varepsilon) \cap \Pi_\pi$ to be zero. Furthermore, since $\mathcal{M}_\varepsilon$ (respectively $\mathcal{N}_\varepsilon$) are C^1 exponentially close to $W^u(\gamma_\varepsilon)$ at $z = \pi$ (respectively $W^s(\gamma_\varepsilon)$), and, by construction, $\mathcal{N}_\varepsilon \cap \Pi_\pi$ is the reflection of $\mathcal{N}_\varepsilon \cap \Pi_\pi$ about the q axis, we also must have $\mathcal{M}_\varepsilon \cap \mathcal{N}_\varepsilon \cap \Pi_\pi$ to have zero p coordinate. In other words, the orbit $q_0(z)$ has $p_0(z = \pi) = 0$, which, by the above-mentioned symmetry, means that $q_0(z)$ is periodic.

Acknowledgments. We thank Steve Weibel for writing the numerical shooting code used in the research and in making the figures. C.S. gratefully acknowledges financial support in the form of a CONACyT-Fulbright-IIE Fellowship. T.K. was funded in part by a grant from the National Science Foundation # DMS-9307074.

References

[1] S.B. Angenent, J. Mallet-Paret, and L.A. Peletier. Stable transition layers in semilinear boundary value problems. *J. Diff. Eq.*, 67:212–242, 1987.

[2] G. Carpenter. A geometric approach to singular perturbation problems with applications to nerve impulse equations. *J. Diff. Eq.*, 23:335–367, 1977.

[3] N. Fenichel. Geometrical singular perturbation theory for ordinary differential equations. *J. Diff. Eq.*, 31:53–98, 1979.

[4] A. Fonda and M. Ramos. Large-amplitude subharmonic oscillations for scalar second-order differential equations with asymmetric nonlinearities. *J. Diff. Eq.*, 109:354–372, 1994.

[5] S. P. Hastings and J. B. McLeod. Periodic solutions of a forced second order differential equation. *J. Nonlinear Science*, 1:225–245, 1991.

[6] S. P. Hastings and J. B. McLeod. Chaotic motion of a pendulum with oscillatory forcing. *Am. Math. Monthly*, 100:563–572, 1993.

[7] C. K. R. T. Jones. Geometric singular perturbation theory. *C.I.M.E. Lecture Series*, 1994.

[8] C. K. R. T. Jones, T. Kaper, and N. Kopell. Tracking invariant manifolds up to exponentially small errors. *SIAM J. Math. Anal.*, 1993. to appear.

[9] C. K. R. T. Jones and N. Kopell. Tracking invariant manifolds with differential forms in singularly perturbed systems. *J. Diff. Eq.*, 108:64–88, 1994.

[10] C. K. R. T. Jones, N. Kopell, and R. Langer. *Construction of the FitzHugh-Nagumo pulse using differential forms*, volume 37 of *IMA Volumes in Mathematics and its Applications.* Springer Verlag, New York, 1991.

[11] T. J. Kaper and C. Soto-Treviño. Geometric theory for periodic orbits in singularly-perturbed systems. in preparation.

[12] T.J. Kaper and G. Kovačič. Multi-bump orbits homoclinic to resonance bands. 1993. preprint.

[13] T.J. Kaper and G. Kovačič. A geometric criterion for adiabatic chaos. *J. Math. Phys.*, 35(3), 1994.

[14] W. L. Kath. Slowly varying phase planes and boundary-layer theory. *Stud. Appl. Math.*, 72:221–239, 1985.

[15] H.L. Kurland and M. Levi. *Transversal heteroclinic intersections in slowly varying systems*, pages 29–38. SIAM, Philadelphia, 1986.

[16] X. B. Lin. Shadowing lemma and singularly-perturbed boundary-value problems. *SIAM J. Appl. Math.*, 49(1):26–54, 1989.

[17] J. Mawhin and M. Willem. *Critical point theory and Hamiltonian systems.* Springer Verlag, New York, 1988.

[18] H. Ockendon, J.R. Ockendon, and A.D. Johnson. Resonant sloshing in shallow water. *J. Fluid Mech.*, 167:465–479, 1986.

[19] P. Rabinowitz. Minimax methods in critical point theory with applications to differential equations. Providence, R.I., 1986. CBMS Reg. Conf. 65, Amer. Math., Soc.

[20] C. Robinson. Sustained resonance for a nonlinear system with slowly varying coefficients. *SIAM J. Math. Anal.*, 14(5):847–860, 1983.

[21] S. K. Tin, N. Kopell, and C.K.R.T. Jones. Invariant manifolds and singularly perturbed boundary value problems. 1995. to appear in SIAM J. Num. Anal.

[22] M. Willem. *Periodic solutions of differential equations with periodic nonlinearities*, pages 157–179. Pitman Research Notes in Mathematics Series. Longman Scientific & Technical, 1992.

Dynamics of large-scale bed forms in coastal seas

H.E. de SWART[1] and S.J.M.H. HULSCHER[2]
[1]*Institute for Marine & Atmospheric Research, University of Utrecht, Princetonplein 5, 3584CC Utrecht, The Netherlands*;
[2]*Delft Hydraulics, P.O. Box 152, 8300 Emmeloord, The Netherlands.*

Abstract

The formation of offshore sand ridges and sand waves in coastal seas is considered. The growth of these bed forms can be described as an instability mechanism due to tide-topography interactions. First extensions of the depth-integrated model proposed by Huthnance (1982) are studied. The most preferred bed forms resemble sand ridges of which the crests are rotated counterclockwise (on the Northern Hemisphere) with respect to the principal current direction. This orientation can become clockwise, as is observed for shore-face connected ridges, by allowing for different directions of depth-averaged currents and sediment transport. Finally a model is analyzed which includes the effect of secondary circulations in the vertical plane on the sediment transport. As a result both sand ridges and sand waves are found. Conditions are discussed for which a weakly nonlinear analysis can be carried out.

1 Introduction

The bottom of many coastal seas and estuaries exhibits patches of rhythmic patterns with a broad spectrum of spatial scales. An example of small-scale undulations are the ripples which can often be observed at sandy beaches during low-tide conditions. They have a characteristic wave-length of 10 cm and a typical amplitude of 1 cm. Larger-scale phenomena close to the beach are discussed by e.g. Lippman and Holman (1990).

In this paper we focus on offshore sand waves and sand ridges which both have horizontal length scales much larger than the local water depth. The sand ridges have wave-lenghts in the order of 5 km, amplitudes vary between 1 and 15m and their crests are rotated slightly cyclonically (counterclockwise on the Northern Hemisphere) with respect to the principal direction of the tidal currents. Furthermore, historical data indicate that the ridges hardly propagate (Van der Meene, 1994). In figure 1 the location of various sand ridge systems in the North Sea is shown. Sand waves have smaller wave-lengths (usually a few hundreds of meters), their crests are perpendicular to the tidal current direction and they propagate several meters per year. Figure 2 shows sand waves in the southern part of the North Sea detected with an echosounder.

Sand ridges and sand waves are often observed simultaneously, see Huntley et al. (1993). The profiles of these bed forms can be highly asymmetric, where the steepest slope of the sand waves indicates the direction of the mean sediment transport. Understanding the behaviour of these large-scale morphological patterns is relevant

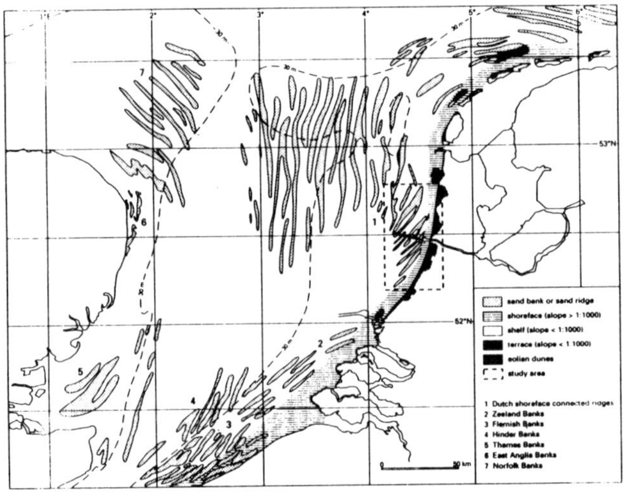

Figure 1: Location of sand ridges in the North Sea (from van der Meene, 1994).

for the management of coastal zones, the offshore industry (trenching of pipe-lines), navigation, etc..

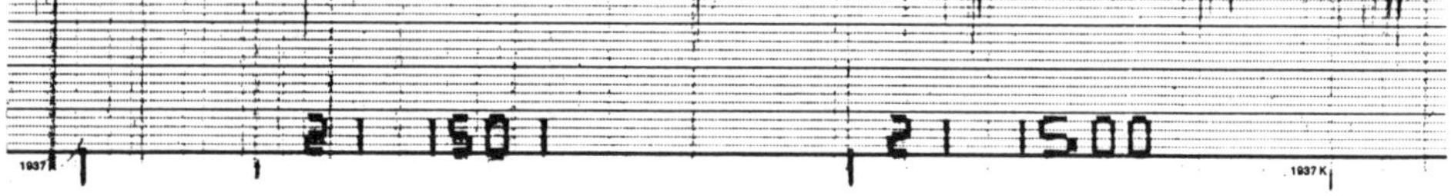

Figure 2: Observation of sand waves in the North Sea (15 km west of The Hague) using an echosounder.

A simple model for the formation of sand ridges was developed by Huthnance (1982) and modifications were studied by De Vriend (1990) and Hulscher et al. (1993). Here the water motions are described by depth-integrated shallow water equations and forcing is due to tides. The model is supplemented with a parameterization for the sediment transport (assumed to be of bedload type) and an evolution equation for the erodible bottom. Huthnance studied the linear stability properties of a basic state, describing a tidal current over a flat bottom, with respect to arbitrary bed from perturbations. He found that the largest feedback between water motions and bed forms is obtained for perturbations having the characteristic wave-lengths and crest orientations corresponding to sand ridges. The physical mechanism of instability is that, for bedforms with their crests turned cyclonically with respect to the principal tidal current direction, Coriolis and bottom frictional forces generate anticyclonic residual horizontal circulations around the bank, see Zimmerman (1981) and figure 3a. They cause the actual flow velocities on the upstream side of the bank to be larger than those on the down-stream side. Since larger flow velocities imply larger sediment transport rates there will be a convergence of sediment transport at the crest, both during the ebb and the flood period. In case the bank crests are turned anticyclonically with

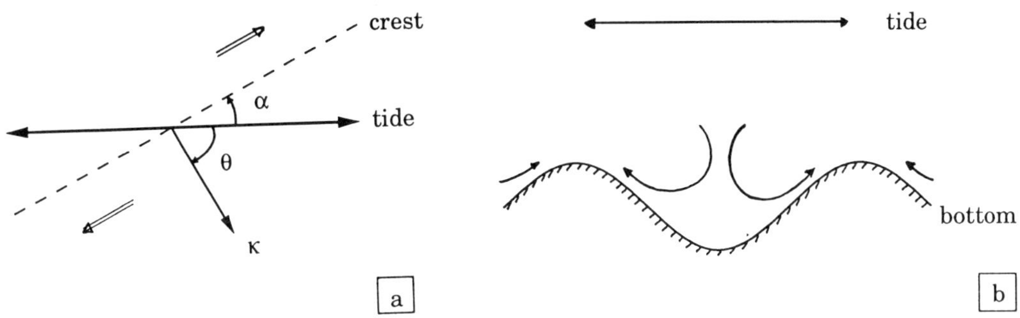

Figure 3: a. Horizontal residual circulations (indicated by arrows) near sand ridges on the Northern Hemisphere (top view) b. Vertical residual circulations near sand waves (side view). For further explanation see text.

respect to the tidal current direction the residual circulations induced by Coriolis and bottom frictional forces are of opposite sign and the instability mechanism will be weaker. Thus earth rotation effects explain the preferred orientation of the banks, but they are not necessary to understand their formation. The reason that sand ridges have the largest growth rates is that for larger-scale bed forms residual circulations are weaker, whereas for smaller scale bed forms the instability mechanism is counteracted by a diffusion process related to the preferred down-slope direction of the sediment transport.

A model for sand waves is discussed in Fredsoe and Deigaard (1992) which is a synthesis of earlier work by these authors, see the references in this book. They consider fully developed sand waves, generated by steady currents, which are assumed to be form-preserving and which propagate in the downstream direction. Using a bed load transport parameterization and a given shear stress distribution in terms of the local water depth they are able to compute the shape, length and height of the sand wave. Alternative models are proposed by Hulscher et al. (1993) and Hulscher (1995). Here it is assumed that the sand waves are driven by oscillating (tidal) currents and the feedback between bed forms and water motions is explicitly computed. An important conclusion is that the presence of sand waves is related to secondary circulations in the vertical plane. As can be seen from figure 3b the latter cause an increase of the currents moving upslope (and thus of sediment transport rates). Similarly, there is a corresponding decrease in the flow velocity and sediment transport on the downstream side. Consequently, both during ebb and flood conditions there will be a convergence of the sediment flux at the crest of the bed form.

A limitation of these studies is that either only a linear stability analysis is carried out, whence no information on the amplitude behaviour is obtained, or a parameterized description of the water motion is used. The stability studies indicate that for realistic parameter values there is a broad spectrum of unstable modes. This implies that a strongly nonlinear analysis is required to describe the amplitude behaviour. On the other hand the observed sand ridges and sand waves all have a well defined length scale with only moderate spatial and temporal modulations which are characteristics of weakly nonlinear phenomena. Therefore the aim of the present paper is to investigate whether conditions can be found for which a weakly nonlinear analysis of sand waves and sand ridges can be carried out.

In section 2 the equations of motion for the currents, sediment transport and bottom evolution are presented and in section 3 the stability concepts are discussed. Use is made of the averaging method to obtain a relatively simple system of partial differential equations. In section 4 a more extensive analysis of depth-averaged models of the type proposed by Huthnance (1982) will be presented. This includes elliptical tides, threshold of sediment movement and lateral mixing. The basic assumption of this two-dimensional model is that direction of the sediment transport coincides with that of the depth-averaged tidal current. However, this is not correct because Coriolis forces (due to earth rotation) cause the veering of the horizontal velocity field over the vertical. This effect is parameterically included in the extended two-dimensional model of section 5. As a consequence bed forms can be found which have their crests rotated anticyclonically (clockwise on the Northern Hemisphere) with respect to the principal tidal current direction. This might be important to understand the characteristics of the shore-face connected ridges, such as those observed near the Dutch Coast, see figure 1.

Finally in section 6 the effect of secondary circulations, which occur in the vertical plane, on the dynamics of bed forms is investigated. These circulations are the principal driving mechanism for sand waves. It is shown how the effect of secondary currents can be incorporated in a depth-integrated model. The result will be referred to as the parameterized three-dimensional model and it extends the one discussed in Hulscher et al. (1993).

2 Model formulation

2.1 Equations of motion

In order to describe the dynamics of morphological patterns in coastal seas a model is required which describes both the water motions, the sediment transport and the evolution of the bed. As sand ridges and sand waves have horizontal length scales which are large compared to the local water depth a model based on the shallow water equations can be used. The water motions are assumed to be driven by tidal forces, hence the currents will be oscillatory with a characteristic amplitude U and frequency σ. The coupling between water motions and bed form changes is described by a bottom evolution equation, which follows from mass conservation of the sediment, together with a parameterization of the sediment transport. The domain is an open coastal sea, hence horizontal boundaries are absent. Only a brief outline of the derivation of the model will be presented, more details can be found in Hulscher et al. (1993).

It is assumed that a characteristic length scale of the bed forms is the tidal excursion length $l_e = U/\sigma$, which is much smaller than the tidal wave-length. The equations of motion are made dimensionless by introducing time scale $1/\sigma$, length scale l_e and velocity scale U. Furthermore the bottom elevations are scaled with the undisturbed water depth H and U^2/g is a scale for the free surface variations. Finally it is assumed that the morphological time scale is much larger than the tidal period. This implies that the averaging method can be applied such that the bottom topography may be considered as time independent on the fast time scale whereas its evolution on the slow time scale is determined by tidally averaged rather than instantaneous sediment fluxes. The resulting nondimensional model is

$$\frac{\partial \vec{u}}{\partial t} + \left(\vec{u} \cdot \vec{\nabla}\right) \vec{u} + f \vec{e}_z \times \vec{u} = -\vec{\nabla}\zeta - \frac{\vec{\tau}}{1-h} + A\nabla^2\vec{u}, \tag{1}$$

$$\vec{\nabla}\left[(1-h)\vec{u}\right] = 0, \tag{2}$$

$$\frac{\partial h}{\partial T} + \vec{\nabla} \cdot \langle \vec{S} \rangle = 0. \tag{3}$$

Here t and T are fast and slow time coordinates, respectively, $\langle \cdot \rangle$ denotes an average over a tidal period and $\vec{\nabla}$ is a vector differential operator with components $\partial/\partial x$ and $\partial/\partial y$, where x and y are horizontal coordinates. Furthermore $\vec{u}$ is the depth-averaged velocity field (components u and v), ζ the elevation of the free surface and h the elevation of the bed with respect to the rest level. Finally $\vec{e}_z$ is a unity vector in the vertical direction, $\vec{\tau}$ is the bottom shear stress (scaled by $\rho\, H\, U\, \sigma$, where ρ is the density of the sea water) and $\vec{S}$ is the volumetric sediment flux. The momentum equations (1) contain two parameters: the Coriolis number f (ratio of Coriolis parameter and tidal frequency σ) and the diffusion coefficient A (ratio of tidal period and horizontal diffusion time scale). Their characteristic values at midlatitudes are $f \sim 0.8$ and $A \sim 0 - 1$.

2.2 Bed shear stress and sediment transport

In this paper the following parameterizations for the volumetric sediment flux $\vec{S}$ and the bed shear stress $\vec{\tau}$ will be studied:

$$\vec{S} = \mu\, H(|\vec{\tau}| - \tau_c)\, \left[|\vec{\tau}| - \tau_c\right]^b \left[\frac{\vec{\tau}}{|\vec{\tau}|} - \lambda \vec{\nabla} h\right], \tag{4}$$

$$\vec{\tau} = R_1\, \vec{u}\,(t - \phi_1) + R_2\, \vec{e}_z \times \vec{u}\,(t - \phi_2), \tag{5}$$

where H is the Heaviside function. Note that the sediment flux is assumed to depend on local flow variables only. This is a good approximation for material which is mainly transported as bed load such that the particles keep contact with the bed. The transport increases with increasing intensities of the bed shear stress. Furthermore, the flux is mainly in the direction of the bed shear stress, but there is a bed slope correction term involved which measures the effect that the sediment transport has a preferred down-hill component. Below a critical shear stress τ_c there is no transport. Many parameterizations of this type are discussed in Van Rijn (1993). In the present context realistic values of the morphological parameters are $\mu \sim 1$ (this is a scaling parameter, to be discussed later), $2 < b < 6, \tau_c \sim 0.1, \lambda \sim 0.01 - 0.02$.

The bed shear stress is expressed as a linear combination of the depth-averaged velocity $\vec{u}$ and the horizontal velocity field perpendicular to $\vec{u}$. Moreover phase lags ϕ_1, ϕ_2 are introduced. The motivation for these choices is as follows. First of all turbulence theory and observations indicate that bed shear stresses should be represented by a quadratic friction law. In many idealized models for tidal flow they are replaced by linear terms where the coefficients are chosen such that the average dissipation during a tidal cycle remains unchanged, see e.g. Zimmerman (1982). The reason that equation (5) contains two terms on its right-hand side is that in general the direction of the bed shear stress will not coincide with that of the depth-averaged velocity. In the present setting this is primarily due to earth rotation effects which cause the veering of the horizontal velocity field over the vertical (Ekman spiral effect, cf. Prandle (1982)). Finally the bed shear stress will not have the same phase as the depth-averaged velocity field due to the presence of internal friction in the fluid. The coefficients R_1, R_2, ϕ_1, ϕ_2 (phases in rad), which are all of order 1, can be computed from a three-dimensional shallow water model.

In this paper the following cases will be investigated:

* $R_1 = r =$ constant, $\mu = r^{-b}, R_2 = \phi_1 = \phi_2 = 0$.
* $R_1 = r_1, \phi_1 = \varphi_1, R_2 = r_2, \phi_2 = \varphi_2$ and all variables are constant.
* R_1, ϕ_1, R_2 and ϕ_2 depend on the flow state.

The first case represents the two-dimensional depth-averaged model of Huthnance (1982) a.o., in which the directions of sediment transport and principal tidal currents coincide. The second is the extended two-dimensional model which accounts for Ekman veering effects. The final case results in the parameterized three-dimensional model which also incorporates the effects of secondary currents in the vertical plane on the sediment transport.

3 Stability analysis

We will now investigate the stability properties of a simple basic state of the model (1)-(3) with the parameterizations (4) and (5). As can be seen there exist basic states of the type

$$\vec{u} = \vec{u}_0(t), \qquad \vec{\nabla}\zeta = \vec{\nabla}\zeta_0(t), \qquad h = h_0 \equiv 0. \tag{6}$$

They describe a spatially uniform tidal flow over a flat bottom. Next small bed form perturbations are introduced and we investigate whether there is a positive feedback between the bed forms and the water motions. Thus we substitute in the equations of motion

$$\vec{u} = \vec{u}_0 + \vec{u}', \qquad \vec{\nabla}\zeta = \vec{\nabla}\zeta_0 + \vec{\nabla}\zeta', \qquad h = h', \tag{7}$$

where the perturbations are indicated by accents and are assumed to be small. After linearization we obtain

$$\frac{\partial \vec{u}'}{\partial t} + (\vec{u}_0 \cdot \vec{\nabla})\vec{u}' + f\,\vec{e}_z \times \vec{u}' = -\vec{\nabla}\zeta' - \vec{\tau}' - \vec{\tau}_0 h' + A\nabla^2\vec{u}', \tag{8}$$

$$\vec{\nabla} \cdot \vec{u}' - (\vec{u}_0 \cdot \vec{\nabla})h' = 0, \tag{9}$$

$$\frac{\partial h'}{\partial T} + \mu\vec{\nabla} \cdot \langle \vec{S}' \rangle = 0, \tag{10}$$

where the perturbed sediment flux for $|\vec{\tau}_0| > \tau_c$ reads

$$\vec{S}' = [|\vec{\tau}_0| - \tau_c]^b \left[\frac{\vec{\tau}'}{|\vec{\tau}_0|} - \lambda\vec{\nabla}h'\right] + \frac{[|\vec{\tau}_0| - \tau_c]^{b-1}}{|\vec{\tau}_0|^2}\left[b - 1 + \frac{\tau_c}{|\vec{\tau}_0|}\right](\vec{\tau}_0 \cdot \vec{\tau}')\,\vec{\tau}_0 \tag{11}$$

and for $|\vec{\tau}_0| < \tau_c$ the transport rate is zero. The perturbed shear stress follows from replacing in equation (5) the variable $\vec{u}$ by $\vec{u}'$. Note however that this expression becomes more complicated if the parameterized three-dimensional model is studied since in that case the parameters R_1, R_2, ϕ_1 and ϕ_2 depend on the flow state.

The pressure terms can be eliminated by cross-differentiation of the momentum equations. This yields the vorticity equation

$$\frac{\partial \eta'}{\partial t} + (\vec{u}_0 \cdot \vec{\nabla})\eta' + f\vec{\nabla} \cdot \vec{u}' = -\vec{e}_z \cdot \left[\vec{\nabla} \times \vec{\tau}'\right] + \vec{\tau}_0 \times \vec{\nabla}h' + A\nabla^2\eta', \tag{12}$$

where

$$\eta' = \vec{e}_z \cdot \left[\vec{\nabla} \times \vec{u}'\right] \tag{13}$$

is the vertical component of the relative vorticity. The equations (9), (10) and (12) allow for spatially periodic solutions, thus

$$\phi'(\vec{x}, t) \equiv (\vec{u}', \eta', h') = \int\!\!\int \tilde{\phi}(\vec{\kappa}, t)\exp(-i\vec{\kappa} \cdot \vec{x})d\vec{\kappa}. \tag{14}$$

This reduces the problem to the analysis of linear algebraic equations and ordinary differential equations in time.

We will now consider a specific basic state for which the velocity components read

$$u_0 = \sin t, \qquad v_0 = \beta\cos t. \tag{15}$$

This represents a simple M_2 (period 12:25 h) elliptical tide. It is assumed that $|\beta| \leq 1$, thus the principal current direction is along the x-axis. Positive (negative) values of

β indicate velocity vectors rotating in a clockwise (counterclockwise) direction. For this basic state the solutions of the spatially Fourier transformed flow equations can be written as Fourier series in time, e.g., the transformed vorticity becomes

$$\tilde{\eta} = \tilde{h} \sum_{n=0}^{\infty} \{\eta_{nc} \cos(nt) + \eta_{ns} \sin(nt)\}, \tag{16}$$

and similar expressions for $\tilde{u}$ and $\tilde{v}$. In practice the Fourier coefficients are computed by truncating the series after an odd harmonic, as explained by De Swart and Zimmerman (1993), and solving the resulting set of linear algebraic equations. From this the components $\tilde{S}_x$ and $\tilde{S}_y$ of the transformed sediment flux can be computed. Averaging these results over a tidal period and substituting the expressions in the bottom evolution equation finally yields

$$\frac{\partial \tilde{h}}{\partial T} = \mu\omega\tilde{h}. \tag{17}$$

Thus from equations (14) and (17) it follows that the bottom topography can be written as $h' \sim \exp(-i\vec{\kappa}\cdot\vec{x} + \mu\omega T)$. In general ω is a complex frequency where its real part denotes the growth rate and its imaginary part the frequency of the perturbation. However, in the present model ω is always real, thus the crests of the bed forms donot move. This is because the equations of motion are invariant to arbitrary rotations of the coordinate axes. Consequently, there is no preferred direction in the model which implies that bed forms cannot propagate. Thus in eq. (17) ω is the growth rate, which depends on the model parameters, the components of the wavevector $\vec{\kappa}$ of the bottom perturbations and the even Fourier components of the transformed vorticity field $\tilde{\eta}$ defined in eq. (16).

4 Two-dimensional models

In this section we briefly discuss the results for a two-dimensional model where we choose $R_1 = r =$ constant, $R_2 = \phi_1 = \phi_2 = 0, \mu = r^{-b}$. As can be traced back from equations (4) and (5) this implies that the bed shear stress coincides with the depth-averaged velocity and that there are no phase lags. In figure 4a a contour plot of the growth rate ω is shown in the wavenumber space for the reference case $f = 0.8, r = 0.6, \beta = 0, \lambda = 0.01, b = 3, \tau_c = 0$ and $A = 0$. This corresponds to a unidirectional M_2 tide with a velocity amplitude of 1 ms^{-1} in a coastal sea at midlatitudes with an undisturbed water depth of 30 m. In this case a dimensionless wavenumber $(k^2+l^2)^{1/2}$ corresponds to a wave-length of approximately 40 km. The figure shows that there is a wide range of bed forms with positive growth rates. The maximum value is attained for a perturbation with a wave-length of approximately 8 km and its crests are rotated cyclonically (counterclockwise on the Northern Hemisphere), by $\sim 30^o$, with respect to the principal tidal current direction. These length scales and preferred orientations can also be observed for most offshore sand ridges, see figure 1.

As was already explained in the introduction the physical mechanism is related to the presence of horizontal residual circulations around the banks, see also figure 3a.

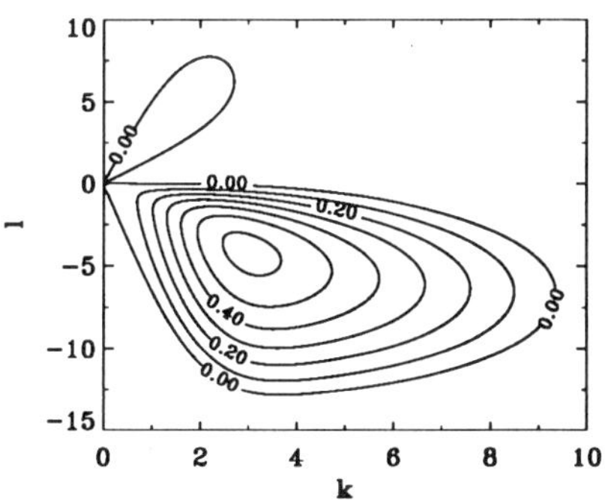

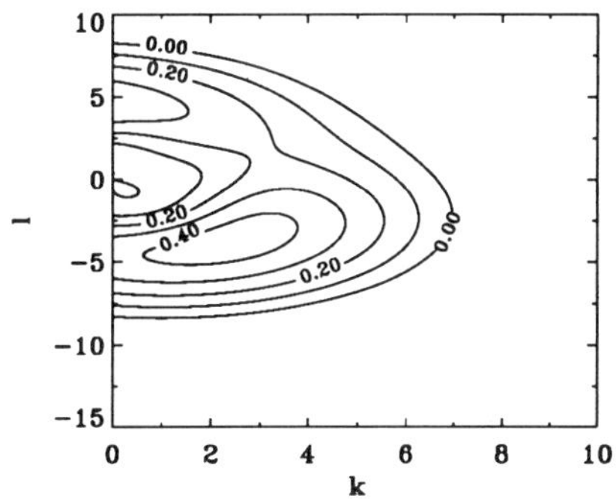

Figure 4: a. Contour plot of the positive growth rates of bed form perturbations for the reference case in the two-dimensional model. b. As a., but for $\beta = -0.8$. For other parameters see text.

These circulations cause a net convergence of sediment transport at the crests of the banks which is most effective for bed forms which have their crests turned cyclonically with respect to the direction of principal tidal currents. This explains the observed asymmetry in the growth rate curves of figure 1 and it can be ascribed to earth rotation effects. The reason that small-scale perturbations (large wavenumbers) are damped is due to the down-slope correction term in the sediment transport (4). When substituted in the bottom evolution equation (3) it can be seen that this mechanism causes diffusion of bed forms which becomes very effective if gradients are large. The figure is qualitatively similar to results presented in earlier studies. The important difference is here that it has been verified that the present solution differs less than 1% from the exact solution. This requires a truncation of the Fourier series in eq. (16) after the third overtide and results in an increase of growth rates by 30%.

We have carried out many sensitivity studies with this model. The results can be summarized as follows:

*Both the incorporation of lateral mixing terms in the vorticity equation (i.e., nonzero A) and a critical shear stress in the sediment flux (i.e., nonzero τ_c) cause a significant reduction of the wavenumber of the preferred bed form and the corresponding maximum growth rate. However, the preferred orientation of the banks is only slightly affected.

*Maximum growth rates increase with increasing values of the parameters f, r and b. This is obvious since increasing Coriolis and bottom frictional forces causes more intense residual circulations around the banks whereas parameter b controls the amount of sediment transport.

*If the ellipticity of the tide is increased the distribution of positive growth rates in wavenumber space becomes more symmetric wheras the preferred wave-length and orientation of the banks are hardly affected, see figure 4b.

*There is no combination of physically realistic model parameters for which a nontrivial selective bed form is obtained. It is a direct consequence of the flow-topography interaction mechanism and can be understood as follows. Consider a bottom perturbation with a fixed orientation α of its crests with respect to the direction of the principal tidal currents and assume a noncircular tide ($|\beta| < 1$ in eq. (15). It appears that the growth rate ω of this bed form is an even function of the wavenumber $|\vec{\kappa}| \equiv \kappa$ and furthermore

$$\omega(\kappa^2) = 0, \qquad \frac{\partial \omega}{\partial \kappa^2} \leq \left.\frac{\partial \omega}{\partial \kappa^2}\right|_{\kappa^2=0}. \tag{18}$$

These results show that mode $\vec{\kappa} = 0$ is neutrally stable and that the spectrum of perturbations with positive growth rates always include the ultralong waves. This is illustrated in figure 5 which shows the maximum growth rate ω_m and corresponding wavenumber κ_m as a function of the bottom friction parameter r for three different values of the Coriolis number f. As can be seen the wavenumber κ_m decreases with decreasing r. For $f = 0$ and $r < 0.06$ all perturbations decay and the ultralong waves are the last to become stable.

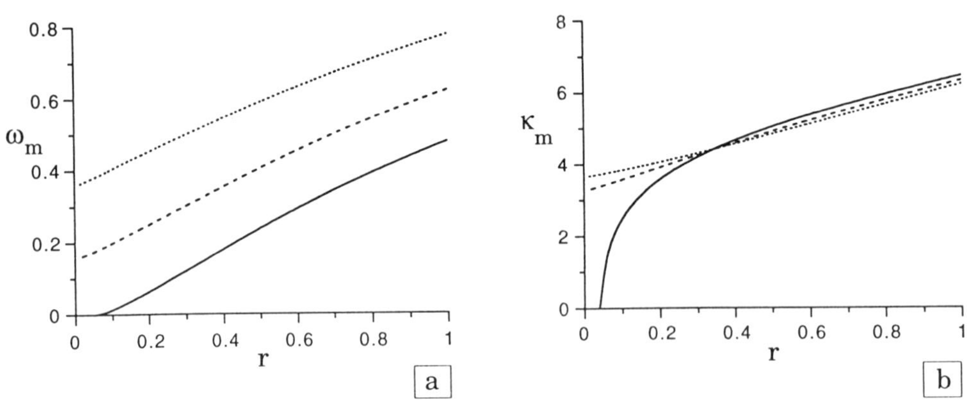

Figure 5: Maximum growth rate ω_m (a) and corresponding wavenumber κ_m (b) as a function of the bottom friction parameter r for $f = 0$ (solid curve), $f = 0.4$ (dashed curve) and $f = 0.8$ (dotted curve).

The only exception occurs for $|\beta| = 1$ which corresponds to a circular tide. There is no longer a principal direction of tidal currents in this degenerate case. The formation of sand ridges is now fully due to asymmetry in the velocity field caused by the basic tide and the first overtide (M_4), thus residual current effects are unimportant. Figure 6 shows the neutral curves $\omega = 0$ in dependence of the wavenumber κ and friction parameter r for $\beta = -1$ and $\beta = 1$. Both curves reach a minimum at nonzero r and κ, which indicates that there are selective modes. Note that the minimum for $\beta = -1$ (cyclonically rotating tidal current vector, on the Northern Hemisphere counterclockwise) is reached for smaller friction parameter than the situation where

$\beta = 1$. This is because a cyclonically polarized tidal current reinforces Coriolis torques thereby causing larger tidal asymmetries and larger growth rates.

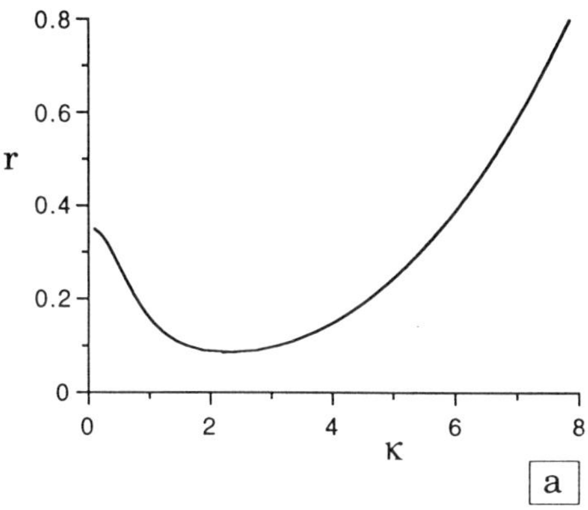

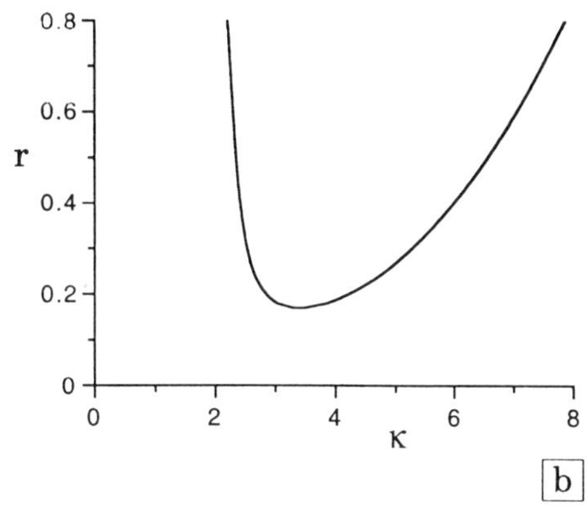

Figure 6: Neutral curves $\omega = 0$ as a function of the friction parameter r and wavenumber κ for a cyclonically (a) and anticyclonically (b) rotating circular tide.

5 An extended two-dimensional model

In this section we investigate the effect of allowing for bed shear stresses which have a direction and phase which differs from those of the depth-averaged current. Thus we consider the case that the coefficients R_1, R_2, ϕ_1 and ϕ_2 are nonzero. Realistic values for these parameters are obtained by identifying the bed shear stress defined in eq. (5) with the bed shear stress computed from the basic state equations

$$\frac{\partial \vec{v}_0}{\partial t} + f\vec{e}_z \times \vec{v}_0 = -\vec{\nabla}\zeta_0 + \frac{1}{2}E\frac{\partial^2 \vec{v}_0}{\partial z^2}, \tag{19}$$

$$\vec{u}_0 \equiv \int_{-1}^{0} \vec{v}_0 dz = \sin t\, \vec{e}_x + \beta \cos t\, \vec{e}_y, \tag{20}$$

$$\frac{\partial \vec{v}_0}{\partial z} = 0 \quad \text{at } z = 0 \quad , \quad \frac{1}{2}E\frac{\partial \vec{v}_0}{\partial z} = s\vec{v}_0 \quad \text{at } z = -1. \tag{21}$$

They describe a basic tidal current over a flat bottom in a three-dimensional shallow water model and they can be solved by straightforward methods, see Prandle (1982). Here $\vec{v}$ is a horizontal velocity field which depends on the vertical coordinate z and $\vec{u}$ is its depth-averaged value. Furthermore parameter E is the Stokes number, which measures the effect of internal friction, and s is the slip parameter which describes the relation between velocity and shear stress at the bottom. The bed shear stress is defined as

$$\vec{\tau} = \frac{1}{2}E\frac{\partial \vec{v}}{\partial z}\Big|_{z=-1} \tag{22}$$

In the experiments we have fixed $b = 3, A = 0$ and $\beta = 0$ (unidirectional tides) and $\lambda = 0.02, \tau_c = 0.1$. Furthermore, $\mu = 1$, so the results are not rescaled. In figure 7 the parameters R_1, R_2, ϕ_1 and ϕ_2 are shown as a function of the Stokes number E for a slip parameter $s = 5$. The latter is a realistic value for tidal flow in a shallow sea. Note that ϕ_1 is always negative, so the depth-averaged flow lags behind the bed shear stress, as is to be expected because of internal friction effects. In this model earth rotation effects

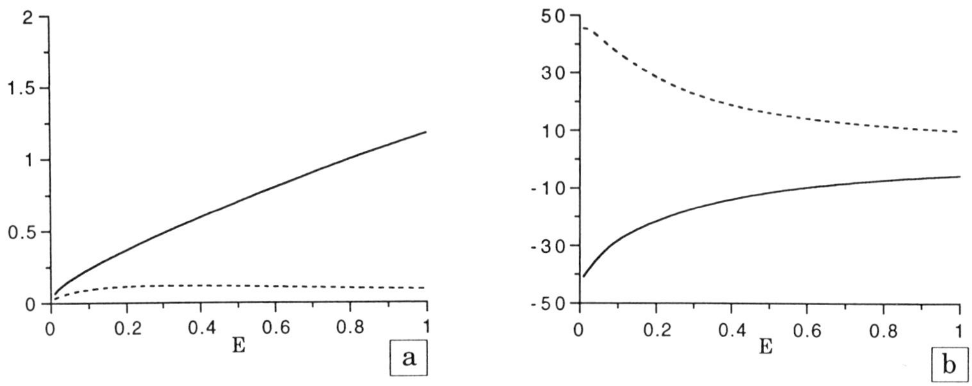

Figure 7: a. Dependence of the parameters R_1 (solid curve) and R_2 (dashed curve) on the Stokes number E for $s = 5, \beta = 0$ and $f = 0.8$. b. As a. but for the parameters ϕ_1 and ϕ_2 (in degrees), respectively.

cause the end points of the bed shear stress vector of the basic state to follow an ellipse during a tidal cycle in a cyclonic direction. Furthermore, the principal axis of this ellipse is rotated anticyclonically (angle ψ) with respect to the direction of the depth-averaged current. If this orientation is large enough (as occurs for small Stokes numbers and large slip parameters) then the preferred bed forms will have their crests more or less parallel to the principal direction of the bed shear stress vector. This can be seen from figure 8 which shows the growth rates for the extended two-dimensional model for $E = 0.25, s = 5$. In this case $R_1 = 0.42, R_2 = 0.12, \phi_1 = -19.27^o, \phi_2 = 25.34^o$ and $\psi = -11.52^o$. The preferred wave-length and crest orientation are 5.0 km and -15.0^o, and thus the crests are rotated anticyclonically with respect to the principal direction of the depth-averaged current. This is fundamentally different from the two-dimensional model discussed in the previous section. The reason for this behaviour is that the preferred bed forms cause large bottom frictional torques, see the second term on the right-hand side of vorticity equation (12). This tidal vorticity is subsequently advected by the depth-averaged current of which the direction in the present model does not coincide with that of the crests. Consequently an intense residual current can develop if the angle $|\psi|$ has large values.

It appears that growth rates are quite small for the present values of the Stokes number and slip parameter. Experiments have demonstrated that they rapidly increase if E and s are increased. Since in that case also the angle ψ between principal bed

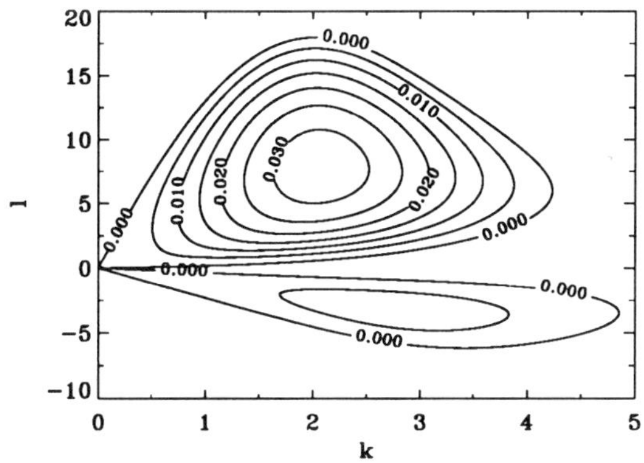

Figure 8: Contour plot of the growth rates in the wavenumber space for the extended two-dimensional model. For parameters see text.

shear stress direction and depth-averaged flow direction decreases the preferred bed forms will have the same characteristics as in the two-dimensional model.

6 A parameterized three-dimensional model

So far we have studied the sediment transport and morphodynamics related to depth-averaged currents. This resulted in preferred bed forms with characteristics similar to observed sand ridges. For the occurrence of smaller scale bottom patterns secondary circulations, which have their main orientation in the vertical plane, are important. Evidence for this can be found in Blondeaux (1990), Vittori and Blondeaux (1990), Fredsoe and Deigaard (1992) and references therein. This knowledge was used in Hulscher et al. (1993) to propose a simple parameterization in a depth-averaged model to account for the effect of secondary currents on the sediment transport. As a result the formation of both sand ridges and sand waves was simulated, where the latter features have wave-lengths of a few hundred meters and their crests are perpendicular to the principal tidal current direction. In this section an extended and more explicit parameterization of the contribution of secondary currents to the sediment transport will be presented. This is motivated by the results of Hulscher (1995) with a three-dimensional shallow water model. The advantage of a parameterized model is that it yields insight in the basic mechanism of sand wave growth whereas also the growth rates can be computed more easily. It is derived from an analysis of the perturbed three-dimensional shallow water equations which read

$$\frac{\partial \vec{v}'}{\partial t} + \left(\vec{v}_0 \cdot \vec{\nabla}\right) \vec{v}' + w' \frac{\partial \vec{v}_0}{\partial z} + f \vec{e}_z \times \vec{v}' = -\vec{\nabla}\zeta' + \frac{1}{2} E \frac{\partial^2 \vec{v}'}{\partial z^2}, \tag{23}$$

$$\vec{\nabla} \cdot \vec{v}' + \frac{\partial w'}{\partial z} = 0, \tag{24}$$

with boundary conditions

$$\frac{\partial \vec{v}'}{\partial z} = 0, \qquad w' = 0 \qquad \text{at} \quad z = 0, \tag{25}$$

$$\frac{\partial \vec{v}'}{\partial z} = \frac{2s}{E}\vec{v}' + h'\left[-\frac{\partial^2 \vec{v}_0}{\partial z^2} + \frac{2s}{E}\frac{\partial \vec{v}_0}{\partial z}\right], \quad w' = \left(\vec{v}_0 \cdot \vec{\nabla}\right) h' \ \text{at} \ z = -1. \tag{26}$$

Here the basic state horizontal velocity field $\vec{v}_0$ and the parameters E (Stokes number) and s (slip parameter) are defined in eq. (19)-(21). Furthermore $\vec{v}'(\vec{x}, z, t)$ is the perturbed horizontal flow field due to the presence of small bed forms whereas w' and ζ' are the perturbed vertical velocity component and free surface elevation, respectively.

The state variables are decomposed in depth-averaged parts and fluctuations (denoted by subscripts s) which describe the secondary flow . We assume that the depth-averaged equations yield results which are already governed by the extended two-dimensional model of the previous section. We now focus on the secondary flow mechanism. This is only significant if advection terms are large, i.e., if the wavenumbers of the bed form perturbations are large. The momentum equations for the secondary flow can thus be approximated by

$$\left(\vec{v}_0 \cdot \vec{\nabla}\right) \vec{v}_s' + w_s' \frac{\partial \vec{v}_0}{\partial z} = -\vec{\nabla}\zeta' + \frac{1}{2}E\frac{\partial^2 \vec{v}_s'}{\partial z^2}, \tag{27}$$

which is a balance between advection, pressure and friction terms. These equations are supplemented by a continuity equation and boundary conditions which follow from (24)-(26) by adding subscripts s to the variables $\vec{v}', w', \zeta'$. Here it is assumed that in the regime where eq. (27) is valid the secondary flow dominates over the depth-averaged flow. These equations include the generation of secondary flow discussed by Zimmerman (1981) and Hulscher et al. (1993). Basically, the internal friction terms cause the generation of tidal vorticity which is subsequently advected by the basic tidal currents. As a result residual secondary circulation cells as sketched in figure 3b will be generated. However, in the present model also vertical advection of basic momentum is incorporated which cannot be neglected using scaling arguments.

This system of equations is analyzed as follows. First the variables are Fourier transformed in the horizontal domain, similar to what was done in section 3, eq. (14). It follows that the intensity of the advection terms in the momentum equations is measured by the wavenumber $\kappa \equiv |\vec{\kappa}|$, which is much larger than the Stokes number E which multiplies the highest-order derivatives. Hence the problem can be analyzed by singular perturbation methods. Here we have only analyzed the case for finite values of the slip parameter, such that the basic velocity field $\vec{v}_0$ at the bottom $z = -1$ is nonzero. In this case the thickness of the boundary layer is $\delta_i = [E/(2\kappa)]^{1/2}$.

The analysis of the boundary layer equations is straightforward because the equations are linear differential equations with constant coefficients in the vertical boundary layer coordinate. In particular the vertical velocity component is constant in this layer and determined by the boundary condition at the bottom. Approximate solutions can then be found by harmonic truncation from which the shear stress $\vec{\tau}'$ can be computed. This is subsequently substituted in eq. (11) for the perturbed sediment transport and thus yields a secondary flow correction on the growth rates of bed forms.

An example of how secondary flow corrections affect the growth rates of bed forms is shown in figure 9 a which is a contour plot of the growth rates in wavenumber space for $E = 1, s = 5$ and other parameters have the same values as in the previous section. In this case the maximum growth rate is $\omega_m = 141.0$ which occurs for $k_m = 131.8, l_m = 12.5$. This describes a preferred bed form with a wave-length of 300 m of which the crests are almost perpendicular to the principal direction of the bed shear stress (crest orientation 84.5° with respect to depth-averaged tidal current direction).

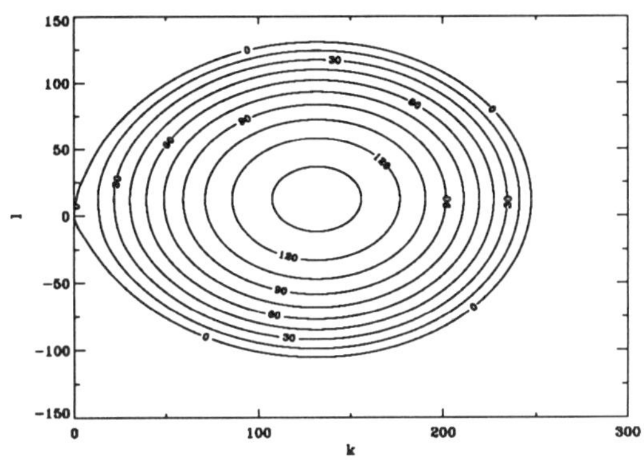

Figure 9: Contour plot of the growth rates in the parameterized three-dimensional model which includes effects of secondary currents, parameters $E = 1, s = 5$.

Note that the growth rates are much larger than those obtained in the models of the previous sections where only sand ridges were obtained. Thus for this choice of the model parameters sand waves dominate over sand ridges. However, if smaller values of s are considered the model predicts preferred bed forms which again resemble sand ridges. For an extensive study, based on a three-dimensional model, see Hulscher (1995).

For the analysis of the nonlinear dynamics it would be helpful if parameters could be selected for which only a narrow spectrum of bed forms with positive growth rates is found. Therefore it was investigated how the maximum growth rate ω_m and the corresponding wavenumbers k_m, l_m depend on the slip parameter s. It appears that they decrease monotonically with descreasing s and for $s < s_c = 0.14$ the basic state (a flat bottom) is stable. At critical conditions again the perturbation with wavenumber $\vec{\kappa} = 0$ is the first to become unstable. Hence this is not a suitable starting point for a weakly nonlinear analysis.

7 Conclusions

In this paper the formation of offshore tidal sand ridges and sand waves has been investigated. It appears that a positive feedback between undulations in the bottom and horizontal residual circulations results in the occurrence of sand ridges, as described by Zimmerman (1981) and Huthnance (1982). From a sensitivity study of the two-dimensional depth-integrated model it is concluded that the crests of these bed forms are always rotated counterclockwise (on the Northern Hemisphere) with respect the principal current direction.

A serious limitation of these two-dimensional models is that the sediment transport is in the direction of the depth-averaged flow. Generally, due to earth rotation effects these directions will differ. This effect has been included in an extended two-dimensional model. Experiments have demonstrated that the most preferred bed forms still resemble sand ridges but for certain parameter values the crests can be rotated clockwise. This result may be helpful for a better understanding of shore-face connected ridges which have this property, see figure 1.

Finally a parameterized three-dimensional model is discussed. It incorporates the effects of secondary circulations, which have their main orientation in the vertical plane, on the sediment transport and morphodynamics. The parameterization is motivated by the results of Hulscher (1995) and extends the one discussed by Hulscher et al. (1993) by including also vertical advection of basic momentum. As a result both the formation of sand ridges and sand waves is simulated. The mechanism of sand wave generation is due to residual secondary circulations and agrees with the description of Zimmerman (1981) and the results of Blondeaux (1990) for a sea ripple model.

All the models investigated in this paper suffer from the problem that no combination of parameter values can be found for which a narrow spectrum of bed forms with finite wave-lenghts exists. It appears that the ultralong waves are always the first modes to be selected if a parameter such as the tidal current amplitude is increased. This is due to the behaviour of the horizontal residual vorticity response. Consequently for realistic parameter values a strong nonlinear analysis is required to yield information on the amplitude behaviour of the ridges. The only exception is for circular tides, such that the bed forms are not generated by residual circulations but rather by tidal asymmetries. However, this turns out to be a degenerate case.

Finally it is important to realize that there are inconsistencies between depth-averaged and three-dimensional shallow water models. It appears that the depth-averaged results obtained from a stability analysis of the three-dimensional model differ from those which follow from the stability analysis of the two-dimensional model, see also the discussion in Zimmerman (1986). Hence for a full understanding of the dynamics of sand ridges and sand waves it is recommended to use the three-dimensional shallow water model, as was done by Hulscher (1995).

Acknowledgements This work has been carried out as a part of the MAST programmes 'G8 Coastal Morphodynamics' and 'CSTAB' . It was funded jointly by Delft

Hydraulics, in the framework of the Netherlands Centre for Coastal Research, and the Commission of the European Communities, Directorate General for Science Research and Development, under contract nrs. MAS2-CT92-0027 and MAS2-CT92-0024-C.

References

Abramowitz, M. and Stegun, I. (1965) *Handbook of mathematical functions.* Dover, New York.

Blondeaux, P. (1990) Sand ripples under sea waves, part 1: ripple formation. *J. Fluid Mech.* **218**, 1-17.

De Swart, H.E. and Zimmerman, J.T.F. (1993) Rectification of the wind-driven ocean circulation on the beta plane. *Geoph. Astrophys. Fluid Dyn.* **71**, 17-41.

De Vriend, H.J. (1990) Morphodynamic processes in shallow tidal seas. In: *Residual currents and long-term transport*, R.T. Cheng (ed.), Springer Verlag, New York, pp. 276-301.

Fredsoe, J. and Deigaard, R. *Mechanics of coastal sediment transport*, World Scientific, Singapore .

Hulscher, S.J.M.H., De Swart, H.E. and De Vriend, H.J. (1993) Generation of offshore tidal sand banks and sand waves. *Cont. Shelf Res.* **13**, 1183-1204.

Hulscher, S.J.M.H. (1995) Sand banks and sand waves in a three-dimensional shallow water model. Preprint, IMAU, Utrecht Univ., subm. for publ..

Huntley, D.A., Huthnance, J.M., Collins, M.B., Liu, C.-L., Nicholls, R.J. and Hewitson, C. (1993) Hydrodynamics and sediment dynamics of North Sea sand waves and sand banks. *Phil. Trans. R. Soc. London* **A343**, 461-474.

Huthnance, J. (1982) On one mechanism forming linear sand banks. *Est. Coastal Shelf Sci.* **14**, 79-99.

Lippman, T.C. and Holman, R.A. (1990) The spatial and temporal variability of sand bar morphology. *J. Geoph. Res.* **95C7**, 11575-11590.

Prandle, D. (1982) The vertical structure of tidal currents. *Geoph. Astrophys. Fluid Dyn.* **22** , 29-48.

Van der Meene, J. (1994) *The shoreface connected ridges along the central Dutch coast*, Neth. Geogr. Studies **174**, KNAG, Utrecht, 256 pp..

Van Rijn, L.C.J. (1993) *Handbook of sediment transport by currents and waves*, Aqua Publ., Amsterdam.

Vittori, G. and Blondeaux, P. (1990) Sand ripples under sea waves, part 2: finite amplitude development. *J. Fluid Mech.* **239**, 23-45.

Zimmerman, J.T.F. (1981) Dynamics, diffusion and geomorphological significance of tidal residual eddies. *Nature* **290**, 549-555.

Zimmerman, J.T.F. (1982) On the Lorentz linearization of a quadratically damped forced oscillator. *Phys. Lett.* **89A**, 123-124.

Zimmerman, J.T.F. (1986) Principal differences between 2D- and vertically averaged 3D-models of topographic tidal rectification. *Lect. Notes Estuar. Coastal Sci.* **16**, Springer Verlag, Berlin, pp. 120-129.